GERIATRIC PSYCHIATRY AND THE LAW

CRITICAL ISSUES IN
AMERICAN PSYCHIATRY AND THE LAW

was edited for

THE AMERICAN ACADEMY OF PSYCHIATRY AND THE LAW
(TRI-STATE CHAPTER)

Volume 1   Edited by Richard Rosner, M.D.

Volume 2   Edited by Richard Rosner, M.D.

Volume 3   GERIATRIC PSYCHIATRY AND THE LAW
Edited by Richard Rosner, M.D., and Harold I. Schwartz, M.D.

GERIATRIC PSYCHIATRY AND THE LAW

Edited by

RICHARD ROSNER, M.D.

Diplomate in Forensic Psychiatry, Clinical Associate Professor,
Department of Psychiatry, School of Medicine, New York University,
and Medical Director, Forensic Psychiatric Clinic for the New York Criminal
and Supreme Courts (First Judicial District), New York, New York

and

HAROLD I. SCHWARTZ, M.D.

Associate Professor of Clinical Psychiatry, Department of Psychiatry,
Mount Sinai School of Medicine, and Chief, Psychiatric Outpatient Services,
and Chief, Program in Psychiatry and Law, Department of Psychiatry,
Beth Israel Medical Center, New York, New York

PLENUM PRESS • NEW YORK AND LONDON

ISBN 0-306-42522-X

To our wives,

M. BERNICE ROSNER, M.S.N.

and

KAREN BLANK, M.D.

whose interests in geriatric psychiatry
have inspired our own.

Contributors

Michael Bell • University of Medicine and Dentistry of New Jersey, Newark, New Jersey

Jon M. Bevilacqua • Deputy Attorney General's Office for Medicaid Fraud Control, New York State Office Building, Room 4A10, Veterans Memorial Highway, Hauppauge, New York

Karen Blank • Department of Psychiatry, Beth Israel Medical Center, New York, New York; Mount Sinai School of Medicine, New York, New York

Harvey Bluestone • Department of Psychiatry, Bronx-Lebanon Hospital Center, Bronx, New York; Albert Einstein College of Medicine, Bronx, New York

Hilda Brewer-Gizzarelli • Forensic Psychiatry Clinic, Bronx-Lebanon Hospital Center, Bronx, New York

Fred B. Charatan • Department of Psychiatry, State University of New York at Stony Brook, Stony Brook, New York

J. Richard Ciccone • Department of Psychiatry, University of Rochester School of Medicine and Dentistry, Rochester, New York

Colleen D. Clements • Department of Psychiatry, University of Rochester School of Medicine and Dentistry, Rochester, New York

Richard Ellison • Department of Psychiatry and Behavioral Sciences, School of Medicine, State University of New York at Stony Brook, Stony Brook, New York; Department of Psychiatry, Veterans Administration Hospital, Northport, New York

Michael L. Freedman • Division of Geriatrics, New York University and Bellevue Medical Centers; Department of Medicine, New York University School of Medicine, New York, New York

Naomi Goldstein • Department of Psychiatry, New York University School of Medicine, New York, New York

Robert Lloyd Goldstein • Department of Psychiatry, The College of Physicians and Surgeons, Columbia University, New York, New York; Member of the New York State Bar

Susan Hamel • Law Offices of Edwin Marger, 6666 Powers Ferry Road, Atlanta, Georgia

Jay E. Kantor • Department of Philosophy, College of Liberal Arts, New York University, New York, New York; Department of Psychiatry, School of Medicine, New York University, New York, New York; Forensic Psychiatry Service, Bellevue Hospital, New York, New York; Franklin Delano Roosevelt Hospital, Veterans Administration, Montrose, New York

Michael Kaufman • Crotona Park Community Mental Health Center, Department of Psychiatry, Bronx-Lebanon Hospital Center, Bronx, New York

Stanley R. Kern • Department of Psychiatry and Mental Health Science, New Jersey Medical School, University of Medicine and Dentistry of New Jersey, Newark, New Jersey; Rutgers Law School, Newark, New Jersey

Jordan I. Kosberg • Department of Gerontology, College of Social and Behavioral Sciences, University of South Florida, Tampa, Florida

Edwin Marger • Attorney in Private Practice, 6666 Powers Ferry Road, Atlanta, Georgia; Member: Georgia, Florida, Washington, DC Bars; Member: Board of Governors, Florida Bar; Trustee: Forensic Science Foundation

Helen L. Morrison • The Evaluation Center, 919 North Michigan Avenue, Suite 3100, Chicago, Illinois; Pediatric Psychiatry Center, Kingwood Hospital, Michigan City, Indiana

John J. Regan • School of Law, Hofstra University, Hempstead, New York

Richard Rosner • Forensic Psychiatry Clinic for the New York Criminal and Supreme Courts (First Judicial Department) of the Department of Mental Health, Mental Retardation and Alcoholism Services of the City of New York; Department of Psychiatry, School of Medicine, New York University, New York, New York

Harold I. Schwartz • Department of Psychiatry, Beth Israel Medical Center; Mount Sinai School of Medicine, New York, New York

Barbara Stanley • Department of Psychology, City University of New York, John Jay College of Criminal Justice, New York, New York; Department of Psychiatry, Columbia University College of Physicians and Surgeons, New York, New York

Michael Stanley • Departments of Psychiatry and Pharmacology, College of Physicians and Surgeons of Columbia University; New York State Psychiatric Institute, New York, New York

Larry H. Strasburger • McLean Hospital, Belmont, Massachusetts; Department of Psychiatry, Beth Israel Hospital, Harvard Medical School, Boston, Massachusetts

Sheldon Travin • Department of Psychiatry, Bronx-Lebanon Hospital Center, Bronx, New York; Albert Einstein College of Medicine, Bronx, New York

Robert Weinstock • Department of Psychiatry, University of California at Los Angeles, Los Angeles, California; Department of Medicine, University of California at Irvine, Irvine, California; Cedars-Sinai Medical Center, Los Angeles, California

Sue S. Welpton • Brookline Council on Aging, 61 Park Street, Brookline, Massachusetts

Foreword

As President of the American Academy of Psychiatry and the Law (AAPL), it is my pleasure to provide this Foreword for the third volume of *Critical Issues in American Psychiatry and the Law*. Through Dr. Richard Rosner's creativity and effort, the first volume of the series was published in 1982. It represented the first major publishing activity of the AAPL other than the *Bulletin* and the *Newsletter*. Dr. Rosner carefully nurtured the project and the second volume, once again addressing a broad range of forensic issues, was published in 1985.

The appearance of the third volume brings two major changes to the series. Dr. Rosner shares editorship of this volume with Harold I. Schwartz, and this is the first of the series to be devoted entirely to one area in forensic psychiatry, the elderly. This is a priority subject. As the baby boom ages and technology for extending life increases, the profile of our population is maturing. The aging of the nation brings with it problems of chronic care, the dementias, the fear of prolonging dying rather than extending meaningful life, the social question of increasing health care costs, and the medical and human question of legal and ethical protection of the elderly patient. This volume could not be more timely.

As Bernice Neugarten has pointed out, we must avoid stereotyping the elderly.[1] Many individuals bring to older age a rich repertoire of interests, abilities, and insights. Among the elderly are the young-old. This group, vigorous and often financially secure, is politically active and as consumers, an important economic force. The old-old as a group face a different existence. John Romano has reminded us that some of the elderly face the four horsemen of aging: bereavement, retirement, isolation, and poverty.[2] Dr. Romano has asked: "What can be done constructively to prevent distress, to identify disability early, to treat it intelligently, and to understand chronic illness and dependency?" Together Rosner and Schwartz have collected 21 contributions in this first text on the elderly and forensic psychiatry. The aging of American society has increased the amount of attention and awareness being given to the overall health needs of the elderly. DRGs, DNR orders, competency, living wills, and the illnesses of the elderly represent important health care problems of the next decade. It is only natural that forensic psychiatry would play an important role in the special concerns of this population.

This volume provides much food for thought for anyone interested or involved in the care of the elderly. The first chapter, written by Dr. Fred Charatan, is an overview of the area of legal issues in geriatric psychiatry. The second section addresses competency and informed consent. The chapter by Drs. Stanley and Stanley provides data on their systematic studies on the capacity to consent to research in three elderly patient groups: those with the cognitive impairments of dementia, those suffering from depression, and those in a normal control group. Philosopher Colleen Clements and I discuss recent legal and ethical trends in consent to treatment, point to problems in the ethical treatment of competent and incompetent elderly patients, and suggest possible solutions. Dr. Robert Weinstock develops a procedure for evaluating competence in geriatric patients and cites case examples from a psychiatric consultation service, which give important clinical confirmation to his analysis. The closely related third section, epigrammatically entitled "End of Life," looks at choosing to die, DNR orders, withholding life support, euthanasia, and determining when a patient can be allowed to die. Dr. Rosner proposes criteria for choosing to die, Dr. Schwartz surveys the evolution of DNR guidelines, Professor Regan looks at life support technology, Attorney Marger looks honestly at euthanasia, and Attorney Bevilacqua tells the New York legal experience.

In the section on diagnostic and clinical factors in the care of the elderly, Dr. Freedman describes the physiology of aging and identifies the major chronic problems of advancing age. Dr. Morrison's chapter on neuropsychiatric assessment of dementia gives a caution in the interpretation of test data obtained from many neuropsychiatric test batteries.

The largest section is devoted to administrative and social policy. Its seven contributions range from Dr. Bluestone's thoughtful and practical description of the operation of a geriatric psychiatric service to Strasburger and Welpton's exploration of divorce among the elderly. The final section focuses on ethnic factors in geriatric and forensic psychiatry. Drs. Bell and Ellison explore questions in psychiatric-legal evaluation of the black elderly, and Drs. Travin and Brewer-Gizzarelli in evaluation of the Hispanic elderly.

The richness of this volume's scope will provide an excellent resource for the forensic psychiatrist, the general psychiatrist, physicians, lawyers, and ethicists working with the elderly and their health care needs. One of AAPL's primary goals is education. AAPL's educational efforts have included the annual meeting, a yearly forensic psychiatry course, the publication of the *Bulletin* and the *Newsletter,* the creation and support of Rappeport fellowships, and the encouragement of research. The ongoing series on *Critical Issues in American Psychiatry and the Law* adds luster to AAPL's efforts to fulfill its mission: to continue the transformation of forensic psychiatry from its alienist roots into its modern day identity as a subspecialty of psychiatry. Drs. Rosner and Schwartz's continuing efforts to encourage scholarship in forensic psychiatry is deserving of special praise. This third volume underlines the important role forensic psychiatry plays in the critical medical and social issues of the day. On behalf of the officers and members of AAPL, I congratulate Drs. Rosner and Schwartz and their contributors for making a meaningful addition to the literature on psychiatry and the law, and to the pressing problem of defining proper care for the elderly patient.

J. Richard Ciccone

REFERENCES

1. Neugarten GL: Time, age, and the life cycle. *American Journal of Psychiatry* 1979; 136: 887–894.
2. Romano J: The modern plague—chronic illness. *Newsletter of the Genesee Valley Psychiatric Association,* 1982; 2:3–5.

Introduction

This is the first volume in this series to be devoted to a single theme, geriatric psychiatry and the law, and that fact is consistent with our belief in the importance of the subject matter. As with the previous volumes, it is our hope that the present text will stimulate constructive thought rather than provide definitive answers.

The core chapters of this volume derive from two special programs presented in January 1986 at the annual meeting of the Tri-State Chapter of The American Academy of Psychiatry and the Law and in February 1986 at the convention of The American Academy of Forensic Sciences. Additional essays were invited to make the presentation more comprehensive for the novice in geriatric forensic psychiatry.

The basic approach to all psychiatric-legal problems in this series has been to analyze them into four component questions:

1. What is the psychiatric-legal issue?
2. What are the legal criteria that determine that issue?
3. What are the relevant clinical data?
4. What is the reasoning process that underlies the psychiatric-legal opinion regarding the issue?

The chapters of this book may be seen as addressing these four components, albeit in some instances a single chapter will encompass one, two, three, or all four. It may help the reader to keep this four-step approach in mind. We have grouped the chapters topically, but they may be understood as well or better if considered in the light of the four questions. This is analogous to medical textbooks being conceptually organized, with equal validity, around types of disease processes or around the organ systems that are affected by those processes. In a similar manner, psychiatric-legal problems may be arranged topically or in accordance with the four component questions; each perspective has its own value.

To exemplify this four stage approach to psychiatric-legal problems, let us imagine that the forensic psychiatrist is called by a colleague
to examine a hospitalized geriatric patient. The first question must be
"What is the psychiatric-legal issue to be evaluated?" It may be an issue
of testamentary capacity, it may be whether the patient is competent to
consent to a proposed surgical procedure, it may be whether a conservator or a guardian is needed to protect the patient's assets or person, it
may be whether the patient is competent to participate in a decision to
refuse life-sustaining treatment, or any of a variety of alternative psychiatric-legal problems.

Depending on what the issue is, the consulting forensic psychiatrist
will have to direct his attention to varying legal criteria. The criteria for
competence to make a will are different from the criteria for competence to consent to surgery. The criteria for being eligible for a guardianship are not the same as the criteria for competence to refuse life-
sustaining medical care. For each issue, a different set of legal criteria
must be considered. Further, as there are 50 states, plus the federal
jurisdiction, there may be 51 sets of legal criteria for each of the many
psychiatric-legal issues, the applicable set depending on the geographic
area in which the patient happens to be located at the given moment of
evaluation.

The relevant clinical data will differ for each different psychiatric-
legal issue. The questions that have to be asked will depend on the legal
criteria that are set forth in the applicable statutes, case law, and administrative codes. There is no such entity as a general psychiatric-legal
evaluation; there is only a specific psychiatric-legal evaluation directed to
a specific set of legal criteria applicable to a specific issue. The fact that a
patient is well or not is often less important than whether the patient is
functionally impaired in a manner specified in the legal criteria that
determine the specific issue; the patient's diagnosis is often irrelevant to
the psychiatric-legal evaluation. Collection of material relevant for
clinical therapeutic purposes may be irrelevant for forensic purposes.

It is extremely important that the forensic psychiatric consultant be
able to explain the reasoning processes that underlie the psychiatric-
legal opinion that he renders. Often that requires that consultant to be
clearer in articulating the steps in his thinking than is ordinarily required for clinical therapeutic psychiatric practice. It is important to
specify exactly what the patient said, how it was said, when it was said, in
what context it was said, rather than presenting a paraphrase or summary of the patient's communications. The raw data upon which the opinion is based must be available for potential judicial scrutiny. In some
instances, it may be valuable to have an exact stenographic record, or an
audiotape recording, or a videotape copy of the interview. The consul-

tant must be prepared to explain why he asked what he did ask, as opposed to asking something else that was not asked. It is important to be ready to explain how one interpreted the phenomena of the interview, and why that interpretation is to be preferred over a possible alternative set of interpretations. Psychiatry is often misperceived by attorneys and laymen alike as "imprecise" and "unscientific" because of failure to clarify the hard data upon which psychiatric-legal opinions are based: the sights, sounds, smells, tactile impressions, and (rarely) tastes that are the experiential foundation of the interview. The forensic psychiatrist's opinion is only as valuable as the data upon which the opinion is based and the quality by means of which he reasoned from the data, in accordance with the specific legal criteria that determine the issue under consideration.

An area of some controversy is whether or not the best physician to call for a geriatric forensic evaluation is a specialist in geriatric psychiatry or a specialist in forensic psychiatry. Ideally, it would be preferable to have a consultant with fine credentials in both psychiatric subspecialties, but such doubly qualified doctors are extremely rare. One of the goals of this volume is to attempt an interim solution to this problem by presenting to geriatric psychiatrists much of the data that they will need to function effectively around forensic problems, as well as to provide to forensic psychiatrists some special clinical considerations relevant to working with geriatric patients.

We have aimed at making each chapter independent, so that the reader can open the book at any point and understand the chapter as a self-contained unit without having to have read the chapters that precede the one being read. As a necessary consequence, there is some overlap in the contents of the individual chapters. We believe that such repetition illustrates the importance of the material that is so emphasized and that it serves a pedagogical function of reinforcing the impress of the data on the reader's memory.

What will not be found here is an introduction to the fundamentals of geriatric psychiatry itself. Our focus is on the interface of geriatric psychiatry and the law, not on the clinical phenomena of mental illness in the elderly.

Because there is a considerable area in which the problems in geriatric forensic psychiatry are the same as in general forensic psychiatry, interested readers may wish to consult volumes one and two in this series to supplement the materials in the present book. Here we have attempted to address some of the psychiatric-legal problems that are unique to the elderly. In the previous volumes we have addressed many of the concerns that are relevant to populations of all ages. The reader will be able to understand the contents of this book without having read the others in this series, but the reader's comprehension of the field will be substantially augmented by study of the companion texts in *Critical Issues in American Psychiatry and the Law.*

RICHARD ROSNER
HAROLD I. SCHWARTZ

Contents

SECTION IV DIAGNOSTIC AND CLINICAL FACTORS

SECTION V ADMINISTRATIVE AND SOCIAL POLICY ISSUES

SECTION VI   ETHNIC CONSIDERATIONS: OVERCOMING POTENTIAL BIAS

I

Fundamentals

1

Geriatric Psychiatry

Legal and Forensic Issues

FRED B. CHARATAN

In any discussion of the legal and forensic aspects of geriatric psychiatry, we should remind ourselves that as members of the legal and medical professions, we do not always operate in a very favorable climate of public opinion. For example, Anton Chekhov, himself a doctor, wrote in "Ivanov"—"Doctors are just the same as lawyers; the only difference is that lawyers merely rob you, whereas doctors rob you and kill you, too." That was in 19th-century Russia, however. Here, in 20th-century America, we practice our professions within a constitutional framework of laws that protect our clients and patients. In no group is this more important than in the aged, for, as Sophocles wisely said, "No man loves life like him that's growing old."

There are some significant features of the relationship of the psychiatrist to the geriatric patient. Usually the psychiatrist is younger, and thus is placed in the symbolic role of a son or daughter, or even grandson or granddaughter. The psychiatrist needs to be sensitive to the possibility of unconscious ageism, a concept first described by Butler.[1] Working with the aged stimulates fears about one's own aging; recalls old conflicts with one's parents; frustrates the doctor seeking large and dramatic therapeutic successes; and may cause anxiety in the doctor whose basic philosophy is that the aged will die soon anyway, and worst of all, during treatment. Sometimes caretakers of the aged defend them-

---

FRED B. CHARATAN • Department of Psychiatry, State University of New York at Stony Brook, Stony Brook, NY 11790.

selves against this anxiety by distancing themselves, by providing mechanistic, impersonal, detached care. They treat the aged with not-so-benign neglect, thereby preserving their own emotional comfort.

COMMITMENT ISSUES

Commitment of a geriatric patient unable or unwilling to enter hospital voluntarily is a recurring problem for the psychiatrist. New York State law provides for a broad range of persons able to make application for admission of a mentally ill person to hospital. They range from family members to public officials. In emergency cases the police have the power (M.H.L. 9.41) to remove a person from the street or dwelling and convey him to an observation ward. This police power—*parens patriae*—is being invoked more frequently nowadays in New York City, where the homeless population includes many mentally ill of all ages.

The experienced psychiatrist knows that occasionally a family member will try to "put away" one of their elderly members for alleged mental illness. This can readily become a self-fulfilling prophecy, for the shock of wrongful incarceration may precipitate mental decline in an old person, and even death. The petition may state that the elderly person has shown progressive mental decline over a period of time. The psychiatrist will need always to maintain a critical approach to the history as given by the relative. In New York, the law states specifically that if the examining physician (psychiatrist) knows that the patient has been under prior treatment, he should try to consult with the professional who treated the patient previously. Unfortunately, this useful provision is often honored more in the breach than in the observance. It is also important to bear in mind that the commitment law states that where the geriatric patient is examined and found to be mentally ill, the fact that he suffers from alcohol or substance abuse does not preclude commitment.

Commitment of a geriatric patient requires a "statement of the facts upon which the allegation of mental illness and need for care and treatment are based." In light of the United States Supreme Court's insistence that involuntary hospitalization should be permitted only when clear and convincing evidence is present, one should establish the facts of mental illness solidly and record them clearly. Jones *et al.*[2] wrote,

> In the course of consultation-liaison with a long-term care unit in a large state
> hospital, we were struck by the frequent commitments of aged persons from
> rest homes and, in particular, by the seeming tenuous fulfillment of legal
> commitment criteria in these cases.

In geriatric patients, symptoms of mental illness fall roughly into two categories—the negative, such as amnesia, cognitive impairment, apathy, depression, and so forth, and the positive, such as delusions, hallucinations, and disordered or destructive behavior. Dangerousness to self or others is less often due to actual or threatened assaultiveness and much more commonly due to wandering or self-neglect. The mentally ill geriatric patient who wanders, hoards, is dirty, unwashed, and unfed is an example of what has sometimes been termed the Diogenes syndrome. Clark *et al.*[3] noted:

> A study of elderly patients (14 men, 16 women) who were admitted to hospital with acute illness and extreme self-neglect, revealed common features which might be called Diogenes Syndrome. All had dirty, untidy homes and a filthy personal appearance about which they showed no shame. Hoarding of rubbish (syllogomania) was sometimes seen. . . . Personality characteristics showed them to be aloof, suspicious, emotionally labile, aggressive, group-dependent, and reality-distorting individuals. It is suggested that this syndrome may be a reaction in late life to stress in a certain type of personality.

For commitment purposes, therefore, dangerousness takes on a different connotation in old age. The constitutional right of refusal of shelter and treatment, together with the morbid desire to freeze or starve to death on the streets of New York, have led to the witticism, "they died with their rights on." Clark and his collaborators stated that one third of their patients had persistently refused offers of help. For psychiatrists, the guiding principle in committing these geriatric patients must always be "the most beneficial alternative" as an essential element of "the least restrictive choice."

Some jurisdictions have developed laws providing for protective services in the case of impaired or disabled adults. The analogy with child protective services is obvious. For example, the Adult Protective Services of the State of Maryland can be used to remedy the plight of the self-neglected old person, as

> it is the declared policy of the State of Maryland that adults who lack the physical or mental capacity to care for their basic daily living needs shall have access and be provided with needed professional services sufficient to protect their health, safety and welfare.

Again, this is the *parens patriae* principle—the state has an obligation to care for the most helpless of its citizens. Yet one or two states, for example, Missouri and Illinois, have quite illogically tried to exclude from the commitment process dementia, or mental decline without psychosis. The hard fact remains that such an elderly person still needs a sheltered environment into which he is placed without his consent, merely to ensure his survival.

Under New York State law, involuntary admission on the certificate of a director of community services or his designee provides broad latitude for admission through the phrase "other conduct demonstrating that he is dangerous to himself." This could be interpreted to include any kind of self-neglect, so long as it is well documented. The same remarks apply to emergency admissions where "likelihood to result in serious harm" is taken to mean "other conduct demonstrating that he is dangerous to himself."

CONSENT TO TREATMENT

Nowadays every psychiatrist should be guided by the doctrine of informed consent. Thus, where the psychiatrist plans to give psychotropic medication or electroshock, he should explain the treatment carefully to the patient, and obtain the patient's signed consent for the recommended treatment. He would do well to document carefully in the chart the fact that he did obtain informed consent, either from the patient, where the patient was competent to give consent, or failing that, consent from the nearest relative, usually spouse or child, in the case of geriatric patients.

Most states have adopted the emergency exception. In any emergency, the patient's consent is implied as a matter of law if a spouse or close relative is not immediately available. In those situations where delay would prevent treatment of the patient and further harm or injure the patient, the consent is again implied. Therefore, the formal consent process can be waived. If the geriatric patient is incompetent, who can consent for him? Case law indicates that the family can speak in behalf of the patient, usually spouse or child.

Recently, a case in Massachusetts, *Rogers v Commissioner of Mental Health*[4] addressed the issue of treatment of the geriatric patient with antipsychotic medications. This restrictive ruling required that only a judge, having found a patient incompetent, may make decisions on the patient's behalf about antipsychotic treatment. But here again, emergencies are the exception. The judge is supposed to use what is called *substituted judgment,* that is, a determination of what the incompetent patient would want, were he competent. Thus, the psychiatrist appears, in Massachusetts at least, to be required to bring the geriatric patient's case to court if he finds the patient clearly incompetent. The judge would then rule via substituted judgment what the patient would want, if he were competent. However, the psychiatrist could still prescribe pro re nata (p.r.n.) medications in emergencies, for example, where the patient was agitated or restless, and behaving in ways that might cause harm to himself. Where there might be some doubt about the degree of dangerousness for which these p.r.n. medications are to be written, a consultation would provide an added safeguard.

Where a geriatric patient refuses treatment, the psychiatrist would need to consider depression as a factor in his decision. The patient might, for example, harbor suicidal wishes; might consider himself unworthy of any offered help; might believe the medications were designed to poison him, etc. The psychiatrist would also want to evaluate the patient's cognitive function, and examine closely the patient's rationale for refusing treatment.

The question of whether to administer electroconvulsive therapy (ECT) or not poses some additional complexities. ECT is often dramatically effective in depression in geriatric patients. A patient may be depressed and yet be competent enough to be able to refuse treatment by ECT after the psychiatrist sought to obtain informed consent from him. The psychiatrist then would have to fall back on treating the patient with antidepressants, which might be less effective. Amnesia after a course of ECT is another problem. Apart from being a troublesome residual symptom in some patients, post-ECT amnesia could interfere significantly with a patient's testimonial capacity in any future legal action.

RESEARCH

There is an increasing interest in research in geriatric psychiatry. Human Subjects Review Committees are the statutory bodies that oversee research in geriatric institutions. The issue of consent is important in light of press reports of exploitation of geriatric patients for drug trials and invasive procedures. The Task Force on Forensic Issues in Geriatric Psychiatry of the A.P.A.'s Council on Aging, chaired by Dr. F. M. Baker,[5] wrote (p. 20):

> Applebaum and Roth[6] have conceptualized consent to research in four hierarchical steps: (1) evidencing a choice; (2) functional understanding of issues; (3) rational manipulation of information; and (4) appreciation of the nature of the situation.

The task force pointed out that the case of Alzheimer's disease is particularly relevant, inasmuch as patients may be aphasic and apraxic and unable to give informed consent. The task force warned that in such cases, the common practice of securing permission of relatives may not hold up in court (Kolata, 1982).[7]

Stanley[8] has made some helpful suggestions in making informed consent more suited to the Alzheimer's patient:

1. Make the consent material more readable.
2. Tailor the consent information to the needs of the patient as suggested in the preamble of the federal regulations for research with humans.

3. Allow patients to review the consent material for a lengthier time prior to assessing competence.
4. Use teaching and review methods coupled with testing.
5. Develop a rapport that promotes a sense of confidence in patients and enables them to ask as many questions as needed.
6. Involve the patient's family members in the consent process. Relatives may use language more familiar to the individual and make the material more understandable.

To summarize, then, in the case of consent issues: if the psychiatrist practices with the degree of care, judgment, and skill that his peers usually exercise under similar circumstances, obtains informed consent that he scrupulously records in the patient's chart; and in the case of incompetent patients, keeps the family informed and obtains informed consent from a close family member, there appears to be little risk of liability. Finally, when consent is not obtained, the emergency exception and the reasonable person standard should shield the psychiatrist from any liability.

WITNESSES AND WILLS: TESTIMONIAL AND TESTAMENTARY CAPACITY

When an old person is called on to testify, questions about his ability to do so may arise. In the absence of any psychosis, such questions usually turn on the ability to recall significant data. The psychiatrist may then be required to give an expert opinion on memory function. Too, the long delay in cases coming to trial may interfere with an elderly witness's ability to recollect events that took place perhaps five to six years ago.

Testamentary capacity is the ability of a geriatric patient to make a valid will. To do this, he must know that he is making a will; know the extent and varieties of his property to be included in the will; and be aware of all those persons related to him who would be regarded as natural beneficiaries of his will. Whenever the competence of the older person making a will might be in doubt, the lawyer setting up the will may ask for a psychiatric evaluation for competency. In such cases it is best to make a very comprehensive examination, including questions and observations directed to the specific issues in the will, for example, details of the property to be bequeathed, the persons who will be the recipients, and so forth. Where the older person is suffering from some impairment due to a psychiatric disorder, he could conceivably come under undue influence of another person, and might be manipulated into making a biased will that could later be challenged. In situations where a presumably valid will is challenged in court, evidence of any impairment found at psychiatric evaluation is obviously of great importance. The psychiatrist who evaluates an old person for testamentary capacity will need to keep the possibility of undue influence in mind.

GUARDIANSHIPS AND CONSERVATORSHIPS

In New York the law states that a committee of an incompetent patient is actually the alter ego of the patient. *Guardianship* refers to control over the patient, whereas *conservatorship* means control of his property. Thus, the committee of the person or guardian makes decisions about the patient's disposition and gives consent for treatment. If the patient recovers and regains his competency, then the court can discharge the committee of the person or guardian.

The court appoints a conservator or committee of the property after someone, usually a relative, has prepared and submitted a petition. The petition must be supported by psychiatric evidence of incompetence of the patient to manage his own affairs. I well recall one case where I examined an elderly man suffering from Alzheimer's disease who was living alone in his own large rambling home on Long Island. I found he had both gross amnesia and a stock and bond portfolio well in excess of one million dollars. Unfortunately, dividend and bond interest coupons had steadily accumulated in his home, for the patient neglected to cash the checks or clip the coupons. The court appointed his daughter, a lawyer, as his conservator, and she had to spend the next several months sorting out his tangled financial affairs. I might add that the IRS was interested in obtaining payment of back taxes from the patient, but waived penalties when presented with clear evidence of incompetence supported by the judge's order for a conservatorship.

A conservatorship protects the estate of the incompetent geriatric patient, and prevents the exploitation of the patient by someone who is unscrupulous and has recognized the inability of the old person to manage his financial affairs. From time to time, one reads in the press about some prominent wealthy elder who has been snared by a predatory younger woman and systematically fleeced. A committee of the property or conservatorship would have prevented such a situation and have protected the natural heirs. Under the law, any committee must submit an annual report to the court that appointed it, and is subject to overall control and direction by the court. Nolan[9] has written a useful article detailing the specifics of a functional assessment of impairments in various areas, such as financial management, activities of daily living, physical functioning, sensory functioning, and social resources and supports.

To quote again from the work of the Task Force on Forensic Issues in Geriatric Psychiatry[5]:

> It is important for the psychiatrist to know which standard is used by his state
> statute of guardianship. The Uniform Probate Code emphasizes lack of cog-
> nitive and communicative ability as a necessary criterion for incompetence.
> Some states require objective evidence of observable behavior as evidence of
> incapacity; others require a diagnosis as to the cause of the socially disap-
> proved behavior. The judge can then decide whether the behaviors con-
> stitute improper care of self and property; whether the improper care is due
> to incapacity; and whether the incapacity is caused by the diagnosis, what is
> termed the causal link.

Therefore, the psychiatrist should document carefully all the evidence
he can collect in favor of functional incapacity from the history, clinical
examination, ancillary tests, and the use of behavioral check lists and
rating scales. The court can discharge a conservator if the patient regains his competence later on.

DEATH AND DYING: THE LIVING WILL

When I was a medical student in London, one of my professors
quoted a couplet that stuck in my mind,

> Thou shalt not kill, nor need'st yet strive
> Officiously to keep alive.

In New York that couplet has been condensed into the code DNR—do
not resuscitate, less poetic, but far more succinct. Lawmakers are now
struggling with the dilemma of how to cope with "endgame," especially
where an aged person becomes incompetent and is no longer able to
express his own wishes regarding life support and heroic or extraordinary methods of treatment. Continuation of extraordinary treatment
measures might, in terminal cases, merely prolong the act of dying;
cessation, however, might expose the physician to malpractice suits.

The living will is a written document prepared by someone while
mentally competent, which specifies the circumstances, were that person
no longer competent to express his or her wishes, that would allow for
cessation of extraordinary or heroic life-saving measures, and would
allow for death to occur in accordance with the natural progression of
the disease. Thirty-five states, but not New York, have enacted living will
legislation. Although the competent patient has the constitutional right
to refuse treatment, the incompetent patient has no such right in New
York. That is why enactment of the living will is needed in New York
State, so that overtreatment of the dying incompetent patient is brought
to an end. The numbers of such patients, especially in nursing homes, is
growing at a rapid pace. Dr. Daniel Cowell of the NIA wrote,[10]

> Of the total population of 23.5 million persons 65 years of age or older in the
> U.S. in 1977, 4.8% resided in nursing homes. Of these, a minimum figure of

> 32% represents those patients who were considered to have some type of
> senile dementia (360,000 patients).

Many of these patients will not be competent to express their wishes regarding extraordinary treatment when their life is drawing to its close. In any living will legislation, statutory immunity for physicians is usually specified, thus freeing the physician from the threat of a malpractice suit for withholding treatment. It seems possible that there are children of some dying geriatric patients who were on bad terms for years, but now resolve to undo their earlier hostility and neglect by directing the attending physicians to engage in furor therapeuticus. If he fails to comply, the family may turn to the malpractice lawyers. Then, too, there are those children of a competent elderly patient in a nursing home who would like the death watch to be as short as possible, and might seek passive euthanasia to that end.

It is important to note that a living will can easily be revoked at any time, by spoken word, in writing, or by a physical act of cancellation. It has also been suggested by the Society for the Right to Die that every living will designate a proxy. In the event of sudden incompetence, for example, a stroke or head injury resulting in coma, the patient then has a surrogate person who is fully conversant with the patient's wishes regarding terminal care, and can relate the patient's requests to the physicians and plan appropriate treatment with them.

The highest court in New York State has ruled that there must be "clear and convincing evidence" that an incompetent hopelessly ill patient would refuse artificial life support before it may be withdrawn or withheld. A living will, recognized by statute, and with immunity for the physicians in attendance, would be the surest way of providing such evidence. The living will would prevent the overtreatment of the dying incompetent patient that, far from extending life, merely prolongs the act of dying.

Wanzer[11] and a number of distinguished colleagues discussed in the *New England Journal of Medicine* the physician's responsibility toward hopelessly ill patients. They stated,

> We believe that a hopelessly ill patient's refusal of life-sustaining treatment is
> not in itself a reason to question the patient's competency, no matter what the
> personal values of the physician or family may be.

They concluded, not surprisingly, that technology competes with compassion, legal precedent lags, and controversy is inevitable. Brown *et al.*[12] studied 44 terminally ill patients and in an article titled appropriately "Is it normal for terminally ill patients to desire death?" found that the majority (34) had never wished death to come early. The 10 patients who had desired death were found to be suffering from clinical depression. Thus, these authors concluded that if their findings can be generalized, patients with terminal illness are no more likely than the general population to wish for premature death.

Finally, in a commentary "The dilemma of 'do not resuscitate' orders," Newman *et al.*[13] point out that where a patient is *non compos mentis* and unable to make an informed decision regarding his medical care, the family is traditionally called on to participate in such decisions, as they are in the best position to express the wishes of an incompetent or unconscious patient. Newman emphasizes that the physician should not write a DNR order if he feels it is inappropriate despite the family's wishes, and recommends consultation with the hospital ethics committee and legal counsel. Newman feels that where a physician cannot accept a family's decision to refuse resuscitation he can refer the patient to a colleague who will be able to accept the patient on those terms. He concludes, "the response to the conflict must be that the patient's interests be given priority over all other considerations, including the advice of counsel."

Our society continues to seek answers from the legal and psychiatric professions in how the last of life should be conducted for competent and incompetent elderly persons. When the laws of our society are in harmony with the ancient laws of Hippocrates and Maimonides, which have guided us so well for so long, we, no less than our geriatric patients, will be the lasting beneficiaries.

REFERENCES

1. Butler RN: *Why Survive? Being Old in America.* New York, Harper & Row, 1975, p 11.
2. Jones LR, Parlour RR, Badger LW: The inappropriate commitment of the aged. *Bull Am Acad Psychiatry Law* 1982; 10:29–38.
3. Clark ANG, Manikar GD, Gray I: Diogenes syndrome: A clinical study of gross neglect in old age. *Lancet* 1975; 1:366–368.
4. Gutheil TG: Medicolegal aspects of geriatric care-Newsletter. *Am Assoc Geriat Psychol* 1985; 7:4.
5. Baker FM, Perr IN, Yesavage JA: *Task Force on Forensic Issues in Geriatric Psychiatry.* Washington, Council on Aging: American Psychiatric Association, 1985.
6. Applebaum PS, Roth LH: Clinical issues in the assessment of competency. *Amer J Psychiatry* 1981; 138:1462–1467.
7. Kolata G: Alzheimer's research poses dilemma. *Science* 1982; 215:47–48.
8. Stanley B: Senile dementia and informed consent. *Behav Sci Law* 1983; 1(4):57–71.
9. Nolan BS: Functional evaluation of the elderly in guardianship proceedings. *Law, Med Health Care*, October 1984, pp. 210–217.
10. Cowell DD: Senile dementia of the Alzheimer's type. *J Am Geriatr. Soc* 1983; 31:61.
11. Wanzer SH, Adelstein SJ, Cranford RE, *et al.*: The physician's responsibility toward hopelessly ill patients. *N Engl J Med* 1984; 310:955–959.

12. Brown JH, Henteleff P, Barakat S, *et al.*: Is it normal for terminally ill patients to desire death? *Am J Psychiatry* 1986; 143:208–211.
13. Newman RG, Meyer K, Mernick M: The dilemma of "Do Not Resuscitate" orders. *NY State J Med* 1986; 86:1.

II

Competency and Informed Consent

2

Competency and Informed Consent in Geriatric Psychiatry

BARBARA STANLEY AND MICHAEL STANLEY

The elderly are a rapidly growing segment of our population—the average life expectancy has increased from 47 years in 1900 to 74 years in 1980.[1] At present, there are more than 25.5 million individuals over the age of 65 in the United States. As a result of medical advances, acute illnesses are no longer the major impediment to health.[2] Instead, illnesses of a chronic nature, such as diabetes, cancer, and hypertension account for much of the medical care given in this country. The elderly, in particular, are prone to developing chronic diseases.

Furthermore, there is great need for research on diseases that afflict the elderly, such as Alzheimer's disease. Little is known about these illnesses and, as our population ages, it will become increasingly important to study and find more effective treatments for these illnesses. As a consequence of greater likelihood of chronic illnesses and also as a result of the increased demand for research, the elderly are placed in the position of having to make frequent decisions about their health care.

Portions of this chapter were previously published in *JAMA*, 1984; 252:1302–1306.

BARBARA STANLEY • Department of Psychology, City University of New York, John Jay College of Criminal Justice, New York, NY 10003; Department of Psychiatry, Columbia University College of Physicians and Surgeons, New York, NY 10032.   **MICHAEL STANLEY** • Departments of Psychiatry and Pharmacology, College of Physicians and Surgeons of Columbia University; New York State Psychiatric Institute, New York, NY 10032.

They are also more likely to be called on to participate in research. At the same time, physicians want to be assured that decisions about treatment and research are based on an adequate understanding of the information relevant to treatment. This desire arises from an increasing awareness of patients' wishes to be well-informed, the physicians' perceived duty to inform patients, and their desire to avoid potential legal problems arising from inadequate information giving. Consequently, it is very important that patients are supplied with relevant information, such as the risks and benefits of a particular treatment; the nature of treatment requirements and regimens; and the alternatives. It is, therefore, vital to physicians from ethical and legal viewpoints to determine whether their elderly patients can make competent decisions (i.e., give informed consent) concerning medical care. If some of these patients are found to be incapable, acceptable procedures for substitute decision making should be considered.

The competency of the aged, as a group, has been the subject of some controversy and opinions regarding the capabilities of the older population diverge widely. On one hand, they are viewed as highly heterogeneous and, therefore, impossible to classify as a distinct group.[3] A contrasting view sees them as quite vulnerable, frail, and dependent.[4,5] Proponents of the latter view often recommend special protective measures, such as legal guardianship for the elderly who are faced with medical treatment decisions.[6]

Because opinions regarding the capabilities of the elderly are so diverse, empirical research on the issue is particularly important. Very few studies[7,8] with the normal elderly and none with the medically ill or psychiatric older patient have investigated their capacity to give informed consent. In studying the normal elderly, the aged were found to have some impairment in their ability to remember consent information.[7] Particular impairment was found in those with lower verbal skills.[8] Poor recall, however, may have little bearing on ability to give informed consent. Competency to consent has been defined most often in three ways: capacity to comprehend relevant information; ability to weigh the benefits and the risks of the proposed procedure; and the capability of reaching a reasonable decision. Recall of information is not usually considered to be a valid criterion because lack of recall may be more indicative of the normal forgetting process rather than of incompetency. Thus these studies shed little light on the elderly's capacity to give informed consent.

Although it does not directly investigate informed consent, research on cognitive functioning, memory, and personality changes in the elderly suggest some areas of possible impairment in the consent capacity of geriatric patients. In the realm of cognitive functioning, decrements in cognitive flexibility, recall and recognition, memory storage and retrieval capabilities, and judgment have been reported.[9–12] However, not all research finds this age-related decrement in memory and other cognitive processes.[13] With respect to personality, changes from an active to a more passive position in relation to the environment and increased introversion and compliance have been noted.[14,15] Changes in cognitive capacities and personality functioning may have an impact on the consent process with elderly patients.

These studies, although not completely consistent, suggest that the elderly may face some special difficulties in giving informed consent. In order to identify more clearly whether geriatric patients have difficulty giving informed consent, to detect what impairments in competency they experience, and under what circumstances these impairments are manifest, we undertook a series of empirical studies examining competency in the elderly population. The findings from two of our studies are to be presented. Our first study examines competency to consent in the elderly population afflicted with chronic medical illnesses and free of diagnosed psychiatric illnesses. We contrasted their capacities with a matched group of young medical patients.[16] This elderly population is frequently a group that liason and consultation psychiatrists are called on to evaluate as a consequence of treatment refusal. The second examines competency in elderly psychiatric patients with cognitive impairment or depression and compares each of these groups with the normal elderly. Thus, we identified three groups of elderly patients who were potentially at risk for having impaired competence. These groups are the elderly medical patients who we chose to compare with young medical patients; the elderly psychiatric patient with cognitive impairment (e.g., Alzheimer's disease); and the elderly psychiatric patient with depression.

In these studies, we addressed capacity to consent to research because the necessity for an informed and voluntary consent is greatest in this case. Therefore, we cannot directly apply our findings to the treatment setting. However, the capacities required to consent to treatment and research have been found to be similar[17] and thus our findings may have implications with regard to treatment consent.

For the elderly medical population, we asked the following questions:

1. Do elderly individuals suffering from physical illnesses expose themselves more often to research procedures in which the risk/benefit ratio is extremely unfavorable? Alternately do they refuse participation in highly beneficial projects more often than young medical patients?

2. Do elderly medical patients comprehend significantly less consent information than young medical patients?

3. Is the quality of reasoning utilized in deciding to participate in research by older medical patients impaired relative to young medical patients?

A parallel set of questions were asked of an elderly psychiatric population. These particular questions were formulated because they operationalized the ways in which competency have been defined legally.

GERIATRIC MEDICAL PATIENTS STUDY

Methods

In this study,[16] as a means of assessing competency, a series of hypothetical research studies were presented to elderly and young medical patients. The research projects, which were written at an eighth grade level,[18] varied according to risk/benefit ratios and whether or not they were directly beneficial to the patients. An interviewer read a series of six research descriptions to the patients. The patients had copies of the descriptions available to them. Competency was assessed directly following the presentation of each description. Measures of intelligence and the level of the patient's attention during the evaluation were also obtained immediately following the competency assessments.

Patient Characteristics

Patients were recruited from three facilities in major metropolitan cities. During the data collection phase, every available inpatient on the general medical units was asked to volunteer. For outpatients, every patient at several medical clinics was approached while the patients were awaiting their appointment. All were diagnosed as having serious medical illnesses, such as diabetes, emphysema, heart failure, or severe hypertension. Heart problems (e.g., hypertension, myocardial infarction), diabetes, and asthma were the most frequent diagnoses in both the elderly and the young. All elderly patients ($N = 39$) were 62 or over ($\bar{X} = 69.2$, $SD = 5.3$). The young medical patients ($N = 41$) ranged from 22 to 44 ($\bar{X} = 33.7$, $SD = 6.6$).

Patients were screened to ensure that their hearing and vision were adequate and were not impaired in a manner that interfered with their full participation in this study. With respect to intelligence and education, no significant differences were found between the elderly and young. The average IQ was in the normal range (young, $\bar{X} = 97.6$, $SD = 14$; elderly, $\bar{X} = 96.4$, $SD = 18$; $t = 1$, $df = 78$, not significant). With regard to education, 37% of the young patients had some education beyond high school; whereas in the elderly group, 27% fell into this category. Again, this is not a statistically significant difference ($\chi2 = 4.0$, $df = 4$, not significant).

Patients were screened for psychiatric illnesses as indicated by a lack of psychiatric treatment in the past 10 years. In addition, they were screened clinically to rule out the presence of severe organic brain syndrome. However, because extensive neuropsychiatric evaluations were beyond the scope of this study, it is possible that neurological or psychiatric conditions in some patients were undetected.

Results and Discussions

Table I reports the results comparing elderly and young medical patients on comprehension of consent information and quality of reasoning employed in reaching the consent decisions. With respect to comprehension, clear differences emerged between the elderly and young. The aged demonstrated poorer understanding of consent information for five of the six research projects. This finding is further strengthened when overall comprehension, derived by summing comprehension scores for all research descriptions, is examined. Elderly medical patients showed significantly poorer comprehension than young patients (elderly, $\bar{X} = 40.8$, $SD = 8.6$; young, $\bar{X} = 48.3$, $SD = 10.4$; $t = 3.14$, $df = 63$, $p < .01$).

With respect to quality of reasoning (Table II), the elderly tended to perform somewhat poorer, demonstrating less rational reasoning then younger patients for two of the research descriptions. However, total quality of reasoning scores for all six descriptions did not show a significant difference between the two groups (elderly, $\bar{X} = 19.3$, $SD = 4.1$; young, $\bar{X} = 18.2$, $SD = 3.6$; $t = 1.26$; $df = 77$, not significant). Table III

Table I. Comprehension Scores for Elderly and Young Medical Patients

	Percentage score[a]		
Comprehension of	Young	Elderly	p value
High risk/low benefit study	74	60	<.01
High risk/low benefit study	72	57	<.01
Low risk/high benefit study	74	64	<.05
Low risk/high benefit study	79	70	<.10
High risk/high benefit study	82	72	<.05
Moderate risk/high benefit study	84	74	<.05
Overall comprehension	75	64	<.01

[a]Percentage of total possible score.

Table II. Reasoning Scores for Elderly and Young Medical Patients

| | Reasoning scores mean ± standard deviations | | |
Quality of reasoning for	Young	Elderly	p value
High risk/low benefit study	3.05 ± 1.20	3.34 ± 1.36	NS
High risk/low benefit study	3.12 ± 1.14	2.56 ± 1.23	<.05
Low risk/high benefit study	2.85 ± 1.32	2.84 ± 1.65	NS
Low risk/high benefit study	3.71 ± 1.66	2.85 ± 1.63	<.05
High risk/high benefit study	3.26 ± 1.06	3.45 ± 0.99	NS
Moderate risk/high benefit study (possible range = 1–7)	3.29 ± 1.27	3.21 ± 1.26	NS
Overall quality of reasoning (possible range = 6–42)	19.3 ± 3.6	18.2 ± 4.1	NS

demonstrates the finding on the "reasonable choice" criteria (i.e., decisions consistent with risk/benefit ratios). Young and elderly patients' agreement to participate in each of the studies is reported. In five of the six studies, there were no significant differences in participation rates. However, the elderly did demonstrate less reasonable decision making on one of the high risk/low benefit studies by agreeing to participate at a significantly higher rate.

Table IV shows comprehension of specific elements contained in consent forms for the four descriptions, with opposing risk/benefit ratios (i.e., risk, benefits, procedures, and purpose of the project). On

Table III. Participation Rate in Hypothetical Studies for Young and Elderly Medical Patients

| | Percentage agreement | | |
Type of study	Young ($N=41$)	Elderly ($N=39$)	p value[a]
High risk/low benefit	25.0	30.6	NS
High risk/low benefit	28.9	51.4	<.05
Low risk/high benefit	82.5	77.8	NS
Low risk/high benefit	80.5	63.2	NS
High risk/high benefit	34.2	28.6	NS
Moderate risk/high benefit	44.7	34.5	NS

[a]Chi-square test was used to determine significant differences.

Table IV. Comprehension Scores of Consent Form Elements for Young and Elderly Medical Patients

	Mean ± standard deviations		
Consent form elements	Young (N=37–40)	Elderly (N=36–37)	p value[a]
Purpose of procedure	11.50 ± 3.31	9.42 ± 2.78	<.01
Procedure	12.54 ± 2.66	10.92 ± 2.24	<.01
Benefits	11.78 ± 3.17	9.51 ± 3.43	<.01
Risks	12.46 ± 3.23	10.45 ± 2.64	<.01

[a] For independent means t test was used for comparisons. Sample size varied as a result of missing data. Possible range = 4–16.

each of these elements, the elderly demonstrated significantly poorer understanding than young patients despite similar outcome on the choice criterion. This finding indicates that the deficits experienced by the geriatric patients are not specific to a single element of consent information (i.e., risks, benefits, alternatives).

The influence of additional factors, such as intelligence and level of attention, on the relationship between age and capacity to consent were also examined. A partial correlation that controlled for the effect of intelligence in the correlation between age and overall comprehension of the research descriptions showed a significant relationship. When the effect of IQ scores were removed, the performance of the elderly remained poor (r_p = −.37; df = 63; p < .01). The partial correlation between age and overall comprehension with the effects of level of attention controlled also showed that a significant relationship continued to exist (r_p = −.38; df = 63; p < .01).

These results present a mixed picture regarding the elderly medical patients' capacity to give informed consent. Both the young and the elderly are more likely to agree to participate in low-risk projects than high-risk procedures and thus tend to make reasonable decisions. The rate of participation for the young compared to the older patients does not differ except in one instance when the elderly agree more often to a risky project. With respect to the quality of reasoning, the elderly show some impairment in this capacity despite the fact that they are able to reach reasonable decisions to the same degree as the young. The greatest differences emerge between the older patients and the younger patients when comprehension scores are examined. The elderly demonstrate poorer comprehension of each of the specific elements of informed consent information, that is, knowledge of risks, benefits, purpose of the project and procedure. Thus as a group, geriatric patients may have some impairment in their competency to give informed consent to research. However, this impairment does not appear to have a significant impact with respect to the quality of their decisions. Thus, although they show poorer comprehension than their younger counterparts, they generally seem to make equally reasonable treatment decisions. It may be that only a basic awareness of relevant information is necessary to make reasonable decisions and it is possible that both of the groups we studied achieved this level of awareness. Thus, the quality of decision making may not be adversely affected until comprehension ability is severely impaired, as in patients who have senile dementia of the Alzheimer's type that has progressed beyond the initial stages.

STUDY OF DEMENTIA AND DEPRESSION IN THE ELDERLY

Methods

In our second study we examined competency in an elderly psychiatric population. In this study we interviewed three elderly groups who were participants in research: a cognitively impaired group suffering from dementia, a depressed group, and a normal elderly sample. In this study, in addition to administering our hypothetical studies, we also examined competency for consent to an actual drug study in which the individuals were participating.

Patient Characteristics

In this study, we assessed the competency of a sample of 12 patients diagnosed as having senile dementia of the Alzheimer's type, 7 elderly patients diagnosed as having major depression, and 14 normal elderly individuals. The small sample size for the depressed group was a result of the type of studies that were ongoing at the time of our date collection. The average age in the dementia group was 69 years ($SD = 6.8$), for the depressed group, the mean age was 68 ($SD = 5.3$), whereas the normals' average age was 70 ($SD = 5.3$). All dementia patients were at least moderately impaired but none was so impaired as to require full-time nursing home care. Despite similar educational backgrounds, the dementia group had lower IQs, as would be expected. The depressed group had significantly higher Hamilton depression ratings ($\bar{X} = 18.4$) than the dementia group ($\bar{X} = 3.7$) or the normal elderly sample ($\bar{X} = 2.1$).

Results and Discussion

For the most part, a comparison of competency variables for the actual drug study will be presented. However, it is also interesting to take a look at differences in agreement rates to our hypothetical projects.

Table V presents the comparison of the cognitively impaired, depressed patients', and normal elderly individual's agreement rate for participation in the series of hypothetical research projects. In examining the findings, there appears to be a tendency for the cognitively impaired group to agree to participate more often in the research studies. This is particularly true for the studies that are described as being high risk. Although these findings are not significant at the .05 level, it may be that the lack of significance is due to the small sample size. These results seem to indicate that dementia patients who have a significant degree of cognitive impairment may not be able to accurately assess risk/benefit ratios in proposed research projects and may need some assistance in making these assessments.

It is noteworthy that several dementia patients reported that they could not decide on their own whether or not to participate. Thus, although not being able to reach a decision, perhaps they retained some protective mechanisms by deferring decision making until they had the opportunity to discuss it with a family member.

In assessing capacity with respect to the drug study, we assessed the participants' competency to consent immediately after they signed the consent form to participate in the drug study. Table VI presents the results from this assessment. As with the elderly medical patients, the

Table V. Comparison of Agreement Rate for Participation in Hypothetical Research Projects by Dementia Patients, Depressed Patients, and Normal Elderly

	Percentage agreement		
	Dementia patients ($N=17$)	Depressed patients ($N=29$)	Normal elderly ($N=11$)
High risk/low benefit study (therapeutic)	60	31	27
High risk/low benefit study	69	59	36
Low risk/high benefit study (therapeutic)	100	86	82
Low risk/high benefit study	81	83	91
High risk/high benefit study (therapeutic)	57	32	27
Moderate risk/high benefit study	64	61	73

Table VI. Competency Scores for Dementia Patients, Depressed Patients, and Normal Volunteers

	Dementia $(N=12)$	Depressed $(N=7)$	Normal $(N=14)$	f test
Overall comprehension (% correct response)	60%	90%	85%	5.32**
Quality of reasoning mean score (Range 1–5)	2.3	3.5	2.9	3.68*
Specific consent form elements (mean score) (range 1–4)				
Risks	2.1	3.1	2.5	1.72
Benefits	2.2	3.1	2.7	1.71
Purpose	2.0	3.0	3.5	8.36**
Method	2.2	3.3	3.1	3.30*

*$p<.05$; **$p<.01$.

dementia patients performance was impaired relative to their comparison group, in this case, the normal elderly. The dementia patients' comprehension of the consent information was significantly poorer than the normal elderly. Further, the quality of reasoning that they employed to reach their decisions also tended to be poorer. For example, when asked why they decided to participate in the drug study, they frequently responded "Because the doctor asked," or "Because my spouse said I should." There was little evidence that they evaluated the pros and cons by themselves. Depressive patients showed no evidence of performing more poorly than normal elderly.

In summary, the preliminary results of this study indicate that senile dementia patients with a moderate degree of illness suffer a significant impairment in competency to consent to research. Further, they appear to have some degree of awareness that their capacity is compromised and tend to turn to others to make decisions on their behalf. We do not yet know if the patients in the early stages of illness will show the same deficits or less impairment.

CONCLUSIONS AND RECOMMENDATIONS

These two studies taken together suggest that elderly medical and elderly psychiatric patients experience particular difficulty in comprehending consent information. Furthermore, the results tend to indicate in a preliminary manner that depression in and of itself does not lead to problems in competency. Thus, whereas depression generally may not present particular problems regarding competency, dementia, even in relatively early (i.e., moderate) phases does. These findings do not mean that every patient diagnosed as having dementia is incompetent nor do they mean that all depressed patients are competent. Instead they serve to identify the groups in which we ought to worry more about competency and possibly be more cautious about evaluating competency. Thus, these results indicate that we ought to be concerned about competency in elderly patients on two counts. First, aging in and of itself leads to some dimunition of competency in medical patients. Second, dementia in the elderly results in further decrements.

These results suggest certain recommendations to follow with the elderly.

In making these recommendations, a division must be made between those patients who are competent and those who are not. No diagnosis is ever presumptive of incompetence. As a result, careful assessment of competency when problems arc suspected should be made.

In addition, when a patient is judged to be of questionable competency (i.e., borderline in competence) efforts should be made to alter the consent process prior to making a determination of incompetence. Simple alterations in procedures can improve the capacity of patients to make health care decisions on their own behalf. These include: (a) making the consent material more readable[18]; (b) tailoring the consent information to the needs of the patient as suggested in the preamble of the federal regulations for research with humans[19]; (c) allowing patients to review the consent material for a lengthier time prior to assessing competence; (d) using teaching and review methods coupled with testing[20,21]; (e) developing a rapport that promotes a sense of confidence in patients and enables them to ask as many questions as needed; and (f) involving the patient's family members in the consent process. Relatives may be able to use language more familiar to the individual and, hence, make the material more understandable. If these efforts at improving competency fail and the patient is judged to be incompetent, some form of substituted judgment must be employed. These substitutes include legal guardianship, durable power of attorney, and the living will. All of these alternatives are at best a workable compromise and the decisions made by these mechanisms may ultimately only partially reflect the patients true wishes.

REFERENCES

1. National Center for Health Statistics. *Health United States*. U.S. Government Printing Office, 1982.

2. President's Commission for the Study of Ethical Problems in Medical and Biomedical Behavioral Research. *Securing access to health care.* U.S. Government Printing Office, 1983.

3. Ostfeld A: Older research subjects: Not homogeneous, not especially vulnerable. *IRB: A Review of Human Subjects Research* 1980; 2:7–8.

4. Vestal R: Physical vulnerability, IRB: *A Review of Human Subjects Research* 1980; 2:5–6.

5. Lawton P: Psychological vulnerability. *IRB: A Review of Human Subjects Research* 1980; 2:5–6.

6. Bernstein J, Nelson K: Medical experimentation in the elderly. *J Am Geriatr Soc* 1975; 23:327–329.

7. Taub HA: Informed consent, memory and age. *Gerontologist* 1980; 20:686–690.

8. Taub HA, Kline GE, Baker MT: The elderly and informed consent: Effects of vocabulary level and corrected feedback. *Exp Aging Res* 1981; 7:137–146.

9. Botwinick J: Intellectual abilities, in Birren JE, Schaie KW (eds): *The Handbook of the Psychology of Aging.* New York, Van Nostrand Reinhold Co, 1977, pp 580–605.

10. Ceci S, Tabor L: Flexibility and memory: Are the elderly really less flexible? *Exp Aging Res* 1981; 7:147–158.

11. Botwinick J: *Aging and Behavior: A Comprehensive Integration of Research Findings.* New York, Springer Publishing Co. Inc., 1978, pp 337–360.

12. Craik F: Aging differences in human memory, in Birren JE, Schaie KW (eds): *The Handbook of the Psychology of Aging.* New York, Van Nostrand Reinhold Co., 1977, pp 384–420.

13. Bromley DB: *The Psychology of Human Aging.* Baltimore, Md, Penguin Books, 1966, pp 215–220.

14. Neugarten BL: Personality and aging, in Birren JE, Schaie KW (eds): *The Handbook of the Psychology of Aging.* New York, Van Nostrand Reinhold Co., 1977, pp 626–649.

15. Chown SM: Personality and Aging, in Schaie KW (ed): *Theory and Methods of Research on Aging.* Morgantown, WVa, West Virginia University Library, 1968, pp 134–157.

16. Stanley B, Guido J, Stanley M, Shortell D: The elderly patient and informed consent: Empirical findings. *JAMA* 1984; 252:1302–1306.

17. Appelbaum P, Roth L: Competency to consent to research: A psychiatric overview. *Arch Gen Psychiatry* 1982; 39:951–958.

18. Flesch R: A new readability yardstick. *J Appl Psychol* 1978; 32:221–233.

19. Final regulations amending basic HHS policy for the protection of human research subjects. *Federal Register* 1981; 46:8366–8392.

20. Miller R, Willner H: The two-part consent form suggestion for promoting free and informed consent. *N Engl J Med* 1974; 290:964–966.

21. Woodward W: Informed consent of volunteers: A direct measurement of comprehension and retention of information. *Clin Res* 1979; 127:248–252.

3

The Elderly, Incompetence, and Treatment Decisions

The New York Experience

J. RICHARD CICCONE AND **COLLEEN D. CLEMENTS**

INTRODUCTION

Laws and regulations are increasingly reshaping the doctor–patient relationship, changing this relationship from convenant to contract[1] to bureaucracy. Mark Siegler sees medicine moving from informal standards of professional ethics (broad and general principles developed over centuries) to a civilly enforced body of law and administrative regulation that will change and control the doctor–patient relationship.[2] It has not been demonstrated that this revamping of the doctor–patient relationship, in the name of protecting patient rights and patient autonomy, serves either societal or individual good. The medical care of the elderly has not escaped these attempts to codify and bureaucratize[3] the doctor–patient relationship. What constitutes good medical care for the incompetent patient who is elderly and suffering from multiple chronic severe conditions? Who is terminal but can be sustained for a short time more? Who is vegetative and capable of long-term survival? These are

J. RICHARD CICCONE AND **COLLEEN D. CLEMENTS** • Department of Psychiatry, University of Rochester School of Medicine and Dentistry, 300 Crittenden Boulevard, Rochester, NY 14642.

now medical, ethical, and legal questions and may soon become social policy and economic questions.

These complex questions have involved psychiatrists doing competency evaluations in conceptual mazes and exposed practicing physicians to legal hazards. This chapter attempts to clarify these questions and develop workable suggestions.

First, we will uncouple the two questions that often arise when caring for the clinically incompetent elderly, terminal, or vegetative patients: (a) What is good medical care? and (b) How can we avoid legal liability? The first question is ethical and clinical, and concerns determining what is good medical care for a particular patient. What is the purpose of the proposed medical intervention or nonintervention? Does the purpose make medical and ethical sense? Can we reasonably expect the desired outcome? What criteria should be used for decision making in this case and who should be involved in the decision making? When asking who should be involved in decision making, the patient's clinical and legal competence becomes a central concern. The second question speaks to legal liability; a concern about protecting the physician and the institution from civil and criminal penalties for a particular clinical decision. It involves the wish of good physicians not to violate the law, and it also involves the realistic desire of physicians and institutions to protect themselves from legal liability. Court orders, ethics committees, committees of the person, durable power of attorney, and statutory law on living wills can be viewed as attempts to minimize legal liability under our increasingly unworkable tort law system.

Having asked these two separate questions, this chapter will examine the intricacies of the competency evaluation and some of the problems typical of a competency question. There are two philosophic issues contained in the competency question. One is an ethical question about the definition of what is good: Is a competent, autonomous choice by the patient the best definition of what is good? A competency evaluation is often done for the express purpose of establishing the patient's informed and free choice and defining that choice as what is good in this particular medical case. This is not as easily answered in the affirmative as recent medical ethics has assumed. There is a long history of philosophy that would question the rationality of such a definition of the good. The second philosophic issue is the status of a competent and an incompetent patient's choice in the eyes of the law. There are limits to a competent individual's civil rights and the police power of the state may be invoked for the common good, for example, when the patient with active TB may be quarantined.

The *Eichner* and *Storar* cases, and the legal decisions and dissents flowing from them, illustrate the questions we have asked and different attempts to answer these questions. The *Eichner* case represents a patient who was once competent in the eyes of the law, and who expressed wishes and views about the treatment and clinical management of vegetative patients. The *Storar* case represents a patient who was never competent, because of severe mental retardation, and therefore never expressed nor could express any wishes and views about the treatment and management of terminal patients. In addition to the potential use of court orders to resolve the questions such patients raise, this chapter will also look at other suggested means for resolving such problems: durable power of attorney, committee of the person, ethics committees, and statutory law on living wills. Finally, we will propose some recommendations for good medical, ethical, and legal management of these difficult cases.

CONSENT: THE POWER TO ACCEPT OR REJECT MEDICAL TREATMENT

Competence is the cornerstone of the patient's legal right to consent to or reject medical treatment. In 1914, Judge Cardozo wrote the often quoted words,

> Every human being of adult years of sound mind has a right to determine what shall be done with his own body; and a surgeon who performs an operation without his patient's consent commits an assault for which he is liable in damages. . . . this is true except in a case of an emergency where the patient is unconscious, and where it is necessary to operate before consent can be obtained.[4]

The legal doctrine of consent to treatment has undergone change in the direction of further supporting patient autonomy and reorganizing the doctor–patient relationship to make it more egalitarian. An adult is presumed to be competent to consent to treatment or to refuse treatment until a judicial finding of incompetency, based on clear and convincing evidence.

The physician, usually a psychiatrist, who is asked to evaluate a patient's competence, must keep two points in mind. First, the physician provides a determination of clinical competence and an opinion regarding legal competence; the physician's opinion regarding legal competence is not legally binding. Second, the consultant should begin by determining who is requesting the consultation and for what purpose.[5] The question is not "Is this patient competent?" but "Is this patient competent for a specific task?"

What follows is the determination of the presence and nature of any psychiatric illness. If a psychiatric illness is present, do the signs and symptoms impair the person's capacity to make the decision to accept or reject treatment? Does the elderly patient's illness impair awareness of the fact that he is being asked to accept or reject treatment? Does the mental illness impair the elderly patient's understanding of the nature of the illness for which treatment is being recommended? Does the elderly patient understand the treatment that is being recommended and the likely side effects of this treatment? Does the elderly patient understand that other treatments are available, including no treatment, and the implications of rejecting the doctor's recommendation?

Competence is not always clearly present or clearly absent. Competence may fluctuate according to a person's environment and according to the complexity of the decision that faces the elderly individual. The mildly demented may be competent to decide simple questions or to consent to nonrisky procedures but have more difficulty when the issues are more complex.

If the elderly individual's clinical incompetence is due to a treatable illness, he may be restored to competence. Therefore, the physician must carefully consider the differential diagnosis of incompetence, which includes an acute psychotic episode, delirium, and major depression. The specific treatment will be dictated by the specific illness or underlying cause of the clinical incompetence. The presence of a full-blown psychosis is not necessary to find incompetency, for example, the nonpsychotically depressed may be seriously impaired by negativistic thinking.

In the situation where the person is rapidly fluctuating from what appears to be a competent to what appears to be an incompetent state, for all practical purposes, the patient is incompetent. The patient who is incompetent for days, has a lucid interval, and then returns to a state of being incompetent is also usually considered incompetent.

WHEN THE PATIENT IS CLINICALLY INCOMPETENT

Traditionally, when an elderly patient is clinically incompetent the physician along with the patient's family members make decisions about what treatment will be given or withheld. This custom has no formal standing in law. In fact, it could lead to significant legal difficulties. What the law currently calls for is documented informed consent before medical treatment can be initiated or withdrawn.

A patient is considered legally competent until someone raises the question. Then, and only then, is the issue joined. In a climate that emphasizes autonomy, those who are charged with "risk management" often advise that formal procedures be followed in an attempt to avoid liability and malpractice suits. In New York State, in nonemergency situations when the elderly patient's capacity to care for himself is questioned, the law currently provides for three procedures: durable power of attorney; conservatorship; and committee of the person. Only the durable power of attorney and the committee of the person have the legal standing to make decisions to accept or reject treatment for the incompetent individual.

Durable Power of Attorney

The statute on durable power of attorney in New York State does not speak to health issues. A durable power of attorney is a power of attorney that, prior to an individual becoming incompetent, contains a specific section that allows for a continuation of the power of attorney if the person becomes incompetent. The New York State Attorney General, however, has issued an opinion that a durable power of attorney may be used, under defined circumstances, to permit an agent to communicate the patient's decision regarding medical treatment. The durable power must state that the patient authorizes the agent to communicate these decisions.[6(p67)]

Conservatorship in New York State

Under New York State law, adults are presumed to be competent for whatever contractual and other decisions are necessary to function in our society. Article 77 of the Mental Hygiene Law provides for the creation of a conservatorship, a mechanism to ensure the financial and personal well-being of an individual that does not require that the person be found incompetent. A petition may be filed by a relative or a friend; the meaning of friend includes corporate bodies, public agencies, or social services officials. If the proposed conservatee is in an institution, the officer in charge of the institution may also file a petition. The petition must contain the reasons for concern about the individual's well-being, the facts demonstrating the need for appointment of a conservator, and the plan the petitioner proposes to "insure the preservation, maintenance, and care of the proposed conservatee's income, assets, and personal well-being, including the provision of necessary personal, and social protective services to the conservatee."[7] When the proposed conservatee is unable to attend a hearing, a guardian ad litem must be appointed unless the court believes that the proposed conservatee's interests are adequately protected by the Mental Health Legal Service or by "counsel chosen by the proposed conservatee." Jury trial is available when any party contests the need for appointment of a conservator. Usually a family member is appointed the conservator, but a non-family member may also be appointed, especially when there are family disagreements. In the case of *Matter of Lyon*,[8] an 84-year-old quadriplegic was confined to a nursing home and had a trust that could be invaded for her medical care. Her son was not appointed conservator because at issue was the extensive and expensive care that was keeping Mrs. Lyon alive. The court felt that it was clear that her son considered her present care to be "squandering" in view of her condition. However, in the opinion of Mrs. Lyon's physician, this care had been life saving. Family members may not always have the best interests of the conservatee in mind—particularly when they stand to inherit the estate that remains after the expense incurred by the conservatee.

Where there is no family member available, and that is not a rare event in our society, the courts have appointed nonprofit corporations as conservator. It is also possible for the department of social services to act as a conservator, even if the conservatee does not receive public assistance.

A conservatorship, although it charges the conservator with looking after the person and supplying their needs, does not empower the conservator with the legal decision-making capacity regarding consent to treatment; such consent requires a committee of the person.

Committee of the Person

Article 78 of the Mental Hygiene Law outlines the jurisdiction, procedure, and purpose for forming a committee of the person and/or of the property. The declaration of incompetence and the appointment of a committee is more far reaching than the formation of a conservatorship. The appointment of a committee is more difficult to get because the court is reluctant to take from a person his civil rights, particularly the right to manage his own affairs. The law requires the person bringing the petition to state the facts showing incompetence and demonstrate why a conservatorship would be inappropriate. When any party disputes the need for a committee, a jury trial is available and the burden of proof falls on the petitioner.

After appointment, the committee is subject to the control of the court and is accountable to it. Section 15, of Article 78, reads "the committee of the person is charged with the responsibility for taking care of the physical needs of the incompetent, furnishing him with such medical and other care as is required, and looking after his health and general welfare."[9]

EMERGENCY CARE

In a medical emergency where the person is unconscious or clinically incompetent, treatment may be given under presumed intent. The Social Services Law 473A provides for a short-term involuntary protective services order (STIPSO).[10] A social service official can act as a petitioner in a special proceeding in supreme court in New York City or in county court elsewhere in the state. The official can bring information to the court that demonstrates that the allegedly endangered adult, as a result of a lack of capacity to understand the situation, is in a life threatening situation or poses imminent risk of serious physical harm to himself or to others. The petitioner is also required to provide a description of the specific protective services needed to remedy that situation. The STIPSO emergency order is limited to 72 hours, with one renewal for 72 hours. It can be instituted simultaneously with a petition for a conservatorship or committeeship or civil commitment for involuntary hospitalization under the mental hygiene laws.

THE ULTIMATE DECISION: DO NOT RESUSCITATE (DNR)

Wanzer *et al.*[11] examined the physician's role in the treatment of the terminally ill adult and in writing do-not-resuscitate orders. Acknowledging the importance of the patient's right to make decisions about his medical treatment, the authors go on to point out that disease, pain, drugs, and a variety of conditions altering the patient's mental state may "severely reduce the patient's capacity for judgment." Therefore, the evaluation of the individual's competence is important in deciding how much to follow the patient's stated wishes. If the patient's capacity to make decisions is impaired, the physician has to turn to family and friends. A surrogate may assist by speaking on behalf of a patient. Where the physician does not have a long-standing relationship with the patient, it may be of special importance to have other means of determining the terminally ill patient's desires. A written statement, the living will, may provide important but not binding evidence of the patient's wishes.

A living will is a written expression of the patient's wish to refuse treatment. Courts consider the living will "persuasive evidence of the incompetent person's intention" and give "great weight" to it in the "substitutive judgment" process. Those in favor of writing the living will into law maintain that (a) it clarifies the law for both physicians and patients; (b) it allows the patient a stronger role in expressing his wishes; (c) it increases a physician's protection from liability, making the physician liable only if the physician fails to act in good faith, that is, deliberately and intentionally acts with malice; and (d) it provides direction for the physician who cannot for any reason comply with the dying patient's written directive—the physician transfers the patient to another physician who is able to comply.[6(pp29–30)]

Even with written documents or spoken wishes, there is no ignoring the physician's central role in the decision-making process. Physicians, through education and experience, have the capacity to diagnose, recommend treatment, and accept responsibility for their recommendations. Part of the training of the physician has been to value life. With modern technology, this life prolonging activity may go on for long periods of time and be fueled by reports of the rare patient who recovered from a similar difficulty. At times physicians may invoke heroic measures because of mistakenly equating the patient's death with their inadequacy. Here as in other circumstances, personal motivation must be examined and therapeutic ambition avoided. Defensive medicine may fuel inappropriate heroic measures in the hopes of avoiding legal liability. An important hurdle to providing humane care for the elderly is the threat of inadequate treatment brought on by current concerns about cost containment. The push to not squander "valuable resources on consumers who are about to check out anyway" may lead to inappropriate DNR orders. The physician must remain mindful of the responsibility to pursue the patient's interests. The patient's best interest may not be the patient's stated choices or opinion. The physician's determination of the patient's best interest is not simply the physician's idiosyncratic opinion, but is based on a professionally developed code or value system that the physician applies to the specific patient.

Many hospitals have established guidelines for do-not-resuscitate policies. Where the patient is competent, these guidelines call for DNR decisions to be made by the patient. If the patient is not competent, the guidelines call for the decision to be reached by consultation with family members or legal guardians. If there is no patient representative available, then the guidelines call for the courts to be approached. Any significant improvement in the patient's condition voids the DNR order and the order can be rescinded any time by any person who contributed to the making of the decision.

LANDMARK CASES

The two important cases in New York State ruling on incompetent vegetative or terminal patients are the *In re Eichner* and *In re Storar* cases.

A detailed look at these decisions may help clarify why physicians find such cases so legally and ethically problematic.

In re Eichner

In the case of *In re Eichner*,[12] Brother Fox was an 83-year-old patient who had been a member of the Society of Mary for 66 years. In 1970, he retired to reside with members of his religious order who ran a high school. He remained active and in 1979 developed a hernia while moving flower tubs. Given his overall good health, a recommendation for an operation was made to which Brother Fox consented. During the operation, he suffered cardiac arrest, went into a coma, required the assistance of a respirator, and was left in a permanent vegetative state. His director, Father Philip Eichner, was informed that there was no possibility of recovery. Two neurosurgeons confirmed that finding. Father Eichner asked the hospital to remove the respirator but the hospital refused to do so. Father Eichner then applied, under Article 78 of the New York State Mental Hygiene Law, to be appointed committee of the person of Brother Fox, with the approval of Brother Fox's surviving relatives (10 nieces and nephews). Father Eichner also asked for permission to remove the respirator. The court appointed a guardian ad litem for Brother Fox and involved the district attorney, who opposed the application. Brother Fox had previously expressed views on the care of the terminally ill when discussing the moral implications of the Karen Ann Quinlan case. He agreed with the termination of extraordinary life support systems when there is no reasonable hope for the individual's recovery. The New York State Supreme Court in Nassau County, pointing to the common-law right to refuse treatment and Brother Fox's previously expressed views, granted the application.

In agreeing with the termination of extraordinary medical care, Brother Fox was within the tradition of Catholic medical ethics, which originated the terms ordinary and extraordinary medical care. As used by Pope Pius XII, these terms have nothing to do with technological sophistication.[13] The term *ordinary* does not mean usual and the term *extraordinary* does not mean highly complex, unusual, or experimental. Ordinary treatments are all medicines, treatments and operations that offer a reasonable hope of benefit and that can be obtained and used without excessive pain or inconvenience. Extraordinary treatments are those that involve excessive pain, expense, or other inconvenience or do not offer a reasonable hope of benefit.

A few months before his operation, Brother Fox repeated to members of his religious order that he did not want extraordinary care. Were he competent, there would be no question that Brother Fox would have a common-law right to decline treatment. Because Brother Fox expressed these wishes prior to becoming incompetent, the supreme court felt that the expressed views on the care of the terminally ill should be respected.

The appellate court heard the *Eichner* case. It modified the lower court decision by adding a mechanism to deal with the patient who had not made his wishes known regarding ordinary treatment while competent. An appropriate person could be appointed to use "substituted judgment" to accept or refuse treatment on behalf of the incompetent individual. The appellate division also established a set of procedures to be followed before future applications of this nature would be granted.

The New York State Court of Appeals found that the supreme court (a lower court in New York State) had ruled properly in respecting Brother Fox's wishes expressed prior to his becoming incompetent. They went on to say that they would not rule on substituted decision making because that issue was not present in this case (Brother Fox had made the decision for himself before becoming incompetent). The court of appeals agreed that the lower courts had established proof of the patient's incompetency and chances of recovery to a clear and convincing standard. The court of appeals rejected the appellate division's suggestion that a mandatory procedure to get approval to stop life-sustaining treatment that would involve physicians, hospital personnel, relatives, and the courts be implemented. The court of appeals held that this suggested procedure come from the legislature.

The court of appeals' reasoning in the *Eichner* case emphasized an individual's common-law right to make treatment decisions. There is no legal requirement for a person to accept every available treatment. The court found that evidence of the patient's wishes need not be in written form. "Clear and convincing" reports of conversations between the patient and family members would qualify. No civil or criminal liability would be incurred by the medical personnel carrying out those wishes because they would be obeying the law that permits patients to make decisions about treatment. More importantly, the court said that there was no requirement for prior court approval although the treating medical team could exercise that option. The court went on to say that any guidance that it provided for future cases was necessarily limited. The court is not empowered to establish procedural rules, but rather has as its role the resolving of issues raised by facts presented in particular cases. The court of appeals commented on the state's right to supercede an individual's decision regarding treatment—there must be a compelling state interest to do so and none was existent in the Eichner case.

In re Storar

In the case of *In re Storar*,[14] John Storar was a 52-year-old retarded, institutionalized individual with a mental age of about a year and a half. His 77-year-old mother lived near the facility and visited him regularly. In July of 1979 he was diagnosed as having transitional cell carcinoma of the bladder and radiation therapy was recommended. The hospital would not provide the radiation treatment without consent of a legal guardian. Mrs. Storar applied to court and was appointed committee of the person (guardian), and with her consent he received radiation. Nine months later, a recurrence of the bladder cancer was diagnosed. It had metastasized to his lungs and probably to his liver and brain as well. The doctors reported that therapy (surgery, chemotherapy, or radiation) was not indicated. Because of bleeding in his bladder from lesions that could not be controlled by cauterization, John Storar became anemic. After initially refusing, Mrs. Storar consented to blood transfusions for John in May of 1980. Approximately a month later, his mother requested that the transfusions be discontinued. The director of the facility, pursuant to Mental Hygiene Law 33.03, sought court authorization to continue the transfusions. In mid-June, Mrs. Storar requested that the transfusions be prohibited. A hearing was scheduled and the court appointed a guardian ad litem and signed a temporary order permitting the transfusions. A month later he had to have his right arm tied down to prevent him from taking out the needle.

In court, there was agreement among the medical experts that John Storar had irreversible cancer of the bladder and an estimated life span of 3 to 6 months. The expert witnesses agreed he was incompetent with an infant's mentality. There was also clear agreement that Storar was continuously bleeding from his bladder. The bleeding led to clots that made urination painful. He also became upset when urinating blood. The frequency of clots and bloody urine would increase following transfusions.

It is interesting to note that the decision of the Supreme Court of Monroe County describes Storar as having frequent attacks of nausea and vomiting, napping frequently, and spending most of his time in his room. He is described as occasionally being able to run for a short distance and walk to the shower, although sometimes he had to lean on someone because of his weakness, which was not relieved by transfusions. The court drew a poignant picture of this man who, with his profound retardation, was not only unable to consent to or refuse treatments, but was also unable to understand his illness and what was happening to him. A strikingly different picture of Storar emerges from the court of appeals' decision. Storar is described as alert and, after transfusion, able to carry out many of his usual activities. He is described as able to take care of himself, feed himself, shower, take walks, and run. The court tells of Storar's mischievious activities, such as stealing cigarette butts and attempting to eat them.

The supreme court ruled, citing Eichner as precedent, that because the blood transfusions were painful and did not offer a reasonable hope of benefit (having no curative purpose, not making him more comfortable), the blood transfusions could be viewed as extraordinary. Pointing out that medical ethics required that the physician focus on making the patient as comfortable as possible, the court held that both medical ethics and traditional Catholic medical ethics not only permitted but supported the termination of extraordinary means of treatment. Because Storar could never have expressed a wish and exercised his right to stop treatment, and because his mother thought that it was in his best interest to stop transfusions, the court applied its substituted judgment. There are potential contradictions here. The court partially based its reasoning on the autonomy of the individual to refuse any treatment, but because continuously incompetent persons could never exercise such autonomy, the court chose to affirm its substituted judgment rather than to affirm the substituted judgment of a committee of the person, in this case Storar's mother. On the basis of clear and convincing evidence, the court established that Mrs. Storar was basing her decision on John Storar's best interest but was unwilling to make such substituted judgment of best interest a legal precedent, relying instead on *re Eichner*'s common-law right to determine treatment. But no one could know what John Storar's preference would be, if he were able to have one. The court is in no better position to know that preference and affirm that right than the committee of the person. Is it more sensible to affirm the court's substituted judgment than to establish the substituted judgment of the committee of the person?

The appellate court affirmed the lower court decision, and did establish the substituted judgment of a committee of the person. It argued that because the terminally ill competent individual has a significant right to refuse the treatment, therefore an incompetent adult has the right to refuse such treatment on the request of a relative and committee (guardian). In dissent, Justice Cardamone said that the court should not decide whether and when to discontinue medical support system for a dying patient. Justice Cardamone concurred that the dilemma was too personal to go beyond the decision of the family or guardian guided by medical advice. When courts try to solve these dilemmas by deciding what is best for the dying patient, they engage in what Justice Cardamone called practical, legal, and moral nonsense. Judges have no special ability to measure the quality of life and deferring to an ethics committee shifts the decision to another unqualified tribunal. Judge Cardamone also argued that Mrs. Storar's substituted judgment could not be enforced by the court, echoing his early sentiments that this was not an issue properly heard in court.

The court of appeals, reversing the decision, pointed out that John Storar had never been competent and it was unrealistic to determine what he would wish. Instead, the court of appeals saw John Storar as an infant, referring to Public Health Law 2504, subdivision two, which allows a parent or guardian to consent to medical treatment on behalf of an infant. However, this law does not permit a parent to deprive a child of life-saving treatment. In order to apply this reasoning to Storar's situation, the court had to divide the threat to his life from cancer from the threat to his life from blood loss. The former was not treatable; the latter was treatable. Thus, the court could hold that there was risk that death would occur without the transfusions, that the transfusions were life-saving. The transfusions were also held to be analogous to food, thus implicitly ordinary. Therefore, the transfusions, according to the court of appeals, did not involve excessive pain, were life saving, and offered a reasonable hope of benefit. They do not speak of the increased sensitivity to pain or the pain caused by blood clots. The majority ruling held that the court might be approached to get the court's view of the propriety of conduct regarding incompetent persons, but emphasized that the procedure was optional. The decision established the court's substituted judgment in this case of an incompetent person who had never expressed a wish, if such a case reached the courts, and in a negative sense. The court would not allow a guardian's substituted judgment for an incompetent patient who had never expressed his choice about life-prolonging treatment in terminal illness. By default, the court's judgment becomes substituted judgment for all such treatments being given if the court's opinion is sought or if the treatment is not experimental. This reinforces Justice Cardamone's view that these decisions are, perhaps, best left to the family and committee of the person to decide.

The court rejected *re Eichner* as precedent for Storar because that case bore only superficial similarities. In *Eichner*, of course, the decision was heavily based on Brother Fox's previously expressed wishes. In *Storar*, there was no way to know what he would have wanted. The Court likened him to an infant, although why he was seen as an infant rather than an incompetent adult is not explained other than to allow use of Public Health Law 2504 as precedent. It appears that the ruling makes it impossible for the family and Committee of an incompetent patient to refuse ordinary treatment or for the court to order cessation of ordinary treatment, unless the incompetent individual has previously expressed such a desire. This places complete emphasis on autonomy and common-law right. The majority's description of transfusions as ordinary treatment is also suspect because of the convoluted reasoning required to separate the hemorrhaging from the cancer when both were intimately connected. Storar was dying from the transitional cell carcinoma, and among the symptoms was hemorrhaging. The dissenting opinions pointed out that the judicial system is unsuited for and has no particular competence in making these decisions. It is also impractical to use the courts because of the time delay usually involved in seeking judicial determinations. Dissenting justices once again preferred involvement of physicians and members of the patient's family, often in consultation with religious leaders, in making such decisions. Justice Jones questioned whether this was ordinary treatment, given that John Storar had inoperable, incurable bladder cancer with a life expectancy of 2 to 6 months. The cancer was causing severe pain, and the blood transfusions were painful, caused apprehension, caused clotting, and added to his increased sensitivity to the pain. The transfusions did not serve to reduce his pain or make him more comfortable or offer reasonable hope of benefit. Therefore, the blood transfusions were extraordinary treatments.

CRITIQUE OF MECHANISMS FOR MAKING TREATMENT DECISIONS FOR INCOMPETENT PATIENTS

The New York State Court of Appeals' ruling in *Eichner* and *Storar* makes medical and social customs insufficient because it does not provide immunity from liability for such family-physician-religious counselor decisions about incompetent patients. The court of appeals has pointed out that it has no special interest in hearing these cases and, in fact, that bringing these cases to court is entirely optional, but that the substituted judgment of a committee of the person, or the agreement on treatment by family and physician are subject to civil and criminal liability. This leaves the physician or the institution to choose between the option of a court ruling or accepting risk for liability after the fact. Instead of being a medical ethics issue, or an issue of proper medical care, this is an issue of liability risk. Is it better to solve this problem by reforming the tort law system, or by putting into place legal mechanisms to deal with providing care or withholding care for the incompetent, terminally ill, or vegetative patient? Some of the legal mechanisms proposed are judicial hearings and court orders, statutory law, ethics committees, and committees of the person. Each has its costs and advantages.

Judicial Mechanisms

The judicial mechanisms for decision making about medical care for the incompetent individual permits a full hearing of the facts, expert testimony, and each adversarial side's day in court. Such mechanisms supply immunity from liability by putting decision making in the court's hands. As has been pointed out, there is a problem with this approach. The court may not be (as some of the dissenting justices pointed out) equipped to deal with these complicated, theological, philosophic, medical, biologic, and sociological issues. Further, there is an implicit contradiction in legal arguments based on autonomy, the civil right of the autonomous individual to have control over his own body and what's to be done to it, being applied to the incompetent individual who never evidenced a choice prior to becoming incompetent. The common-law right to choose what is done to one's own body cannot be the basis for the court's substituted decision because the court has no way of knowing what that individual would have decided if he or she were competent. That leaves either the compelling interest of the state (police power) or the substituted judgment of committees of the person. New York courts have not affirmed the decision-making authority of such committees to determine the best interest of the patient. This leaves only the court. There is little justification for taking this very personal, difficult decision making out of the hands of family members or a guardian. The problem is further exacerbated by the fact that when a petition is brought to the court, it is decided on the basis of either civil liberties or the police powers of the state. There is an additional problem that the court may not be able, in reality, to give a full hearing to the clinical and intimate facts of a medical case. The family or the committee, in concert with the physician and often with theologic consultation, are much closer to the situation, and through their involvement and commitment are much more likely to make a considered, suitable decision. The judicial route does provide some relief from the risk of liability but at the expense of judicial intrusion into the practice of medicine and into the very personal and sensitive doctor–patient relationship that extends to the family and to the guardian.

Statutory Law

Because our litigious society currently has difficulty with established medical practices and easily finds new causes of legal action, there is realistic risk for families and physicians. It may seem desirable to encase this medical custom in legislative protections by having the legislature say that it is both reasonable and legal for family members or guardians to make important decisions in the care of the incompetent patient who is elderly, terminally ill, or vegetative. A statutory law regarding living wills has been passed in a number of states and represents some relief from this difficulty, but also some hazards. One of the hazards in any legislative action is that it is not case specific. Take, for example, the living will. If the goal is to excuse the physician from liability for withdrawing treatment, a law has to specify in advance all those situations where such action would be legal and appropriate. There is no way of knowing or being able to compile such a list covering every possibility. As a result, the list is always narrow and inflexible. Yet there is, of course, a danger of abuse if there are no lists. Perhaps the biggest problem with living wills is their inflexibility, a generic problem with statutory law, followed closely by overattention to following procedures.

Ethics Committees

Ethics committees are composed of people of varying backgrounds who may have no special knowledge, training, or real expertise with which to make these complicated, intimate decisions about treatment. However, they do provide some protection from liability. This advantage is likely to be seized on by physicians and institutions as a way to minimize the evermounting risk of practicing medicine and providing medical care. But it does take the decision function away from family and doctor interaction, either directly or indirectly. Neither the physician nor the family have recourse when they disagree with the ethics committee decision. Although there is little track record or accumulated evidence, it is likely that the problems that occur when families and doctors make decisions will be miniscule compared to the difficulties caused by ethics committees making their decisions at a distance, at a slower pace, and with less involvement. As with most committees, they may generate policies and procedures, create a bureaucracy, and share the inflexibilities of statutory law.

Committee of the Person

Of all the possible solutions, a legislative one that affirms the traditional medical custom of permitting physicians and families to continue making these difficult personal choices seems most realistic. Such a law would not allow liability for ad hoc judgments unless there were evidence of gross negligence and malpractice on the part of the physician. But that is actually what we have right now. We do not need a living will or new statutory law. That leaves us with a committee of the person who can make this decision with the doctor. When family members or physicians disagree, a possibility is the judicial arena that may be best suited for the resolution of intractable conflicts, or the formation of a substituted consent committee that would hear the facts and make a decision that would be binding and free of liability, similar to the New Jersey ombudsman. The issue is finally how much risk the hospital or the physician is comfortable with and willing to take. Because this area is inevitably filled with some risk, and many physicians and hospitals are uncomfortable with assuming any known risk, an alternative is the reformation of the tort law system.

ETHICAL CONSIDERATIONS

The *Eichner* and *Storar* decisions show that the courts are caught up in the same dichotomy that medical ethics has been mired in, the dichotomy between patient autonomy and patient best interest (beneficence). Traditionally, the courts tend to rule in favor of patient autonomy whereas the practice of medicine has emphasized beneficence, doing everything possible to seek the patient's well-being. For the past 15 years, medical ethics has stressed autonomy (what each individual chooses is the good); this has resulted in an ethical nihilism because there can be no general description of what is good. Some have wrapped this nihilism in the political cloth of the democratic process. For example, Englehardt maintains there can be no rational content to ethics in a pluralistic society, but that autonomous moral agents can achieve a democratic consensus.[3]

An emerging fashion in medical ethics is the emphasis on the common good,[15] which in law justifies the police power of the state. This power to enforce positively the public good, more extensive than the *parens patriae* basis for judicially substituted decisions, underlies much public health law. It has been used to argue the state's interest in the viability of its citizens, as seen in laws concerning motorcycle helmets, alcohol regulations, and continuing treatment in the case of Elizabeth Bouvia. The common good is viewed as representing the opposite pole from autonomy. Rather than being viewed as opposite poles, autonomy and the common good can be viewed as levels of good. These levels of good are different and often in conflict. Neither level can be continuously sacrificed without undoing the entire system. General systems theory provides a framework to study the different, interconnected levels of individual and social good and may provide the balance necessary to make ethical, workable decisions.

What is missing from autonomy and the common good is the actual good of the patient, which is neither always autonomy nor always sacrifice for the common good. There is nothing unethical about putting the complex decisions in the doctor's and family's hands, allowing for the flexibility of making difficult decisions in individual cases. When the incompetent patient has no family, the appointment of a committee of the person allows for the decision to be made between that committee and the doctor. Both parties have important responsibilities they must discharge and traditional standards to which they can be held.

CONCLUSION

The *Eichner* decision sets precedent for a guardian making decisions for an incompetent person who was previously competent and expressed his wishes. The *Storar* case, however, does not provide such recourse for the incompetent person who has never been competent or who did not express his wishes. The court expressed its reluctance to see all such cases brought to court but did not affirm the proper arena for deciding them or deal with the liability risk, because it did not wish to make law.

We have looked at judicial decision making and found it wanting for routine cases. We have suggested that only in the case of an incompetent individual whose family members as surrogate decision makers disagree about treatment, or whose physician and family disagree, should the court function reasonably in its judicial role of resolving intractable conflicts. We have found weaknesses in the living will and have highlighted major problems with ethics committees. The committee of the person has limitations in making decisions about withdrawing treatment that will lead to death, and in deciding what constitutes extraordinary and ordinary treatment. Treatment decisions for the elderly individual who may be incompetent, vegetative, or terminal have been complicated by the liability issue, and medical ethics has not clarified the matter.

The legal activism of the 1960s and 1970s appears to be receding in the 1980s in the face of judicial restraint and respect for clinical reality. The Supreme Court in *Youngberg v Romeo*[16] wrote that in determining whether a reasonable standard of care has been met, courts must defer to the judgment exercised by the qualified professional, who is presumed to have acted appropriately. Intrusive laws and burdensome regulations, rather than ameliorating the problem, may make it more difficult to provide humane, compassionate care to elderly patients.

REFERENCES

1. Romano J: How do we transmit the humanistic factors in the practice of medicine? In Warren JV, Trzebiatowski GL (eds): *Medical Education for the 21st Century*. Columbus, Ohio, Ohio State University College of Medicine, 1985.

2. Siegler MA: A physician's perspective on a right to health care. *JAMA* 1980; 244:1591–1596.
3. Engelhardt HT Jr: *The Foundations of Bioethics*, New York, Oxford University Press, 1986.
4. *Schloendorff v The Society of New York Hospital*, 211NY125, 105NE92 (1914).
5. American Psychiatric Association Task Force on Forensic Issues in Geriatric Psychiatry, American Psychiatric Association, Washington, D.C., 1985.
6. *The Physician and the Hopelessly Ill Patient: Legal, Medical, and Ethical Guidelines.* New York, Society for the Right to Die, 1985.
7. New York State Mental Hygiene Law 77.
8. *Matter of Lyon*, 382 NYS2d833 (1976), Aff'd 396 NYS2d 183.
9. New York State Mental Hygiene Law 78.15.
10. Regs 18 NYCRR 454.14
11. Wanzer SH, Adelstein SJ, Cranford RE, *et al.*: The physician's responsibility toward hopelessly ill patient. *N Engl J Med*, 1980; 310:955–959.
12. *In re Eichner*, 438 NYS2d266, 420 NE2d6Y (1981).
13. McCormick RA: Life preservation and the incompetent. In Teichler-Zallen D, Clements CD, (eds) *Science and Morality*. Lexington, Mass, Lexington Books, 1982.
14. *In re Storar*, 438 NYS2d 266.
15. Beauchamp DE: Community: The neglected tradition. *Hastings Center Report*, 1985; 15(6):28–36.
16. *Youngberg v Romeo*, 50 USLW4681 (1982).

4

Informed Consent and Competence Issues in the Elderly

ROBERT WEINSTOCK

Questions regarding competence occur often in the elderly, partly because Alzheimer's disease, one of the most frequent causes of incompetency, primarily afflicts this group. This chapter will focus on competence to handle one's affairs. The main emphasis will be on assessment of competence to give informed consent to accept medical treatment or to refuse medical treatment, including the controversial questions of the refusal of life support systems and life sustaining measures.

Ethical, social, clinical, and legal issues are involved in these assessments, and the questions are often very complex. Significantly, complex as these questions are, and much as they address psychiatric and legal issues and even the application of psychiatry to a legal question, forensic psychiatrists have traditionally played little role in these matters. These matters are addressed usually by general psychiatrists or sometimes even by general physicians, who usually have little awareness or understanding of the complex issues involved. Forensic psychiatrists do, however, have much to contribute.

Competence is, of course, a legal issue; however, emergency situations, court delays, or the reluctance to take the time to go to court, make the psychiatrist the de facto decision maker of last resort in most cases.

ROBERT WEINSTOCK • Department of Psychiatry, University of California at Los Angeles, Los Angeles, CA 90024; Department of Medicine, University of California at Irvine, Irvine, CA 92664; Cedars-Sinai Medical Center, Los Angeles, CA 90048.

Because this is an issue involving the application of psychiatry to a legal question, the forensic psychiatrist has much to offer. Because such cases rarely go to court, knowledge of the relevant legal criteria becomes especially important, as the psychiatrist is the one who ordinarily makes the final decision. It is therefore striking that forensic psychiatrists in general, for the most part, have ignored this entire subject area even on a consultative basis.

In the not too distant past, mental patients were considered automatically incompetent for all purposes merely because they were hospitalized in a mental hospital, even if they were voluntary.[1] Patients were deprived of the right to make contracts, drive an automobile, vote, marry, and so forth. They were considered globally incompetent for all purposes, regardless of the reality. Gradually, with some notable exceptions,[2] most court decisions have changed to reflect better the reality that incompetence for one purpose does not necessarily mean incompetence for another. Legal criteria have come to differ for differing specific competencies. Global incompetences involved in a decision to appoint a guardian, conservator, or committee still exist, with the specific name differing by state. However, the concept of differential specific competencies has been increasingly appreciated and understood. In some states there is the realization that even committed mental patients may occasionally be competent to refuse psychiatric treatment unless a separate assessment of incompetency is made, regardless of whether the patient meets the quite different criteria for civil commitment. In my opinion, the problem with these decisions is that the legally required procedures have been unduly cumbersome and costly, but the facts and issues are inescapable. To ignore them[1] runs the risk of returning to rigid automatic and occasionally invalid rules. What is needed may be a streamlining of procedures and perhaps giving the psychiatrist, patient, and family more opportunity to consider these issues on their own or with the assistance of an institutional review committee.[3,4,5]

One of the problems in assessment of competence is the absence of understanding of the concept of differential competence by physicians. Weinstock[6] showed significant confusion by physicians regarding requests for competence assessments to a psychiatric consultation service. This confusion probably extends to most psychiatrists with no forensic background. In 1968 a survey of attorneys, psychiatrists, and administrators showed severe confusion and lack of information.[7] Almost all, when questioned, believed that there should not be a single criterion for competency applicable to all situations. They believed the criteria should vary according to the function. However, most did not know the existing legal criteria. They also showed little evidence of serious thinking on this matter, as demonstrated by the lack of thoughtful consideration shown by their poorly thought out suggestions. It is doubtful that the situation has improved considerably, even though recent involvement of the law in psychiatry may have heightened awareness slightly.

Most forensic psychiatrists do have familiarity with the concepts of differential and specific competencies. Not all general psychiatrists do. Other civil competencies, of course, include such matters as competence to refuse treatment, contract, handle finances, make a will, marry, divorce, operate a motor vehicle, have a physician's license, hold a job, vote, hold public office, perform judicial duties, practice as an attorney, testify, be a parent, etc. For any conceivable function there needs to be specific functions evaluated. Some are spelled out by statute, others by court decisions or general practice in a community. The criteria differ to some degree from state to state.

Guardianships, conservatorships, and committees assess a person as being globally incompetent for all or almost all purposes to care for his person and/or his estate. Although in the severest of cases a person may be incompetent for all conceivable purposes, the risk of these assessments, although sometimes convenient, is that they encourage clinicians not to think out the specific circumstance but to apply an automatic rule and not even consider the fact that a person may be capable of handling a specific function even if he has been legally assessed as globally incompetent. For example, a patient may be sufficiently competent to decide whether he wishes his leg amputated even if he is not competent to handle his finances. Global assessments also run the risk of being interpreted as "written in stone," and a change in condition such as recovery from a drug-induced delirium or pseudodementia[8] might not always lead to a reevaluation. Their only advantage is the absence of a need to return to court each time an issue arises in a generally incompetent person, an obviously time-saving procedure. However, awareness of the concept of differential competency is important, especially among non-forensic psychiatrists, who often become confused by these complexities and look for an automatic rule as opposed to thinking out the specific circumstances.

LEGAL BACKGROUND—INFORMED CONSENT

Anglo-Saxon law does not provide generally accepted guidelines for establishing criteria for competency to consent to or refuse treatment. Courts in the past used broad rather than narrow criteria and assessed patients as globally competent or incompetent to handle their affairs.[9]

Early cases in England involved not obtaining consent at all. In *Slater v Baker and Stapleton*[10] there was liability because of a deviation from the customary standard that physicians obtained consent before administering treatment. In the early 1900s the law in the United States developed from the common law action for battery in a series of cases involving the absence of any consent prior to treatment.[11,12]

In one case, reference was made to an earlier case in which permission had been granted to operate on the right ear; but the physician, finding an infection in the left ear, operated on it. Despite the success of the operation it was considered a battery.[13] Judge Benjamin Cardozza in 1914 in the case of *Schloendorff v Society of New York Hospital*[14] made his now famous statement that "every human being of adult years and sound mind has a right to determine what shall be done with his own body, and a surgeon who performs an operation without his patient's consent commits an assault for which he is liable in damages." Increased concern about the need for informed consent followed the Nurenberg Code in 1947.[15]

In the late 1950s the standard of the community (the malpractice or negligence standard) developed in those cases where consent was obtained but the consent was inadequate. No consent at all involved a battery; inadequate consent involved negligence or malpractice.[16] The negligence or malpractice standard requires that in addition to inadequate information, there also be damages. Several cases reflected the negligence standard for inadequate consent.[17,18,19] The standard was physician based, that is, based on what competent physicians in the community would do and risks they would consider material.

In the early 1970s the standard shifted to a patient-based reasonable man or a reasonable patient standard,[20] instead of a negligence standard based on the conduct of other physicians. In *Canterbury v Spence*[21] the case involved the failure of a neurosurgeon to disclose a 1% risk of paralysis in a lumbar laminectomy. The patient was paralyzed, and the court noted that a physician must disclose the information a reasonable man would need to make a decision. In *Cobbs v Grant*[22] the test established for adequacy was the patient's need, the materiality to the patient's decision. Complications need to be explained in understandable lay terms. The patient must be told if there is a potential for serious harm or death. However, there is a therapeutic privilege that allows the physician to withhold information that would so upset a patient he could not weigh it dispassionately.

Many jurisdictions adopted relevant statutes in the 1970s. Some considered the need for a procedure different from civil commitment when nonpsychiatric medical treatment was at issue because civil commitment authorized treatment for only psychiatric conditions and was not intended to be relevant to the treatment of nonpsychiatric purely medical conditions. For example, the fact that a patient is a danger to others is totally irrelevant to whether he has the capacity to consent to a surgical operation. Even if a patient is suicidal, he may be willing and competent to consent to unrelated medical treatment. He may even be competent and willing to consent to psychiatric treatment, as in the case of a voluntary psychiatric hospital admission.

Conservators (both mental health and probates) of the person can in California be given the power to consent to medical treatment, but it is not automatic. In 1981 in California, for example, provisions went into effect that allowed for petition to the court for court authorization for medical treatment in the case of an incompetent patient. There is also an emergency exception. California Probate Code 3208 states that a petition may be granted if all of the following apply:

1. There is a medical condition requiring treatment.
2. If untreated, there is a probability it will become life endangering or result in serious threat to the physical health of the patient.
3. The patient is unable to give informed consent.

Section 3210 states that nothing in this section limits action in an emergency for the purpose of:

1. Alleviating severe pain.
2. The patient has a medical condition that, if not immediately diagnosed and treated, will lead to serious disability or death.

A few courts recently have even gone beyond the reasonable patient standard and adopted an individual patient standard. In *Scott v Bradford*[23] the important issue was whether the risks would be material to the making of an intelligent choice by the particular patient. "To the extent the plaintive, given an adequate disclosure, would have declined the proposed treatment and a reasonable person in similar circumstances would have consented, a person's right of self-determination is irrevocably lost." Such a test requires the physician to discuss the proposed procedure with the patient in a sufficiently detailed manner in order to know what would be material to the particular patient's decision. It does not require disclosure of every rare side effect. However, it does potentially lead to the risk of a physician not being able to ascertain everything that would be material to an individual patient's decision, even in an extensive interview, unless specifically so informed by the patient. It may, therefore, be a potentially impractical standard, even though it would certainly be ideal.

Informed consent can also require informing a patient of the risk of *not* undertaking a medical procedure. In one case a physician was considered liable for not letting a patient know that failure to obtain a Pap smear could lead to not detecting a possible cancer.[24]

LIFE-SUPPORT SYSTEMS AND "RIGHT-TO-DIE"

The right to make decisions about accepting treatment also implies a right to refuse that treatment. Since the 1970s relevant decisions have been made regarding these issues in situations where the removal of life support systems was requested. In the Quinlan case in New Jersey[4] the court used the substituted judgment standard in deciding that her guardian and family would know best what her wishes would have been if she were able competently to take into account her noncognitive vegetative existence. They felt that her right to privacy extended to a right to terminate such an existence by natural forces. The court felt that the decision should be left to the family. It also required, however, the physician to determine that there was no reasonable possibility of her "emerging from her present comatose state to a cognitive sapient state" and that a hospital ethics committee should concur with the physician's determination.

In Massachusetts,[25] the substituted judgment standard was also used. The requirement was that an appointed guardian ad litem and the judge should have attempted to ascertain the person's actual interests and preferences, as they did. In ascertaining how the person would have decided, the present and future incompetency of the individual was one of the factors to be taken into account. The court allowed termination of life support systems but required that the question be submitted to the court for hearing.

The President's Commission for the Study of Ethical Problems in Medicine and Biomedical and Behavioral Research[26] also used the substitute judgment standard in these cases by saying that the "decisions of surrogates should, when possible, attempt to replicate the ones that the patient would make if capable of doing so." If there is no relevant evidence, the "best interests" standard should then be applied. Guidelines developed by the Los Angeles County Medical and Bar Associations[27] give recommendations that when surrogate decision makers agree, a responsible physician may go along with their decision and not necessarily be required to take the case to court, and the physician could designate the surrogates if none has been so appointed or designated. If they disagree, the physician should maintain life-sustaining treatment until the disagreement is resolved or the court appoints a conservator to decide. The patient's desires should take priority if known. Otherwise the best interests standard would apply.

In California the concept of proportionality[28] has been utilized by the courts in terms of benefits to be gained versus the burdens caused. Proportionate treatment is

> that which, in the view of the patient, has at least a reasonable chance of providing benefits to the patient, which benefits outweigh the burdens attendant to treatment. Thus, even if a proposed course of treatment might be extremely painful or intrusive, it would still be proportionate treatment if the prognosis was for complete cure or significant improvement in the patient's condition. On the other hand, a treatment course that is only minimally painful or intrusive may nonetheless be considered disproportionate to the potential benefits if the prognosis is virtually hopeless for any significant improvement in condition.

The court stated that a focal point of the decision should be the "reasonable possibility of a return to a cognitive and sapient life" or "a remission of symptoms enabling a return to a normal functioning integrative existence." The court required the physicians to determine the diagnosis and prognosis "under the generally accepted standards of medical practice in the community and, whenever possible, the patient himself should then be the ultimate decision maker." The court also stated that "if the treating physicians have determined that continued use of a respirator is useless, then they may decide to discontinue it without fear of civil or criminal liability."

In addition, in the Bartling case[29] in California it was decided that any adult who has a terminal illness, or a serious illness that is probably incurable but has not been diagnosed as terminal, and who is capable of giving informed consent may have life-sustaining treatment discontinued even if withdrawal of such treatment will surely hasten his death. The court also gave incompetent patients the same rights as competent ones to refuse medical treatment, that is, by means of substituted judgment.

The Los Angeles County Guidelines[30] state that "all life-sustaining interventions, including nutrition and hydration, are legally equivalent. It is legally possible for the caregiver to withhold or withdraw any or all of them." They recommend, however, that decisions be made jointly by members of the health care team, family, and other persons, and that the hygiene and dignity of the patient be maintained. The guidelines encourage the physician and family to make decisions on their own in California in situations where the courts have allowed such decisions, without the necessity of always going to court. This situation is obviously different in other states that have not gone so far as California. In those states one should always take these situations to court, and I would also recommend it for cases at all questionable in California.

Another related and complex issue is that of DNR (do not resuscitate) or no-code. Such decisions should also be made by competent patients or by substitute decision makers if the patient is incompetent. Ideally, the healthy person should make a living will or appoint someone to have a durable power of attorney if he should become incompetent in the future in those states that allow it. However, few persons have done so. In reality these decisions are often made unofficially solely by medical staff. Such an event is, of course, legally very questionable and risky. However, further discussion is beyond the scope of this chapter.

SUBSTITUTE DECISION MAKERS

Although courts theoretically make a decision about a person's competence and who the substitute decision maker should be, the psychiatrist often in reality does so. Emergency situations, delays with court, or even the reluctance to go to court, and the fact that suits are rare in these circumstances if one considers the relevant facts and weighs in good faith the needs of the patient against a respect for autonomy, encourage this situation. Psychiatrists in fact generally make decisions about competency; and the psychiatrist, along with the primary care physician, generally decides who the substitute decision maker should be.

In circumstances in which there is a life-threatening emergency situation and the patient is unable to give consent, the doctrine of "implied consent" assumes that the patient would be presumed to want treatment and that he would have given consent if able.[31] Such circumstances include that where the patient is unconcious or in shock. This doctrine can at times be extended to the incompetent patient when there is no substitute decision maker to be found and the need for urgent treatment clear. Courts have commented on who a reasonable substitute decision maker might be, so there are guidelines for situations in which the physician must choose. In California a court decided that one should consider immediate family members or nonfamily friends who are in the best position to know the person's feelings and desires about treatment, would be most affected by the decision, are concerned for the patient's comfort and welfare and have shown an interest in the patient by visits or inquiries to the patient's doctor or hospital staff.[32]

The standards used in proxy decision making are either the best interests or substituted judgment standards. These standards originated in circumstances where family disputes were being resolved or issues regarding control of an incompetent's property or gifts from his estate arose.[5] The best interests standard requires an attempt to make an objective assessment of what will promote an incompetent person's welfare without regard for the patient's previously stated wishes or preferences. It is, however, not always easy to determine what is in a person's best interest, and the decision maker's preferences and prejudices may inevitably influence the decision. The standard also ignores a fundamental value in informed consent—the right to individual self-determination. The substituted judgment standard requires the decision maker to ascertain what the incompetent person would have decided if competent. The substituted judgment standard allows an incompetent person to make foolish judgments not in his self-interest much like a competent person could make. An example of the application of the substituted judgment standard is allowing a Jehovah's Witness to refuse blood transfusions even if it would result in death.[33] In cases where no previous preference has been expressed, relevant factors include previous general expressions, religious beliefs, effect on the family and how that would have affected the decision, probability of side effects, the benefit, and the prognosis.[5] In those cases where there is no way to determine how the patient himself would have decided, the best interests standard could be the only possible one to apply.

LAWS REGARDING GUARDIANSHIPS, CONSERVATORSHIPS, AND COMMITTEES

Caring for the well-being and the estate of a person who is unable to care for himself is based on the parens patriae power of government and has a history dating back to the Twelves Tables of Rome.[34] Although convenient and automatic once a person is declared incompetent for all purposes, there is the danger of overlooking the fact that a person may be de facto competent for a specific purpose, even if he is de jure incompetent.[35] Except in cases of severe organic brain syndromes, it is rare that one can truly make a judgment that a person is incompetent for all purposes. Even a psychotic patient may be competent for matters not affected by his psychosis, as indicated in a study by Weinstock.[36] The name for global incompetency assessments granted by the courts differs from state to state. It may be called a guardianship or a committee. In states such as California, conservatorships that ostensibly are limited guardianships may also in reality refer to global assessments of incompetency.[37] The criteria are often vague and refer to one being generally unable to care for oneself or being unable to provide for food, clothing, or shelter.[38] In order to assess such functioning truly, basic living skills must be assessed, such as the ability to shop in a store, handle finances, etc. If global incompetency is being considered, or something that essentially is no different, it is especially important to do a functional assessment under differing circumstances, an assessment that is done infrequently.[37] In California either a mental health conservator or a probate conservator of the person (as opposed to the estate) can be given the power to consent to medical treatment, but it is not automatic.[39] However, a probate conservator of the person cannot admit a patient to a mental hospital, but a mental health conservator can. It is interesting that differential competency to consent to medical treatment is recognized by the law for conservatorships in California, even though the person is otherwise considered globally incompetent. In my experience, practice differs in how the conservatorship laws are interpreted, even in differing counties in California. For example, mental illness is not defined in the California statutes for mental health holds in a manner similar to short-term commitment and conservatorships. In my experience, dementia is generally not considered a mental illness in Los Angeles County. Therefore, one cannot place a demented patient on a mental health hold or conservatorship in Los Angeles, but must resort to a probate conservatorship of the person. In Orange County, however, dementia is considered a mental illness so mental health holds and conservatorships are possible. Criteria, of course, also differ from state to state.

Criteria and practice for relatively global assessments of incompetency as are required for guardianships, conservatorships, and committees differ according to the state. However, most such procedures involve declaring a patient incompetent for all or almost all conceivable purposes. It is, therefore, especially important for the psychiatrist to keep in mind that even though a person may have been legally adjudicated as incompetent, he may be competent for a specific purpose nonetheless. In such a situation it is incumbent on the psychiatrist to state that fact, bring it to the court's attention again, and give the patient the opportunity to make his own competent decision.

CONCEPT OF INFORMED CONSENT

The three main variables for informed consent are that there (a) be provision of information; (b) the patient be competent; (c) there be understanding (or appreciation).[40] A precondition for consent to occur is that it be voluntary, that is, free from unfair persuasions and undercurrents. The requirements for disclosure are that there be discussion of the risks, benefits, alternatives, and nature of treatment. In my experience, the alternatives are, however, rarely presented to a patient. According to the report of the President's Commission,[41] only 14% of physicians surveyed saw provision of alternatives as important to informed consent. What types of alternatives are needed is also unclear. State statutes that confront this issue make reference to reasonable and recognized alternatives, viable alternatives, or medically acceptable alternatives.[42]

State statutes also differ regarding risk. Some states do not require informing a patient of risks inherent in a procedure. Others oblige the physician to do so. Some states require informing a patient of common, usual, or most frequent risks. Others do not require it if the risks are too commonly known. Some do not require informing a patient of minor, infrequent risks. Another requires only significant and substantial risks, and it is unclear whether the reference is to frequency or severity.[42] It therefore is clear that it is necessary to refer to the statute in one's own state, should such a statute exist. Otherwise, the requirements are unclear and probably should be determined by the reasonable patient or individual patient standard discussed earlier.

Exceptions to the need to get informed consent are: (a) an emergency; (b) incompetency; (c) waiver; and (d) therapeutic privilege.[43] The emergency exception in my experience is sometimes overlooked when rules are followed too rigidly by frightened medical personnel. I have encountered situations in which nursing personnel, for example, have refused to act even when it could place a patient's life in danger unless a psychiatrist was called to declare the patient incompetent or legal proceedings started, even when such actions were dangerous to the patient's welfare. In my opinion, this behavior is a result of an excessive fear of the law by medical personnel and an overlegalistic interpretation. Forensic psychiatrists could go far in dispelling these fears, including the fact that a refusal to act in an emergency could first place that person at risk for negligence and lead to the possibility of a successful malpractice suit. It is certainly better medical practice, and in my opinion usually safer legally as well, to save a patient's life when the situation is unclear and there is no time to go through legal formalities.

Incompetency is the issue that most often involves the psychiatrist in informed consent matters. Unfortunately, considering the complexity of the issues entailed, it is rare that forensic psychiatrists become involved even in a consultative capacity. Incompetency has in statutes been variously defined as lack of capacity to consent to treatment, lack of capacity to be informed about treatment, and adjudicated incompetency prior to the need for care.[44]

Waiver means that a patient may decline to be informed and say he will do whatever the doctor thinks best, and that he would rather not be burdened with the details.

Therapeutic privilege refers to the privilege described in *Cobbs v Grant*[22] for the withholding of information that would so upset a patient he could not weigh it dispassionately. This exception should not be relied on too readily, however. There should be good evidence for it.

In considering what test of competency to use, it is important to use a realistic test and not expect the patient to understand all the details or to take a course in medicine or surgery. Studies have, moreover, shown in normal people that most of what is presented to them is neither understood nor remembered in circumstances of consent. Robinson[45] showed poor retention in all categories of informed consent in cardiac patients tested postoperatively. Casseleth[46] tested cancer patients one day after they had given informed consent to chemotherapy, radiation, or surgery. Only 60% actually understood the nature and purpose of the treatment. Only 55% were able to list even one potential complication. Only 40% actually had read the consent forms carefully. In spite of the theory of informed consent, most felt that the consent forms were to protect the physicians' rights. They did, however, believe that the forms were worthwhile and comprehensible, but 31%, after completing them, thought official documents were out of place in the clinical setting. It was the researchers' impression that most relied on their physician, and that trust in the physician was the factor most patients preferred to rely on.

Because even unquestionably competent patients did not understand much of the information presented to them and the procedures to which they consented, it is important not to expect more from a slightly impaired individual. In addition, many consent forms are impossible to comprehend. In a study of the readability of consent forms,[47] few of them measured up. Most required a graduate school level education to comprehend. It is therefore clear that many normal patients will not understand the information in a consent form. Another study[48] showed that consent forms must be brief to maximize comprehension and intelligent decision making. In my opinion, the required degree of understanding should be determined by the risks and benefits of a proposed procedure.

TESTS FOR COMPETENCY TO GIVE INFORMED CONSENT

Psychiatrists generally become involved in the matter of informed consent for nonpsychiatric medical treatment when asked to evaluate competency. Although informed consent is mentioned in many court decisions and many state statutes, the actual test for competent informed consent is rarely defined. Roth[49] summarized some of the alternative tests that have been used to assess competency. They are (a) evidencing a choice; (b) the choice was reasonable; (c) the choice was based on "rational" reasons; (d) there was a generalized ability to understand; (e) the patient actually understood.

The actual test depended on the circumstances and depended on weighing the risks versus the benefits of a particular procedure.

Appelbaum and Roth[50] listed a hierarchy of tests to be used in assessing competency to consent to research. In my opinion, such a hierarchy is also useful in determining competency to consent to non-psychiatric medical procedures. The tests in order of increasing stringency generally following Appelbaum are:

1. Choosing—the most basic test.
2. Understanding—the most widely accepted test. Either the ability to understand the specific contemplated procedure or actual understanding as demonstrative of such ability can be tested. This test can be subdivided according to the degree of understanding required. In more complex situations, more understanding should be required than in less complex situations. I find the generalized ability to understand less useful except insofar as it is used to confirm the specific ability to understand a specific procedure.
3. Reasoning—the ability to manipulate information rationally.
4. Appreciating—the most demanding test. It includes affective as well as cognitive components. It requires full emotional appreciation. It is the ability to consider the relevance of facts to the situation and take crucial data into consideration.

Some examples of conditions in which these tests might lead to an incompetency assessment are:

1. Choosing—situations in which the patient is mute, manic, with a profound thought disorder, exhibiting catatonic excitement
2. Understanding—dementia with inability to remember over time; delirius (with need to retest when stable), poor intelligence, poor education
3. Ability to reason—delusions or hallucinations, cognitive delusions in bipolar disorder, looseness of associations, delusions of hopelessness, futility, nihilism, ambivalence (as in an extreme obsessive-compulsive)
4. Appreciating—depression, mania, grandiosity, hopelessness, helplessness, marked denial, severe phobia, panic, anger, agitation, impulsivity

Roth proposed the idea of considering the risks and benefits of a specific procedure in determining the threshold for a test for competency. I would like to propose the use of an actual risk/benefit ratio in determining the threshold for competency. Therefore, the risk/benefit ratio can be high either if the numerator is high or the denominator low (see Table I). An example of the functioning of Table I would be a very risky procedure where the risks are high and the benefits low. The risk/benefit ratio is therefore high. There is a presumption of incompetency if the patient accepts this procedure unless there is a very understandable explanation. In other words, one must be "out of one's mind" to accept. Similarly, if there is a safe life-saving procedure where the risks are low and the benefits high, the risk/benefit ratio is low. There is a presumption of incompetency if one refuses unless there are understandable reasons for doing so. One must ordinarily be "out of one's mind" to refuse. Understandable or recognizable reasons are sufficient even if not wise. For example, the patient may wish to wait to be treated by his own physician even though he knows the risk of waiting.

In some situations the risks and benefits are inextricably entangled with the patient's subjective experience. If the patient has terminal cancer, it is important to make sure first that the patient is not suffering from a major depression or a temporary adjustment order with depression. If he is, the depression should be treated and any life prolonging treatments continued, because the risks would be seen as low to moderate and the benefits high. There would be a fairly high threshold for competency if the patient refused and such a depressed patient would not meet this threshold. He would be unable to appreciate the situation and perhaps even unable to reason if he had delusions of hopelessness. However, if the patient had reached a conclusion over a period of time, had expressed even before becoming severely ill that he would not wish life prolonged after a certain point, his decision might be competent. He might be able to give rational reasons for refusal, such as refusing che-

Table I. Competency Threshold Table

Risk/benefit ratio		Patient consents	Patient refuses
High	Risk high/ benefit low	High threshold to be competent	Low threshold to be competent
Moderate	Risk low/ benefit low —or— Risk high/ benefit high	Moderate threshold	Moderate threshold
Low	Risk low/ benefit high	Low threshold to be competent	High threshold to be competent

motherapy for cancer because it would prolong the pain or make him lose his hair, or that he had lived a good life and would wish to die comfortably. In that situation, such a procedure as chemotherapy might be seen to have high risks (such as prolongation of pain, loss of hair) and low benefits (continuation of life with increasing pain). Such an eventuality might then lead to a high risk/benefit ratio and there would be a low test for competency if a patient refuses. If life support systems were refused in such a circumstance, there would be a possibly high risk/benefit ratio if the person saw living on life support systems as painful and very unpleasant. The use of a threshold for competency is consistent with the concept that every patient should be permitted to make his own decisions, if at all possible, unless his decision is too irrational. If it is too irrational, a more careful testing is done and a more stringent test is employed. Such a procedure is consistent with the values of our society and balances the competing concepts of autonomy and paternalism.

Another example of the use of the table is with psychiatric treatments. Voluntary psychiatric patients are generally held to the lowest test of competency, that is, consent. Such a procedure results from seeing the risks as low to moderate and the benefits high. Therefore, there is a fairly low ratio. If the patient consents, the threshold for competency is low (perhaps too low) because we would ordinarily require some understanding in a competency test for even the simplest safe medical procedure. Perhaps such a situation parallels the tendency of physicians to accept frequently the signature of a patient who agrees to a low risk/high benefit procedure without doing a competency assessment. There is a tendency to believe that any patient who accepts the "good" judgment of the physician must be competent. If one were to use a more realistic test, probably some test of understanding should be used for voluntary psychiatric patients that at least requires them to know that they are voluntarily agreeing to psychiatric treatment in a psychiatric hospital. Otherwise, they should probably be committed because they are unable to give informed consent, even if they are willing to sign a paper. In my experience, psychiatrists generally allow patients to sign a voluntary admission paper without even questioning the issue of competency. This issue is relevant because studies have shown little awareness of the meaning of voluntary admission regardless of standard and also little understanding of voluntary consent forms among psychiatric inpatients.[51,52,53]

To consider such patients as having given informed consent to treatment either because they signed an admission form or even a specific form for psychotropic medications (as used in California) is probably misleading unless they have been evaluated as competent to make this decision. Otherwise, commitment or a psychiatric hold would probably be more appropriate with the patient also evaluated as incompetent to consent or refuse psychotropic medication. However, it is important to note that considering the fairly low risk/benefit ratio, the threshold for competency could be low if the patient consents. Similarly, the studies that show voluntary psychiatric patients do not understand the forms they signed may not be that different from those that show that medical patients frequently do not understand their forms. The relevant test in my opinion would be an understanding test with the information explained by the psychiatrist and an attempt made to make sure the patient understood.

A test somewhat similar to the risk/benefit ratio standard (Table I) has been suggested by Drane,[54] who proposed a sliding scale for competency. As the risk increases, the stringency of the standard increases. The lowest standard is one in which the proposed procedure is not dangerous, and in the patient's interest. The second standard is one in which the illness is chronic, the procedure more dangerous and of less benefit. The highest standard is for decisions that are very dangerous and fly in the face of both public and professional rationality. Balancing the values of autonomy and paternalism is the cornerstone of a good competency assessment. For the highest standard, competency requires the capacity to appreciate the nature and consequences of the decision. Appreciation requires the highest degree of understanding, more than just the medical details of the illness and treatment. For appreciation, personal reasons need not be publicly accepted but must not be purely private, idiosyncratic, or incoherent. The person must be able to give reasons and the reasons may derive from a philosophical view or religious beliefs shared by even a small minority.

Appreciation is thus the most stringent, highest standard as recommended by myself, Appelbaum, and Drane. Discussion of appreciation as involving affective, as well as cognitive, understanding originated in discussions of the insanity defense.[55,56,57] An example of the use of such a test is a patient with depression who is so discouraged that he feels there is no hope or that whatever he does will make no difference. He has a local lesion that looks like a melanoma, but he refuses to have it removed because there "is no point in doing so." In my opinion, this is a low risk/high benefit procedure and the threshold test for a patient refusing is high. In my opinion, he would not meet the threshold for competency, even if he was able to understand cognitively the risks and benefits of the proposed procedure and could explain the reasons of his physicians, but refused because it was pointless. He would, in my opinion, be unable to appreciate (on an affective basis) the risks and benefits of the procedure and might even have his reasoning affected by delusions of hopelessness, and would therefore be incompetent.

OTHER ISSUES TO CONSIDER

It is important to remember that the autonomy of a patient must be respected. The patient's wishes should be overridden only if there are compelling reasons that suggest that the patient is not fully capable of weighing the risks and benefits of a procedure and his wishes should be overridden for his own benefit. One might even use a substituted judgment or a "thank you theory"[58] type of standard that suggests that the patient himself would have wanted the procedure if he was not incapacitated by a psychiatric problem. However, competent patients who truly understand the risks and benefits have a right to use poor judgment and make unwise decisions. The patient's wishes should ordinarily be respected if there is some possibility that his way of doing things might work, even if the decision is clearly against good medical judgment. An example would be a patient with a gastrointestinal bleed who requires a work up as to cause. The patient, however, refuses, says he feels fine and did well for months the last time this happened. He has business affairs to handle, whereas the nursing staff says he probably wishes to go out and drink alcohol. However, he knows the risks and knows he could have a massive gastrointentinal bleed and die. He may be showing some emotional denial, but it is possible that he will do well like last time. Considering the risk/benefit ratio, a fairly high test, the ability to reason test seems appropriate. He is able to reason, knows what is involved but says he prefers to handle his affairs and will come back, perhaps in a week or two, to complete his medical work up on an outpatient basis. He is making a very unwise decision but would be appropriately assessed, in my opinion, as competent to make that decision.

Psychodynamic factors are also important in consent and refusal,[59] both as an aspect of understanding the patient's decision and also as leverage in attempting to change a patient's mind when he is believed to be making an unwise, yet competent, decision.

The psychiatrist, forensic or otherwise, should attempt to understand how a patient's fears compromise understanding and address those fears therapeutically. In unclear situations such an attempt will sometimes help clarify the problem and even help in the assessment of a patient's competency. It is also important to understand which defenses are mobilized. For example, if a patient has a need to appear self-sufficient and healthy to compensate for feelings of inadequacy, it might help to compliment the patient on the strength he has shown in being of support to his family and/or people at work, but say that failure to get needed treatment would not be the best way to look after things because he would not be there when others need him if he did not take care of himself. One might also show awareness that it is difficult for someone as active as he to stay bedridden for a period of time in the hospital and have others care for him. If a patient is using denial because a relative had had a similar problem that eventually was diagnosed as cancer, it could help to clarify any ways in which the patient's situation might be different and in which there would be good reasons to suspect that he would have a good prognosis with treatment. Often patients are not really told about a procedure,[60] and it is critical to make sure an adequate explanation has been given. If one evaluates a patient and discovers he does not understand a procedure, the evaluator does not know whether information was provided and whether it was done in simple language a patient can understand.

I find it useful in doubtful situations, where the psychiatrist does not fully understand the procedure, to jointly discuss the situation with the patient and the medical physician. Sometimes this is important in mild dementia where the psychiatrist does not know whether the patient was given a full clear explanation and forgot it, or whether an understandable explanation never truly was given. In situations where there is delirium and the patient's mental state is changing, it is useful if there is time to wait until the patient's mental condition stabilizes and the underlying cause of the delirium treated. Otherwise, it may be necessary to call the patient temporarily incompetent and emphasize the need for reevaluation in the future because of the changing mental status picture. Such a statement is important if the psychiatrist does not continue to be involved. In my experience, there is otherwise the danger that the original competency evaluation may be considered as if "written in stone" and a reevaluation never requested. The concept of being very careful about competency assessment when a perceived high-risk procedure is being contemplated (regardless of whether the perception is true) is demonstrated by the California statute[61] regarding ECT. The statute requires that even if a patient is voluntary and consenting, there is a need in cases of ECT to get a second medical opinion to document the patient's actual competency to consent to the treatment. Roth[59] also suggests involving family and friends. Others can be important, firstly in supplying data, because the patient can be a poor informant. Also, people the patient trusts and who talk the same language, coming from the same class or race, can often be helpful in persuading a patient to make a wise decision. The psychiatrist's responsibility should not end with an assessment of competency if the patient is making an unwise yet competent decision, but family and friends whom the patient trusts more than his physician can be very helpful in calming his fears and anxieties and persuading him to make a reasonable decision.

In addition, extra efforts can be helpful with the elderly in assisting them to understand what is involved in a procedure and enabling them to give informed consent.[62] Careful explanation and provision of information can be especially important. Even if there are some cognitive deficits, such additional effort can sometimes enable an older person to give informed consent.

PROCEDURE FOR EVALUATION

I recommend the following procedure for assessing competency to give informed consent to or to refuse medical treatment. As with most competency assessments the emphasis should, in my opinion, be on the specific function to be tested and not on a general test for competency that may not be relevant, is inefficient and unnecessarily time-consuming, can confuse some clinicians, and may not even test the specific function in question. Consistent with this procedure, a competency decision should not be based on a diagnosis or even a general psychological test for competency because the specific function in question may not be tested or could be statistically lost in the midst of many irrelevant items. Moreover, an emphasis on a general or global test of competency as opposed to a specific test of competency contributes to confusion. Such a procedure would support what I consider to be an inappropriate idea that the best way to assess a patient's competency for one purpose is to assess it for a totally different purpose.[63] The procedure I recommend is the following:

1. Determine the threshold for competency by considering the risk/benefit ratio and referring to Table I.

2. Evaluate the patient's understanding of the specific procedure being contemplated or the patient's ability to understand. The degree of understanding required should be determined by the risk/benefit ratio and resulting threshold. Mere consent or refusal in and of itself should not be taken as an adequate competency test unless there is no reason even to question the patient's competency. If no adequate explanation of the procedure has been given to the patient, the patient's capacity to understand could be evaluated. However, it would probably be more useful, more accurate, and more relevant to evaluate the patient's actual understanding after making sure an adequate explanation was given (often by being present).

3. Depending on the threshold decided on the basis of the risk/benefit ratio of the procedure, it would be important to test the patient's reasoning ability and the basis for his decision. It also might be necessary to evaluate the patient's full appreciation (emotional and cognitive) of the situation.

4. A mental status exam should then be performed and a history taken in order to determine whether the findings are consistent with the competency assessment, and the deficits and diagnosis discovered of the type that might produce the problems found in the competency assessment. If these are consistent, a relatively brief history and mental status examination may be sufficient.

5. If the situation is unclear or a very high test of competency is being utilized, one should do a very thorough mental status exam and history and then return to the specific issue and procedure in detail in order to evaluate the patient's specific abilities very carefully.

Other issues of importance include being aware of the ethical need to inform the patient of the purpose of a competency evaluation.[64] In addition, in unclear or questionable situations involving very high threshold competency tests, or those open to challenge because family members are disagreeing, it would be useful to take the case to court if at all possible. Emergency considerations could, however, preclude such an option if a delay might be dangerous, life threatening, or very painful for a patient. Assessment could be very controversial in situations requiring a high competency test or in which family members disagree, and it therefore would be helpful, if at all possible, to ask a court to make the determination, because unlike the psychiatrist, a judge is ordinarily immune from any potential liability for making controversial decisions in such situations.

RESEARCH ON COMPETENCY TO GIVE INFORMED CONSENT IN THE GERIATRIC POPULATION AND WITH OTHERS

Even though competency to give informed consent is a prominent issue in the geriatric population, there is little research in this area. Stanley studied the reasoning ability[62] of geriatric patients to evaluate possible research protocols. Patients over age 62 showed poorer understanding of consent information for five out of six projects and significantly poorer overall comprehension. There was significantly poorer understanding than younger patients of the purpose of the procedure, the nature of the procedure, the risks, and the benefits. Significantly, however, the elderly patients made reasonable decisions, nonetheless, in five out of six studies. The one study in which they did not make a reasonable decision was a high risk/low benefit study in which the elderly agreed to participate at a significantly higher rate than younger patients. Such an outcome suggests that elderly patients may not need to understand fully all the details in order to make a reasonable decision.

Weinstock,[36] in a study of competency to give informed consent to medical procedures, found that the only patients in a Veterans Administration population who were found to be incompetent to give informed consent were those with dementia. Patients with other psychiatric problems were found to be competent during the 6-month period studied. These findings suggest the possibility that psychotic patients are more competent than is generally believed to make decisions about treatment that is unrelated to their psychiatric disorder. Another possible explanation is that psychotic patients tend to follow passively their doctors' wishes and that under such circumstances assessments are not requested by the physician, or alternatively a low threshold for competency may be required under many circumstances. Such findings also suggest that the capacity to have a full understanding of a procedure is not necessary for a patient to be competent and to be able to reach a competent decision. These findings give added basis to the caution in using either a psychiatric diagnosis or a general competency assessment in evaluating the ability to make a specific decision.

Reisberg distinguished between the phases[65] of dementia and at which point deficits became sufficiently severe that patients became incompetent to give informed consent regarding treatment. In the earliest phase, the "forgetfulness phase," the deficit is primarily subjective, the patient is aware of it and has trouble recalling names and places. The next phase is the "confusional phase," in which the patient has trouble with recall of recent events, decreased concentration, and trouble recalling many appropriate words, but still has a spared vocabulary. In the early part of this phase insight is maintained, but it diminishes in the latter part. In the "dementia phase" the individual can no longer survive on his own and needs assistance for basic activities. Through the early confusional phase patients can make reasonable decisions and are especially interested in participating in studies that could help cognitive abilities. In the late confusional phase, however, judgment is compromised as well as insight. Even though one might still try to get consent from the patient, it would be advisable at this stage also to get consent from the spouse or that of the next of kin.

In addition to the previously mentioned studies, Weinstock[36] evaluated requests to a psychiatric consultation service at a VA hospital for patients over the age of 60. Of the 66 competency requests during a 6-month period, 27 were for patients over the age of 60. Interestingly, however, of these 27 consults, 20 were at least partially for an evaluation of ability to give informed consent. That means that in the patients over 60, only seven consults were for all other types of competency, including conservatorships, competency to handle VA money, and so forth. This finding gives added credence to the impression that competency to give informed consent is an important issue for the elderly.

Lidz[43] did a study of how informed consent operates in practice in a psychiatric hospital, that is, in reality as opposed to in theory. Regarding information, they found that patients were typically given information after consent was given and not before for the purpose of aiding in carrying out the treatment plan and not to aid in the decision. The information came from many people. Disclosures were brief and incomplete by physicians, and there was poor disclosure of alternatives. Regarding understanding, it was typically incomplete, technically limited, and frequently idiosyncratic. In reference to the decision made, most believed it was up to the doctor; consent forms were treated as an insignificant ritual, and rather than present alternatives to a patient, staff felt committed to help patients make the *right* decision. They concluded that truly informed consent did not exist in the observed settings. Tancredi[66] also suggested that informed consent has hardly achieved its promise and that there is only a facade of patient involvement.

CLINICAL CASE HISTORIES

The following are case histories from a psychiatric consultation service at a VA Hospital. Consults were from medical and surgical wards. The risk/benefit ratio as discussed in Table I is utilized to help decide the threshold for competency to give informed consent. The appropriate test is then utilized to assess the patient's competency to give informed consent to the procedure in question.

CASE HISTORY WITH MODERATE-HIGH RISK/BENEFIT LOW-MODERATE; RISK/BENEFIT RATIO HIGH—LOW THRESHOLD TO REFUSE

A 65-year-old man with a severe peripheral vascular disease and diabetes was evaluated for competency to give informed consent. He had multiple cerebrovascular accidents with residual hemiparesis and expressive aphasia and was admitted to a medical ward with congestive heart failure and pneumonia. He was refusing permission to amputate his leg, saying that he would rather die than be crippled further and would rather take a chance on gangrene developing. At first he refused to communicate but later did communicate when informed about the purpose of the interview. All questions had to be posed so that he could answer YES or NO. He was eating during the interview and showed no signs of psychomotor retardation although he did become tearful on occasion. There was no past history of depression.

According to the medical resident, the patient's family had no interest in him and requested "no code." He appeared to understand the treatments and what was contemplated and why, but merely wished to be left alone and not further disfigured, did not care whether he lived or died, but wanted an end to his discomfort. He appeared to understand the risks and benefits of treatment, not to be suffering from a temporary situational depression, having been in this state of mind for some time. His depression appeared reality based and not to involve an unrealistic appraisal of his situation. He was evaluated as competent to refuse treatment.

CASE HISTORY WITH LOW-MODERATE RISK/BENEFIT MODERATE-HIGH; RISK/BENEFIT RATIO LOW—LOW THRESHOLD FOR COMPETENCY TO CONSENT

A 65-year-old man with cancer of the bladder was referred from urology for a competency evaluation. He had refused a total cystectomy recommended by his physicians. He had signs of a mild dementia possibly secondary to alcoholism. He knew the year but was off by one month on the date. He knew the name of the hospital and the general area, but forgot the city. He knew the President and could do serial 7's. He recalled two out of three objects after 3 minutes. It appeared that he had not understood the alternatives open to him. We explained that the best prognosis was with surgery and that radiation was possible but less than ideal. After a full explanation he consented to surgery and was able to remember our discussion 30 minutes later. He was assessed competent to give informed consent. This case is also an example of the need to make sure that adequate information was supplied to the patient and that it was explained in terms that the patient could understand.

CASE HISTORY WITH LOW RISK/BENEFIT HIGH; RISK/BENEFIT RATIO LOW—HIGH THRESHOLD TO REFUSE

A 76-year-old man with chronic renal failure, who now has a signicant electrolyte imbalance and may require dialysis as well as catheterization, was referred for evaluation. On interview with the patient he was uncooperative, hostile, and childlike. He would stick out his tongue in a playfully hostile manner and smile in a teasingly inappropriate way. We attempted to explain the seriousness of the situation to him, including the fact that he would die without treatment. Although he knew the procedure being contemplated, he could not understand the danger and felt that all he needed was some water to make his kidneys work. He did not seem capable of appreciating the life-threatening aspect of his situation or of weighing the risks and benefits. He was oriented to person, knew the month but was a year ahead in the year, knew the city but was incorrect in giving the hospital name. He could remember one out of three objects after 5 minutes. He could do calculations but made some serious errors in the Bender-Gestalt test.

Because of the high threshold for competency necessary to refuse this low risk/high benefit procedure, he was assessed incompetent to refuse treatment, even though he may have been competent for other purposes.

He was unable to appreciate the situation and had some difficulty in understanding secondary to dementia in the middle confusional phase. The consent of the family was obtained and the medical situation judged by the internists as sufficiently urgent not to allow for time to go to court.

CASE HISTORY WITH LOW RISK/BENEFIT LOW-MODERATE; RISK/BENEFIT RATIO MODERATE—MODERATE THRESHOLD TO REFUSE

An 80-year-old man in a nursing home with end stage renal disease was evaluated. He had multiple medical problems including adenocarcinoma of the prostate, arteriosclerotic vascular disease, and basilar artery disease. There was no evidence of depression or psychiatric disorder in the past. The patient had been on hemodialysis in the last few years. Consultation was requested because the patient was now asking to be taken off the dialysis program.

Mental status exam revealed a frail, elderly man lying quietly in bed. His calm was punctuated by occasional agonal groans. The patient extended his hand in greeting, maintained eye contact, and had an appropriate smile. Speech was soft yet coherent. Thought processes were goal directed without loosening. There was at times a preoccupation with the pain and misery of illness and his current life situation. He did not express self-destructive intent; however, he asked, "Is there a point to continuing?" The patient's affect was appropriate to thought with some lifting of mood when he spoke of other subjects. He did not have sleep problems. He did have some loss of appetite and weight loss. He was oriented, sensorium was clear without fluctuation in a lethargic, fatigued state. General intellectual functions were slowed yet intact. Discussion with his family revealed a gradual decision on the part of the patient to forgo treatment rather than an acute response to any particular stress. The patient demonstrated an understanding of his current medical situation and the facts relevant to his case. He was able to understand the likely outcome of continued dialysis or interruption of treatment. He appeared to have depression, but his mood seemed appropriate to his situation. It did not appear to be a major depression or a temporary adjustment disorder with depression, both of which would have been amenable to treatment. He appeared to be aware of his terminal state and pain. He was, in our opinion, competent to refuse treatment. He met both the understanding and the ability to reason tests. In fact, he also seemed to appreciate what his situation involved.

CASE HISTORY WITH HIGH RISK/BENEFIT LOW; RISK/BENEFIT RATIO HIGH—HIGH THRESHOLD TO CONSENT

An 88-year-old man was found to have a large left frontal mass. He also had Parkinson's disease, peptic ulcerative disease, and a history of a cerebrovascular accident. Consultation was requested by the surgery resident prior to surgery to remove the frontal tumor because of their concern about his competence, considering the operative risk. This request was made even though the patient consented to surgery. The patient showed no ability to understand his current medical condition. He was not able to appreciate the risks of surgery in a man his age with his medical problems, even after they were carefully explained to him. He believed he would be fine without being able to supply any evidence to support his optimism. He seemed unable to appreciate the dangers or even to repeat them to us although we explained them to him. He was oriented to person and place but was off by about 2 weeks on the date. Recent and remote memory showed some problems but only minor ones. In my opinion, he had a mild dementia interfering with his ability to appreciate the risks of surgery for a man his age. He was also having some trouble in reasoning and understanding secondary to denial and a dementia at the early confusional phase. In my opinion, he was not competent to give informed consent to the procedure.

Mental illness can, of course, also be a reason for incompetency to consent to medical treatment. However, the fact that someone is committed for psychiatric treatment (or on a psychiatric hold in California) should not, in my opinion, mean that he could or should be automatically treated for an unrelated medical problem. Such is the law in California; although some jurisdictions, such as Orange County, sometimes blur this distinction. It is my impression that practice differs throughout the country as it does even within the State of California. It should be automatic only if the consent or refusal directly relates to the reason for the commitment, such as part of a suicide plan. Otherwise, the question should be handled as a competency question even if the reason for incompetency is mental illness. Situations in which the person is mentally ill and gravely disabled could probably be handled either by a procedure such as a mental health conservatorship (where the conservator is given the power to consent to medical treatment) or by a declaration of incompetency to give informed consent. The situation should be handled much like other forms of incompetency, and the mental health hold or commitment reserved for the psychiatric treatment. As can be seen, the specific legal maneuvers can be complex and differ by state.

CASE HISTORY WITH LOW RISK/BENEFIT HIGH; RISK/BENEFIT RATIO LOW—HIGH THRESHOLD TO REFUSE

A 68-year-old depressed woman refused antibiotics for a respiratory infection because she was discouraged, did not want needles, and did not care if she died since everything was empty and hopeless anyway. She did not actively wish to kill herself but was merely refusing treatment because she did not want painful needles. She intellectually understood the risks and benefits and could repeat them, but she did not appreciate the situation because of her depression. Her hopelessness reached delusional proportions and interfered with her ability to reason, so she did not even meet that test. Her husband died 2 months ago and she had vegetative signs of decreased appetite and early morning awakening. She was willing to take antidepressants since they helped in a similar episode ten years earlier, so she was not refusing psychiatric treatment and was willing to be a voluntary psychiatric patient. However, she should in my opinion, be considered as incompetent to refuse her medical treatment.

CASE HISTORY WITH LOW-MODERATE RISK/BENEFIT HIGH; RISK/BENEFIT RATIO LOW—HIGH THRESHOLD TO REFUSE

A nongeriatric patient, incompetent because of psychosis, was a man refusing needed abdominal surgery for an intestinal obstruction. The patient refused this procedure because he had a delusion that he had an extra penis in his abdomen that was responsible for all his difficulties. He would not sign a consent form unless the surgeon agreed on paper to remove the extra penis. The surgeon was understandably reluctant to include such a strange statement in the consent form. The patient had been diagnosed as chronic schizophrenic and maintained on Haldol. He was otherwise doing fine and would be willing to increase his Haldol if recommended. He therefore was willing to be a voluntary psychiatric patient. However, his reasoning ability was hampered by his delusion. His understanding was also slightly affected because although he did understand what the doctors believed the necessary procedure should be, he himself believed a different procedure should be involved and that the "extra penis" that was not removed in a previous surgery was what was truly responsible for his troubles.

Considering the relatively high threshold for competency to refuse this procedure, and because he was refusing the procedure the surgeons were contemplating, he did not meet the reasoning test and not even the understanding test, and was therefore incompetent to refuse this procedure. In spite of being incompetent for making a decision to refuse this procedure, he might very well be competent to decide about another procedure about which he had no delusional ideas, even if the same threshold of competency existed.

CONCLUSIONS

Informed consent is an important issue in dealing with the elderly. It comes most in question in the late confusional phase. Competency is best assessed by looking at the specific issue in question. Even in cases where someone is generally incompetent or de jure incompetent by being on a conservatorship or guardianship, it is important to be alert for the fact that patients may be competent for the specific purpose in question. Informed consent represents a balance between patient autonomy and beneficence or paternalism. The threshold for competency is best assessed by a risk/benefit ratio assessment.

Even if it is an ideal not always realized in practice, competency to give informed consent represents a balance between competing forces and values in our society. It is important to keep the necessity of balancing competing values in mind when doing an assessment. Utilizing only one value or principle or looking for an absolute rule will lead to difficulty. The physician should try to do what is best for a patient but should keep in mind that a competent patient has a right to make his own decisions, even wrong ones. A useful concept is that of requiring the patient to have "recognizable reasons"[67] even though they may not be fully rational reasons.

I would also recommend streamlining of procedures and giving more authority to the physician and family as opposed to the courts. Courts do not always understand medicine and are rarely available on an emergency basis and take time and money. In my opinion, it would be an error to deal with the problem by throwing out the whole concept of informed consent and specific competencies, and reverting back to judgments of global incompetency or automatic judgments of incompetency for one purpose when a person is found incompetent for another. Informed consent is an important concept. It involves allowing a competent patient to make his own decisions on the basis of information provided by the physician. There is and should be a sharing of decision making. Physicians should be encouraged to present alternatives to competent patients (or their families), even though a preference could be stated by the physician to them.

Unfortunately, confusion is widespread. In my opinion, this confusion is not the result of confusion in the concepts. The problem is that most physicians and psychiatrists do not understand legal concepts and forensic psychiatrists have traditionally avoided this area. In my opinion, forensic psychiatrists need to become aware of criteria for competency to give informed consent and need to play either a direct role or an educative role and help make informed consent and specific competencies. In addition, courts need to streamline procedures and give more autonomy to physicians and psychiatrists, perhaps with the assistance of institutional ethics committees to help in education and also in difficult cases. Courts do not have the flexibility and responsivity to be involved in most cases. They should, in my opinion, reserve their role for the complex and controversial cases where there is no time pressure and not the routine ones. Either the patient or the physician should be able to request a court hearing. In questionable cases physicians should try to keep patients alive until the courts decide. Efforts should also be made by physicians to insure that DRGs (Diagnostic Related Groups) and other cost-saving measures are not used to cause patients to die who would not themselves wish to do so. Forensic psychiatrists should increase their visibility in what is a complex matter with many ethical ramifications. They need to educate themselves and play a role in educating the general psychiatrist about the many facets presented by these complex, and at times very difficult, questions.

REFERENCES

1. Stone A: The right to refuse treatment. Why psychiatrists should and can make it work. *Arch Gen Psychiatry* 1981; 38:359.
2. *Stensvad v Reivitz*, No. 84-C-383-S (W.D. Wis Jan 14, 1985).
3. *Parham v J.R.*, 99 S.Ct. 2493 (1979).
4. *In re Quinlan*, 70 N.J. 10, 355 A.2d 647 cert denied 429 US 922 (1976).
5. Solnick PB: Proxy consent for the incompetent terminally ill adult patient. *J Leg Med* 1985; 6:1–49.
6. Weinstock R, Copelan R, Bagheri A: Physicians' confusion demonstrated by competency requests 1985; *J Forensic Sci* 30:37–43.
7. Allen R, Feister E, Weihofen H: *Mental Impairment and Legal Incompetency*. Englewood Cliffs, N.J., Prentice Hall Inc, 1968, p 251.
8. Wells CE: Pseudodementia. *Am J Psychiatry* 1979; 136:895–900.
9. Roth L: Competency to consent or to refuse treatment, in Grinspoon L (ed): *Psychiatry 1982 Annual Review*. Washington, American Psychiatric Press, 1982, p 353.
10. *Slater v Baker and Stapleton*, 2 Wils 359, 95 Engl Rep 860 (KB1767).
11. *Pratt v Davis*, 224 Ill 300, 79 NE 562 (1906).
12. *Rolater v Strain*, 39 Okla 572, 137 p96 (1913).
13. *Mohr v Williams*, 95 Minn 261, 104 NW 12, 14 (1905).
14. *Schloendorff v Society of New York Hospital*, 211 NY 125, 105 NE 92 (1914).
15. Lebacqz K, Levine RJ: Respect for persons and informed consent to participate in research. *Clinical Research* 1977; 25:101–107.
16. Rosoff AJ: *Informed Consent, A Guide for Health Care Providers*. Rockville, Md, Aspen Systems Corp, 1981, p 58.
17. *Salgo v Stanford Univ Board of Trustees*, 154 Cal App 2d 560, 317 P 2d 170 (1957).
18. *Natanson v Kline*, 183 Kan 193, 409–410, 350 P 2d 1093, 1106 rehearing denied 187 Kan 1986, 354 P 2d 670 (1960).
19. *Mitchell v Robinson*, 334 SW 2d 11 (Mo 1960) affirmed after retrial 360 SW 673 (Mo 1962).
20. Rosoff AJ: Informed Consent, A Guide for Health Care Providers. Rockville, Md, Aspen Systems Corp, 1981, p 38.
21. *Canterbury v Spence*, 462 F 2d 772, 783–788 (DC Cir 1972).
22. *Cobbs v Grant*, 104 Cal Rptr 505, 502 P 2d 1 (S Ct en banc 1972).
23. *Scott v Bradford*, 606 P 2d 554 (Okla 1979) rehearing denied (1980).
24. *Truman v Thomas*, 27 Cal 3d 285, 611 P 2d 902, 165 Cal Rptr 308 (1980).
25. *Superintendant of Belchertown State School v Saikewicz*, 373 Mass 728, 752–753, 370 NE 2d 417, 431 (1977).

26. *Deciding to Forego Life-Sustaining Treatment,* 3, 5 U.S. Government Printing Office, 1983.
27. *Principles and Guidelines Concerning the Foregoing of Life-Sustaining Treatment for Adult Patients,* Los Angeles County Bar and Medical Associations, 1985 and 1986, p 1.
28. *Barber v Superior Court,* 147 Cal App 3d 1017–1020 (1983).
29. *Bartling v Superior Court,* 163 Cal App 3d 186 (1984).
30. *Principles and Guidelines concerning the Foregoing of Life-Sustaining Treatment for Adult Patients,* Los Angeles County Bar and Medical Associations, 1985 and 1986, p 4.
31. Jonsen A, Siegler M, Winslade W: *Clinical Ethics.* New York, MacMillan, 1982, p 89.
32. *Barber v Superior Court,* 163 Cal App 3d 1006, 1021 fn 2 (1983).
33. *In re estate of Brooks,* 32 Ill 2d 361, 205 NE 2d 435 (1965).
34. Gutheil T, Appelbaum P: *Clinical Handbook of Psychiatry and the Law.* New York, McGraw Hill, 1982, p 221.
35. Meisel A: The exceptions to the informed consent doctrine: Striking a balance between competing values in medical decisionmaking. *Wisconsin Law Rev* 1979; 58:413–488.
36. Weinstock R, Copelan R, Bagheri A: Competence to give informed consent for medical procedures. *Bull Am Acad Psychiatry Law* 1984; 12:117–125.
37. Morris G: Conservatorship for the "gravely disabled". California's non-declaration of non-independence *Int J Law Psychiatry* 1978; 1:395–426.
38. California Welfare and Institutions Code 5000 ff.
39. California Probate Code Section 1801.
40. Meisel A, Roth L, Lidz C: Toward a model of the legal doctrine of informed consent. *Am J Psychiatry* 1977; 134:285.
41. *President's Commission for the Study of Ethical Problems in Medicine and Biomedical and Behavior Research: Making Health Care Decisions: The ethical and legal implications of informed consent in the patient–practitioner relationship* 1982; 18:67.
42. Andrews LB: Informed consent statutes and the decision making process. *J Leg Med* 1984; 5:196–197.
43. Lidz C, Meisel A, Zerubavel E, *et al.: Informed Consent, A Study of Decision-making in Psychiatry.* New York, The Guilford Press, 1984.
44. Andrews LB: Informed consent statues and the decision making process. *J Leg Med* 1984; 5:209.
45. Robinson G, Mercu A: Informed consent: Recall by patients tested postoperatively. *Ann Thorac Surg* 1976; 22:209–212.
46. Casseleth B, Zupkis R, Sutton-Smith K, *et al.:* Informed consent—Why are its goals imperfectly realized? *N Engl J Med* 1980; 302:896–900.
47. Grundner T: On the readability of surgical consent forms. *N Engl J Med* 1980; 302:900–902.
48. Epstein L, Lasagna L: Obtaining consent-form or substance? *Arch Intern Med* 1969; 123:682–688.
49. Roth L, Meisel A., Lidz C: Tests of competency to consent to treatment. *Am J Psychiatry* 1977; 134:279–284.
50. Appelbaum P, Roth L: Competency to consent to research: a psychological overview. *Arch Gen Psychiatry* 1982; 39:951–958.
51. Appelbaum P, Mirkin S, Bateman A: Empirical assessments of competency to consent to psychiatric hospitalization. *Am J Psychiatry* 1981; 138:1170–1176.
52. Olin GB, Olin HS: Informed consent in voluntary mental hospital admissions. *Am J Psychiatry* 1975; 132:938–941.
53. Klatte EW, Liscomb WR, Rozynko CV, *et al.:* Changing the legal status of mental hospital patients. *Hosp Community Psychiatry* 1969; 20:199–202.

54. Drane JH: Competency to give informed consent-A model for making clinical assessments. *JAMA* 1984; 252:925–927.
55. Goldstein AS: *The Insanity Defense.* New Haven, Conn, Yale Univ Press, 1967.
56. Zilboorg G: Misconceptions of legal psychiatry. *Am J Orthopsychiatry* 1939; 9:540, 553.
57. Diamond B: Criminal responsibility of the mentally ill. *Stanford Law Rev* 1961; 14:62.
58. Stone A: *Mental Health and Law: A System in Transition.* Washington, National Institute of Mental Health, 1975.
59. Roth L: Competency to consent or to refuse treatment, in Grinspoon L (ed): *Psychiatry 1982 Annual Review.* Washington, American Psychiatric Press, 1982, pp 355–356.
60. Ashley M, Seszak R, Roth L: Legislating human rights-informed consent and the mental health procedures act. *Bull Am Acad Psychiatry and Law* 1980; 8:133–151.
61. California Welfare and Institutions Code Section 5326.7.
62. Stanley B, Stanley M, Pomara N: Informed consent and geriatric patients, in Stanley B (ed): *Geriatric Psychiatry: Ethical and Legal Issues.* Washington, American Psychiatric Press, 1985.
63. Abernathy V: Compassion, control and decisions about competency. *Am J Psychiatry* 1984; 141:53–58.
64. *Principles of Medical Ethics with Annotations Especially Applicable to Psychiatry.* Washington, American Psychiatric Association, 1982, pp 5–7
65. Reisberg B, Gordon B, McCarthy M, *et al.:* Insight and denial accompanying progressive cognitive decline in normal aging and Alzheimer's disease, in Stanley B (ed): *Geriatric Psychiatry: Ethical and Legal Issues.* Washington, American Psychiatric Press, 1985.
66. Tancredi L: Competency for informed consent—conceptual limits of empirical data. *Int J Law Psychiatry* 1982; 5:51–63.
67. Freedman B: Competence, marginal and otherwise, concepts and ethics. *Int J Law Psychiatry* 1981; 4:53–72.

III

The End of Life

5

Psychiatric Assessment of Competence to Choose to Die

Proposed Criteria

RICHARD ROSNER

This chapter explores the emotion and value laden area of patients rights to refuse medical treatment, with the knowledge that such refusal may be tantamount to a decision to hasten their personal death. In general there are four major areas in which such issues will arise. (a) Patient decisions regarding whether to authorize extraordinary treatments, so-called heroic therapy regarding their condition. (b) Patient decisions regarding charted "do not resuscitate" orders, that is, orders directing the medical staff not to use cardio-pulmonary resuscitation in the events of a cardiac arrest. (c) Patient decisions to authorize physicians to use whatever level of narcotic/sedative medications may be neccessary to control the patient's pain, with the understanding that the administration of such large doses of medications may prematurely hasten the patient's demise. (d) Patient decisions to seek discharge from the hospital, against medical advice, with the understanding that they will go home to die in a familiar setting, rather than to be subjected to the loneliness and pain of inefficacious treatment in a hospital or nursing home.

RICHARD ROSNER • Forensic Psychiatry Clinic for the New York Criminal and Supreme Courts (First Judicial Department) of the Department of Mental Health, Mental Retardation and Alcoholism Services of the City of New York; Department of Psychiatry, School of Medicine, New York University, New York, NY 10016.

CASE HISTORY A

"Don't torture me any more!" the old man screamed as the intern shooed out his family and asked for the defibrillation equipment. The man was 78 years of age, he had been discharged only 2 weeks earlier from the urological surgery department of a university hospital, being told that his prostate was not yet "ripe" for surgery. He was also told, though he did not share that data with his family, that he had "some heart problem" and that he should "see a doctor for it." One day prior to this admission, he had felt unable to urinate and called his urologist; it was recommended that he re-present himself at the university hospital's emergency room for urological evaluation and possible admission. At the emergency room, he was diagnosed as having an acute myocardial infarction. As there were no available beds in the university hospital's cardiac care center, he was transferred to the emergency room of a nearby municipal hospital. After several hours of waiting (the city was in the grip of a mighty snowstorm and the staff cardiologists had to battle weather and traffic to arrive at all), he was transferred to the cardiac care center of the municipal hospital. Within minutes the still half-frozen team of cardiology fellows and residents had attached the myriad EKG monitoring leads, the nasal oxygen tube, the intravenous fluid lines, the urinary catheter, the central venous pressure subclavian vein line, and the usual panoply of medical technology and techniques had been deployed. No one stopped to explain what these tubes and machines were for, nor inquired whether the patient wanted them. Scant moments later, they were needed to control an episode of ventricular fibrillation. Notwithstanding appropriate medications, electrical treatments were needed repeatedly. The patient complained of the pain of the process, complained that he never asked for the kind of treatment to which he was being subjected. He said, "I'm an old man. I'm going to die soon anyway. The treatment is worse than the sickness. Don't torture me any more!" He was ignored.

CASE HISTORY B

The doctor took the patient's family out into the hallway to speak with them out of the patient's hearing. He explained that the patient was gravely ill, with multiple ailments, and that continued life would surely be both painful and of unpredictably short duration. He wanted the family's advice regarding whether to list the patient as a "no code" case, that is, to write orders for the patient not to be resuscitated should he sustain a cardiac arrest. After due consideration, the family indicated that such a decision would have to be made by the patient. The doctor questioned whether or not the patient was competent to participate in such a complex decisional process.

CASE HISTORY C

The patient was suffering from disseminated metastatic carcinoma and was in excruciating pain. His internist expressed concern that administration of narcotic and sedative medications sufficient to control the patient's discomfort would run the risk of respiratory depression, perhaps respiratory arrest. The patient wanted relief from pain at any risk; his family wanted the patient kept alive as long as possible. The internist requested a psychiatric-legal consultation regarding the patient's competence to authorize the use of enough medicine to control his pain, even at the risk of hastening the time of his inevitable death.

CASE HISTORY D

A young woman, the unwed mother of four infant children, was advised that she had a rhabdomyosarcoma that required the immediate amputation of her left leg and subsequent chemotherapy. She was reluctantly advised that the cancer had spread and that, even with chemotherapy, her prognosis was poor. She indicated that she wanted to go home, to care for her children "while I still have a leg to stand on," and that she would rather have a short time with her family as "a whole person" than a somewhat longer time with them as "a cripple." Psychiatric-legal consultation was sought as to whether or not the patient was competent to discharge herself from the hospital against medical advice.

These relatively common case examples are meant to demonstrate the four general categories noted earlier. The point is that it is not rare for physicians to encounter problems regarding the competence of patients to engage in actions that entail rejection of proposed medical/surgical treatments or to request pain relief such that the patient's requests, if granted, might lead to the patient's death.

Although the laws regarding patients' rights in such matters vary from state to state, in New York that matter was addressed definitively by Judge Cardozo's decision that every individual "of adult years and sound mind has a right to determine what shall be done with his body."[1] The same decision also affirmed such competent adult patients' rights to control the course of their medical treatment (the point being reaffirmed in *In re Storar*).[2] In *Storar,* the New York Court recognized that a patient's right to determine the course of his medical treatment was paramount to what might otherwise by the doctor's obligation to provide medical care, and that the right of a competent adult to refuse medical treatment must be honored, even though the recommended treatment may be beneficial or even necessary to preserve the patient's life.

The critical matter thus becomes the assessment of whether or not a given patient is possessed of "sound mind," the determination of the patient's competence. However, the determination will itself depend on the criteria that are used for evaluate the patient's competence (as well, of course, on the patient's psychiatric status at the time of the evaluation).

The focus of this discussion has to be distinguished from related, but separate, clinical situations. We are not concerned with cases of legally incompetent minors, nor of legally adjudicated incompetents. We are not concerned with cases of persons who are seriously ill but who are reasonably expect to survive. We are here concerned with adults who are expected not to survive, that is, persons who are estimated to have a 1% or less chance of survival. Further, we are not concerned with persons who erroneously believe that their chances for life are 1% or less, we are concerned with adults who correctly believe that their illnesses are terminal. The issue is not the possibility that the specific treating physician may be wrong in his prognosis, rather we are interested in those cases in which the physician's grim prognosis is confirmed by reputable colleagues and the data all support (by clear and convincing evidence, or beyond a reasonable doubt) that the patient has a 1% or less chance for avoiding death due to his current illness. Thus, we are interested in the limiting extreme case of essentially certain death by illness of an adult patient who has not been legally judged to be incompetent. These cases may be few, but they are important.

To contrast the cases we are concerned with by comparison with the more familiar instances, we will briefly review three seemingly related, but distinguishable, decisions: *In re Karen Quinlan*,[3] *Superintendent of Belchertown State School v Saikewicz*,[4] and *Eichner v Dillon*.[5]

In the *Quinlan* case, the father of a comatose patient petitioned the New Jersey Court for permission to have the patient disconnected from a respirator when "the strong likelihood is that death would follow soon after its removal." The Court held that

> upon the concurrence of the guardian and family of Karen, should the responsible attending physicians conclude that there is no reasonable possibility of Karen's even emerging from her present comatose condition to a cognitive, sapient state and that the life-support apparatus now being administered to Karen should be discontinued, they shall consult with the hospital "Ethics Committee" or like body of the institution in which Karen is then hospitalized. If that consultative body agrees that there is no reasonable possibility of Karen's ever emerging from her present comatose condition to a cognitive, sapient state, the present life-support system may be withdrawn and said action shall be without any civil or criminal liability therefore on the part of any participant, whether guardian, physician, hospital or others.

In the *Saikewicz* case, a resident of a Massachusetts State school for the mentally retarded suffered from acute myeloblastic moncytic leukemia and the Court was petitioned to appoint a guardian with the authority to refuse to permit Saikewicz to receive chemotherapy. The Court opined that "the decision in cases such as this should be that which would be made by the incompetent person, if that person were competent."

In the case of *Eichner v Dillon*, Brother Joseph Charles Fox, at age

83, suffered brain damage as a result of a cardiac arrest during the course of an operation to correct an inguinal hernia. The Court was petitioned for authority to discontinue the life-support system that was sustaining Brother Fox. The Court held that

> the physicians attending the patient must first certify that he is terminally ill and in an irreversible, permanent or chronic vegetative coma, and that the prospects of his regaining cognitive brain function are extremely remote. Thereafter, the person to whom such certification is made . . . may present the prognosis to any appropriate hospital committee. . . . that committee shall either confirm or reject the prognosis. . . . Upon confirmation of the prognosis, the person who secured it may commence a proceeding . . . for appointment as the Committee of the incompetent, and for permission to have the life-sustaining measures withdrawn. The Attorney General and the appropriate District Attorney shall be given notice of the proceeding. . . . Additionally, a guardian ad litem shall be appointed to assure that the interests of the patient are indeed protected by a neutral and detached party wholly free of self-interest.
>
> Where this procedure is complied with, and where the court concludes . . . that the extraordinary life-sustaining measures should be discontinued, no participant—either medical or lay—shall be subject to criminal or civil liability as a result of the termination of such life-sustaining measures.

These court decisions address patients who have already been adjudicated incompetent and, thus, are tangential to our inquiry. Karen Quinlan and Brother Joseph Fox were physically unable to respond to questions regarding their medical treatment; they were comatose. Joseph Saikewicz was regarded as so mentally retarded as to be incompetent to participate in treatment decisions. Our interest, however, is in patients who have not yet been adjudicated incompetent, who are not comatose, and who are not mentally retarded.

In the assessment of such cases, the forensic psychiatrist is advised to employ a four-step process by posing the following questions.

1. What is the psychiatric-legal issue?
2. What are the criteria used to determine the specific issue?
3. What are the relevant clinical data?
4. What is the reasoning process that underlies the psychiatric-legal opinion?

It is necessary to clarify the psychiatric-legal issue or issues to be addressed because, as indicated in the earlier examples, there is a variety of seemingly similar issues and precision is important. If the issue is the competence to refuse treatment of a minor, that is different from the competence of a mentally retarded person; it is not the same as the competence of a comatose person, nor is it the same as the competence of an adult facing a terminal illness. It is not just that the psychiatric techniques used in the evaluation of a minor, a retarded person, a co-

matose person, and an adult are different, it is that the laws (both statutes and case law) differ for these different competency issues. Each
psychiatric-legal issue must be approached in the context of the specific
laws governing it. The clinician is well advised to seek the assistance of
competent legal counsel to ascertain the exact psychiatric-legal issue and
the exact laws that pertain to that issue.

The criteria used to determine the specific issue cry for clarification.
This matter of criteria is understandable by way of analogy. If someone
were to ask a psychiatrist if a given patient had a diagnosis of bipolar
disorder, manic, without psychotic features, the psychiatrist would understand that he would have to compare the patient with the clinical
criteria in DSM-III.[6] If the patient matched the criteria, the answer
would be affirmative; if the patient did not match the criteria, the answer would be negative. Analogously, if someone were to ask a psychiatrist if a given patient was competent to refuse recommended medical
treatment that was deemed essential to that patient's life, the psychiatrist
should understand that he should compare the patient with the relevant
legal criteria.

Using yet another analogy, Loren Roth, Alan Meisel, and Charles
Lidz set forth five categories of tests for competency regarding that
specific psychiatric-legal issue, five different criteria that might be used
to assess competence to consent to treatment: "(1) evidencing a choice,
(2) 'reasonable' outcome of choice, (3) choice based on 'rational' reasons,
(4) ability to understand, and (5) actual understanding."[7] Further, they
explained that

> although in theory competency is an independent variable that determines
> whether or not the patient's decision to accept or refuse treatment is to be
> honored, in practice it seems to be dependent on the interplay of two other
> variables, the risk/benefit ratio of treatment and the valence of the patient's
> decision, i.e. whether he or she consents to or refuses treatment.[7]

Thus, the criteria used to determine competence to consent to treatment
vary depending on the context within which that competence must be
determined. The clinician should be aware of the likelihood that competence to refuse treatment is an analagously complex determination.

The relevant clinical data will depend on the criteria used to determine the psychiatric-legal issue. For example, if the issue is competence
to have made a valid last will and testament, then the relevant data will
address the patient's past state of mind at the time that the will was
signed. Alternatively, if the issue is competence to make a valid contract
currently, then the relevant data will address the patient's present state
of mind. What the examining psychiatrist must ask the patient, what the
patient must demonstrate in his answers, are governed by the exact legal
criteria that will be used to decide the issue in question.

The explication of the reasoning process that the physician has employed to come to his conclusion regarding the patient's competence is crucial to establishing the validity of his conclusion. This matter is of particular importance in assessment of competence to refuse life-sustaining treatments because of the possibility of the physician's own emotional reactions distorting his judgments. If the patient is very much older than the physician, there is the risk that the patient will be viewed as a quasi-parental figure. The doctor may be influenced by what he thinks would be best for his own parent, if that parent were situated similarly to the patient. The doctor may be at risk for succumbing to a wish to rescue this quasi-parent, at any cost, as he might wish to rescue his own parent. Less happily, the doctor may be at risk for experiencing towards the patient some of the hostile feelings he had towards his own parent, so that he may be emotionally biased towards favoring the patient's death by early termination of life-sustaining treatments. These biases are difficult enough to cope with when they are conscious, they are even more dangerous when they operate outside of the doctor's awareness. One of the few ways of trying to cope with such affective intrusions into the physician's judgments is to insist on a clear statement of the reasoning processes that sustain the judgments. The same type of problems arise if the patient and the physician are close in age: the doctor may be influenced by what he feels he would want to do for himself if he were in the patient's place, whereas his concern should be what the patient wants to do and whether the patient is competent to make that decision. The protection of the patient's autonomy, the integrity of the medical profession and the interests of society require that the doctor explain how he arrived at his conclusion regarding the patient's competence to refuse life-sustaining care.

Having outlined some of the problems in the assessment of the competence of patient's to choose to die, it may be appropriate to suggest some approaches to solutions. The assessment should be conducted with due consideration of the clinical context in which the patient exists. For the sake of clarity, we have tried to focus on those instances in which death is essentially certain, that is, the likelihood of survival is 1% or less. In such instances, there is no valid consideration of risk/benefit ratios of further treatment: there is no substantial likelihood of benefit, there is only aggravation of risk with concomitant pain and suffering. The only variables left to consider are the duration and the intensity of the pain and suffering. We would suggest that increased duration and increased suffering should require a lowering of the criteria used to determine competence, whereas shorter duration and lower suffering should require a raising of the criteria of competence (see Figure 1).

Whereas this context-related use of criteria may be readily under-

	Dying long	Dying short
Pain high	Low criteria	Middle criteria
Pain low	Middle criteria	High criteria

Figure 1. The criteria to choose to die.

stood, it still leaves the criteria themselves to be stated. We suggest that the criteria elaborated in the previously cited article by Roth, Meisel, and Lidz can be applied appropriately in the determination of competence to choose to die. The high criteria of actual understanding might be required when pain is low and the wait to die is short. The low criteria of evidencing a choice might be required when pain is high and the wait to die is long. The middle criteria of ability to understand or choice based on rational reasons or reasonable outcome of choice might be used when the pain is high and the wait for death is short or when the pain is low and the wait for death is long. Figure 2 may show this more clearly.

There is no way to avoid the decision of one's own fundamental values in such matters. If one believes that life should be preserved at any cost, then one should make clear one's value system to the patient and, perhaps, to the patient's family. The patient should have the option of choosing another physician to attend him, or another psychiatrist to consult on the matter of his competence to select death as an opportunity to avoid protracted suffering. To participate in such evaluations of patients' competence to choose to die, without revealing one's conscious biases, is not to provide the patient with essential information; it is not fair and it may even be dishonest. Even if one decides not to participate in such evaluations, one is defining one's values regarding the issue. It is not that a doctor cannot make a forthright attempt to prevent his values from intruding on his technical skills, it is that the doctor must know his values (must acknowledge them to himself) if he is to be able to prevent them from unduly influencing his decisions. These are two separate issues: the doctor should know his values to decide if they can

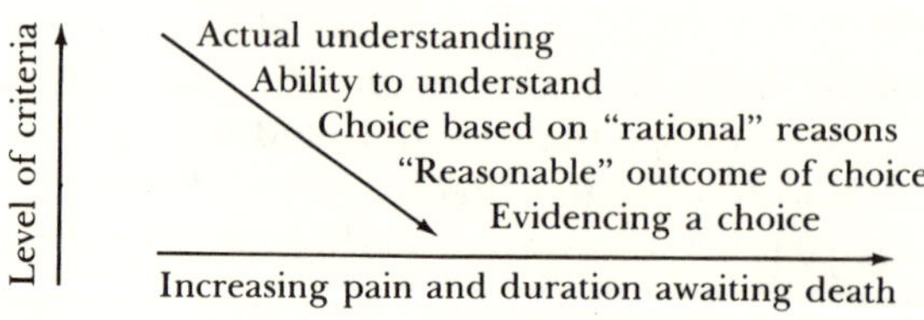

Figure 2. As the intensity of the pain and the duration of the wait for death (suffering) increase, the criteria used to determine a patient's competence to choose to die decreases.

be kept from interfering with his task to assess the patient's competence, and the patient should know the doctor's values to decide if another doctor should be asked to take on the case. It may be incumbent on some physicians to disqualify themselves from participation in the assessment of patients' competence to choose death.

The thrust of our argument is that precision is needed in the determination of competence to refuse life-sustaining treatments. We believe that the doctor must be aware of the potential of being biased by his own emotions and by his own values; that such an awareness of the potential for bias is one of the few techniques to avoid succumbing to such biases. We believe in the right of a competent patient to refuse to accept medical treatments essential to sustain his life. The critical matter is to be clear on which criteria will be used to determine if the patient is or is not competent. We have suggested that such a criterion should be related to the context in which the patient must decide; that the criterion should be lowered when the intensity and duration of anticipated pain and suffering are increased and that it should be raised when the intensity and duration of the pain and suffering are decreased. We have recommended that the criteria used in the determination of a patient's competence to consent to treatment should also be used in the determination of a patient's competence to choose to die. It is hoped that our discussion will contribute to further clarification of the process of evaluation of competence to refuse life-sustaining treatments.

REFERENCES

1. *Schloendorff v Society of New York Hospital*, 211 NY 125, 105 NE 92 (1914).
2. *In re Storar*, 52 NY 2d 363, 420 NE 2d 64, 438 NYS 2d 255 (1981).
3. *In re Quinlan*, 70 NJ 10, 355 A.2D 647 cert denied 429 US 922 (1976).
4. *Superintendent of Belchertown State School v Saikewicz*, 370 NE 2d 417 Mass (1977).
5. *Eichner v Dillon*, 426 NYS 2d 517 (App. Div. 1980).
6. *Diagnostic and Statistical Manual of Mental Disorders*, ed 3. Washington, American Psychiatric Association, 1980.
7. Roth L, Meisel A, Lidz C: Test of competency to consent to treatment. *Am J Psychiatry* 1977; 134:279–284.

6

Do-Not-Resuscitate Orders

The Impact of Guidelines on Clinical Practice

HAROLD I. SCHWARTZ

As the ability to resuscitate, to bring back the dead, has evolved from Elisha's biblical account of mouth to mouth resuscitation[1] to today's complex of intubation, mechanical respiration, medication, and monitors, so have evolved the complexities of who will be resuscitated and when and to what degree. Good medical practice, ethical and economic considerations, and the specter of liability are compelling medical practitioners to examine how these decisions are made.

It is noteworthy that the term *code* is most commonly used for a resuscitation effort. What is a code? To the intern and resident it is a frantic few minutes in which his or her skill, presence of mind, and endurance will be maximally challenged and a patient will live or die. To the patient, family and friends it may be the last hope—or, an unwanted, unwarranted invasive intrusion into life's most solitary moment. The dictionary defines a code as a system of secret writing in which letters, figures, and the like are arbitrarily given certain meanings. In this case the term has two specific uses. It appears first in the process of calling a resuscitation team, as in "code blue" or "code zero" announced over the public address system. It also refers to the coded designations that will guide the team's resuscitative efforts. Some hospitals assign letters, *A* through *E* for example, to signal the extent of resuscitation felt to be appropriate. The use of the term *code* highlights the degree to which

HAROLD I. SCHWARTZ • Department of Psychiatry, Beth Israel Medical Center; Mount Sinai School of Medicine, New York, NY 10003.

91

resuscitation decisions and the orders which emanate from them have historically been shrouded in secrecy. At one New York hospital, for instance, a patient in danger of cardiac arrest, until recently, would have been assigned one of three code statuses; a full code, meaning an all out effort, a floor code, or NTBR—not to be resuscitated. The term *floor code*—something between an all out effort and no effort at all, had no precise meaning. It was defined in each case by discussion, nuance, or gesture—and it was usually noted in pencil, somewhere other than in the chart or in the order book.

A number of important clincial, ethical, and medicolegal considerations surrounding the assignment of resuscitation status and the issuing of do-not-resuscitate (DNR) orders have begun to be addressed in recent years by hospitals, physicians, legislators, and other policy-making groups. These include the degree of patient and family participation in the decision, including mechanisms of substituted judgment for incompetent patients, ultimate responsibility for the assignment of code status and communication of these decisions to the staff through written orders and chart notes, and mechanisms for resolutions of disputes between physicians, patients, and families when differences of opinion arise about resuscitation status. This chapter will review the very brief history of the development of policy guidelines for the issuing of DNR orders. It will go on to review the few studies that have examined the impact of such policies on clinical practice. Finally, it will examine in greater depth a recent study that points to significant differences between the guidelines recommended by a 1983 presidential commission and the actual attitudes and perceived practices of clinicians. Evidence is accumulating that guidelines and hospital policies may have less impact on clinical practice than policy makers may wish.

THE DEVELOPMENT OF POLICY GUIDELINES

It was only in 1974 that the American Heart Association and the National Academy of Sciences began the process of attempting to establish uniform standards for the use of cardiopulmonary resuscitation (CPR) and the writing of DNR orders.[2] These standards were updated in 1980.[3] They require the following:

1. DNR orders should be written in the order book for benefit of all staff.
2. The patient's family should understand and agree with the decision although the family's opinion need not be controlling.

3. Physicians and the entire medical team should be in general agreement with the decision.
4. Confirmatory written opinions by appropriate consultants may facilitate this agreement.

Though written in 1980 these guidelines can properly be thought of as primitive. Nowhere do they mention the wishes of the competent patient.

In September, 1982, the Medical Society of the State of New York issued an advisory set of guidelines for all hospitals and medical personnel in the state.[4] These guidelines can be summarized as follows:

1. The attending physician should determine the appropriateness of a DNR order for any given patient.
2. When a patient is capable of making his own judgments the DNR decision should be reached consensually by the patient and physician.
3. When the patient is not capable of making his own decision, the decision should be reached after consultation between the appropriate family member(s) and the physician.
4. If a patient disagrees, or in the case of a patient incapable of an appropriate decision, the family member(s) disagree, *a DNR order should not be written.*
5. The attending physician shall write formal DNR orders and document the facts and considerations relevant to this decision in the chart.
6. The attending physician shall ensure that appropriate staff members are aware of the DNR order and its meaning.
7. The DNR order shall be subject to review at any time by all concerned parties on a regular basis and may be rescinded at any time.

These guidelines are clearly an advance. They place the patient in a central position in the decision-making process and offer guidelines for how to proceed when there are differing opinions. Medical societies in a number of other states have followed suit and published guidelines.

It should be emphasized that the medical society guidelines are only advisory. There is no statute in New York (and in most other jurisdictions) bearing directly on issues such as the use of family members for substituted judgment when patients are incompetent. There is as yet no clear legal consensus on this and other issues as basic as the legality of DNR orders themselves.

Finally, in March of 1983, the President's Commission for the Study of Ethical Problems in Medicine and Biomedical and Behavioral Research issued a report entitled *Deciding to Forego Life-Sustaining Treatment*.[5] This report was the culmination of a massive effort to review the ethical, medical, and legal issues involved in such treatment decisions. The President's Commission issued its own set of guidelines, which are summarized below:

1. A competent and informed patient or an incompetent patient's surrogate is entitled to decide with the attending physician.
2. When cardiac arrest is likely, a patient (or a surrogate) should usually be informed and offered the chance specifically to decide for or against resuscitation.
3. DNR orders should be written in the chart.
4. When a physician and a competent patient disagree, ultimately the physician must follow the patient's decision or transfer responsibility to another physician.
5. Incompetent patients should have surrogates (family).
6. If no family is available, judicial appointment of surrogates is necessary only for controversial cases.
7. If the surrogate and physician disagree, consultation, review by an institutional committee, and, if necessary, judicial review should be sought.
8. Accrediting agencies should require hospitals to have explicit policies.
9. Health care professionals should develop skills necessary to help patients and families make ethically justified decisions about resuscitation.

Note the further emphasis on procedures for substituted judgment and the handling of disputes. The addition of a requirement that physicians develop skills necessary to help patients and families with ethical dilemmas is another advance. Table I reviews the evolution of key issues in these three sets of guidelines.

THE IMPACT OF DNR POLICIES

The need for clear and effective policy guidelines on the issue of determining resuscitation status is clear from ethical and clinical points of view. Though the results of outcome studies vary, with survival rates ranging from less than 5% to greater than 20%,[6] there is no doubt that overall survival of CPR is low. Although there is ongoing debate about the role of factors such as age, the factor most clearly correlated to outcome appears to be the prognosis of the patient's condition prior to

Table I. The Evolution of DNR Guidelines

	American Heart Association 1980	NY State Medical Society 1982	President's commission 1983
Competent patient decides	−	+	+
Incompetent patient's family decides	±	+	+
Orders should be written	±	+	+
Encourages staff discussion	+	+	+
Provides guidelines for disagreement	−	+	+
Requires explicit policies	−	−	+
Requires development of skills in ethics	−	−	+

cardiac arrest.[7,8] Cardiopulmonary resuscitation is a massively invasive procedure that has many serious complications that can reduce the quality of life for survivors.[9] The futile application of CPR to patients with no meaningful chance of recovery may be devasting to the patient and family and comes at great economic and psychological cost.[10] The growing concern that the poor survival rate of CPR reflects in part its indiscriminate use with inappropriate patients has led hospitals to implement their own DNR guidelines. As a result, official hospital guidelines are no longer uncommon.

Several studies have reviewed the impact of hospital DNR policies on actual practice. McPhail *et al.*[11] reported on the results of the implementation of a DNR policy at McMaster University Medical Centre. A preliminary survey there revealed no uniform practice with regard to the writing of DNR orders or the documentation of resuscitation decisions in the chart. House staff and nurses on duty at the time of a cardiac arrest would often have to decide about a CPR without the benefit of written orders or knowledge of the attending physician's intentions. In January 1979, a DNR policy was adopted that required that physicians, after determining the patient's condition to be irreversible, discuss the relevant facts with the patient in order to determine the patient's wishes with regard to CPR. The physician was to obtain the wishes of family when patients were deemed imcompetent. When the patient refused CPR, a DNR order was to be written in the chart and the relevant considerations documented in the chart. One year following implementation of this policy a survey questionnaire was sent to a sample of house staff, nurses, and attending physicians regarding their awareness of the policy and reactions to it. Though a finding that only 58% of those surveyed had read and understood the policy pointed to the need for improved dissemination of hospital policy, a large majority of those familiar with the policy felt it had been useful. Over two thirds felt that the DNR decision and its documentation had been clarified and almost three quarters felt that the policy had spurred more discussion about DNR decisions between staff. The study made no attempt to verify that the subjective responses of the respondants actually correlated with changed practices.

Bedell and Delbanco[12] reported the results of a study at Beth Israel Hospital (Boston) that did attempt to examine the actual DNR practices of the medical staff in an institution with established DNR policy guidelines. They interviewed patients who survived CPR and their attending physicians as well as the house staff responsible for their care with an eye toward determining the degree to which doctors were actually discussing resuscitation in advance with their seriously ill patients. They sought to determine the degree to which physicians reached conclusions about the use of CPR without discussing it with their patients, or their patients' families and the correlation between the physician's estimate of what the patient wanted and the patient's own statement of his wishes. The results of their study point to a startling discrepancy between what physicians state they ought to do and what they actually do in practice. Only 19 of the 154 patients surveyed had discussed resuscitation with their doctor before suffering cardiopulmonary arrest though 68% of their physicians had formed an opinion about their patient's wishes regarding resuscitation. Only 33% of the families had been consulted. Although 93% of private attendings and 100% of house officers believed that patients should "at least sometimes" be included in discussion about the resuscitation decision, only 10% of those who believed such discussions to be proper actually held such discussions (to be fair we must note that the cardiopulmonary arrest was unanticipated in the majority of cases). Of the 25 patients who had survived CPR, 8 stated unequivocally that they had not wished to be resuscitated and did not wish to be in the future. Of the 16 physicians caring for these 8 patients all but one believed the patient wished to be resuscitated.

Comments by study physicians helped to explain the findings. Some believed that responsibility for resuscitation decisions is the doctor's alone and that patients will initiate these discussions if they wish or will communicate their wishes nonverbally. It was felt that such discussions may be harmful to patients and may unnecessarily burden family members with guilt. Physicians expressed their concern that writing a DNR order would result in neglect of other aspects of the patient's care. Others complained that the procedure to be followed in writing DNR orders was too cumbersome. Several physicians expressed disbelief that anyone would refuse resuscitation. Obviously, merely having a DNR policy in place does not address problems that physicians may experience in implementing it.

In a recently published study,[13] Zimmerman *et al.* examined the DNR policies in 19 intensive care units (ICUs) in 13 hospitals. Each hospital had written DNR policies based on published guidelines. Of the 5,030 admissions to intensive care units that they examined, 313 (6.2%) had DNR orders written while in the ICU. However, the proportion of patients in each ICU with DNR orders varied widely from as low as 0.4% to 13.5%. These variations in the frequency of use of DNR orders could not be explained by factors such as diagnosis and severity of illness, nor were they explained by the type of hospital. Similarly, wide variability in the timing of the writing of such orders during hospitalization could not be explained by clinical variables. A closer examination of 80 DNR orders in one hospital revealed that the patient had participated in the DNR decision in only 11 cases (14%). The absence of patient participation was explained by coma or reduced consciousness in many, but not all, cases.

The results of this study illustrate that, though written guidelines for DNR orders are increasingly common, the actual use of written DNR orders varies widely between institutions. The physician remains the dominant individual in the initiation of DNR orders and variations in their use seem to be more related to differences in physician treatment decisions than to any clinical or social variables.

It is increasingly clear that the establishment of guidelines within an institution is not necessarily followed by changes in DNR practice. Many factors may interfere with the implementation of DNR policies. In the study by McPhail *et al.*, the majority of clinicians had not heard of the hospital's guidelines. Obviously, adequate dissemination and education about DNR guidelines is a necessary first step to ensure implementation. In the study by Bedell and Delbanco, clinicians' attitudes inhibited the practice of uniform DNR policies. Doctors in that study expressed concern that discussions of DNR decisions would be harmful to patients and burdensome to families and that patients who wished to participate in DNR decisions would make their wishes known on their own initiative. Still more revealing were statements of disbelief that anyone would refuse resuscitation. Such comments illustrate the degree to which the question of DNR touches on critical psychological issues that influence the clinician's conduct. Fear of death and the wish for omnipotence and control are factors that may stand in the way of the physician's ability to share decision making with the patient and family.

With these issues in mind, Perry *et al.*[14] surveyed a group of medical professionals to asses their perceptions of how resuscitation status was determined in their hospital and their opinions on how it should be determined for patients and for themselves.

The results of this questionnaire survey point to considerable discrepancies between how these clinicians believe DNR decisions should be handled and how they believe they are actually handled, as well as marked differences in how they feel about DNR decisions for themselves as compared to their patients. The majority of those surveyed felt that the DNR decision was made by the doctor (only 1% believed the patient actually makes the decision). Only 40% of respondents believed the decision should be the patient's. In a statistically significant contrast, 61% would want to make the decision for themselves. Although most respondents felt the patient was ignorant of his own resuscitation status, most would want such information if they were ill. The majority of respondents felt that the medical staff is poorly informed about patients' resuscitation status, whereas they endorsed the notion that it should be documented in the chart. Of note, when presented clinical vignettes the respondents frequently chose a resuscitation status that differed from the expressed wishes of competent informed patients.

The authors noted significant differences between the attitudes and beliefs of these clinicians and the guidelines of the presidential commission that we discussed earlier. They attribute this difference in part to physicians' belief that very ill and anxious patients, though competent in the broadest legal sense, are psychologically regressed and impeded in decision-making capacity and as a consequence prefer to delegate responsibility to their physicians. These medical professionals appeared to be uncomfortable discussing resuscitation status with patients and the authors emphasize that, in general, physicians lack training in discussing such sensitive issues with patients and families. In noting the contrast between the respondents stated wish to decide their own resuscitation status were they ill with their reluctance to allow patients to decide for themselves, the authors discussed three points that have important bearing on physician behavior. First, when viewing the question from a position of health, clinicians may have difficulty projecting themselves into the sick role. Second, being medically informed, they may feel themselves more capable of making such decisions than the lay public. Third, medical professionals, as a group, may be characterologically less inclined than others to delegate important decisions. This trait, which may be critical to adequate performance as a physician in many situations, would obviously impair the physicians ability to engage in shared decision making.

CONCLUSION

The shrouds of secrecy around DNR decisions are falling away. In the evolution of guidelines two key principles have emerged:

1. DNR decisions should be reached consensually by the attending

physician and the patient or the patient's family or other surrogate when necessary. This principle is firmly rooted in the doctrine of informed consent.

2. DNR orders should be written and the reasoning behind them should be documented in the chart.

Though the establishment of hospital guidelines for DNR decisions is critical, such guidelines have failed to resolve many of the problems in DNR practice. This failure can be traced to two factors, the lack of regulating legislation and psychological conflicts the physician faces that may prevent him from confronting the DNR issue directly.

Most states do not yet have legislation regarding DNR orders. In the absence of appropriate regulations, hospital guidelines can not adequately protect clinicians from situations that may lead to liability. For example, in New York, a state with no statute legitimizing or regulating DNR orders, the State Medical Society has recommended that when the patient is not capable of making his own decision, family members should be consulted. Though several hospitals have adopted guidelines based on this recommendation, the issue of DNR orders for incompetent patients remains controversial. Criteria for determinations of competency remain vague and in New York, as in many states, there is no statute outlining mechanisms for substituted judgment. Though relying on the judgment of family seems reasonable, it is not supported in statutory law and can lead to challenges on the basis of conflict of interest. The need for statutory regulation is clear. In New York the Governor's Task Force on Life and the Law is currently formulating recommendations for such legislation.

The second obstacle to improvement of DNR decision-making practices is the psychological conflicts that physicians themselves face. Although inertia and resistance to change may explain a small part of the problem, there are compelling psychological forces that play on physicians, as indeed on all of us, when facing questions of life and death. The imposition of hospital guidelines and even legislation cannot by themselves change behaviors that are so deeply rooted. We have seen these factors at play in the studies by Bedell and Delbanco and by Perry *et al.* Though increasingly viewed as paternalistic, and contrary to the aims of informed consent, the physicians' perogative to decide that discussion of an issue will be harmful to patients has a long history. Many physicians believe that DNR discussions will be harmful and that, if patients do not bring the subject up, they do not want to talk about it. It may well be that physicians are responding to the psychological regression that so frequently accompanies severe illness, even in seemingly competent patients. The doctrine of informed consent may indeed place an unwarranted burden on the patient in this condition. Of course, the physician's own difficulties acknowledging his failure to prevent death and the vulnerability he may experience in talking about death are also important issues. Most physicians have never received any special training in working with families. Many are wary of negotiating extreme emotional reactions in family members and, fearing that DNR discussions will provoke such reactions, may avoid these discussions altogether.

Hospitals cannot eliminate these problems with guidelines. Nor can states legislate them away. If the actual practice of DNR decision making and order writing is to improve in accordance with evolving guidelines, clinicians must be trained to help families and patients make these difficult decisions. To do this effectively they must be sensitized to the ways in which their own attitudes and anxieties influence their behavior. This process should start in medical school and be considered as much a part of continuing medical education as training in the life support measures that raise the question of DNR orders.

REFERENCES

1. Mistein BB: *Cardiac Arrest and Resuscitation.* Llyod-Luke Medical Books, London, 1963.
2. National Conference Steering Committee: Standards for cardiopulmonary resuscitation (CPR) and emergency cardiac care (ECC), *JAMA* 1974; 227:837, 864.
3. *Standards and Guidelines for Cardiopulmonary Resuscitation (CPR) and Emergency Cardiac Care (ECC). JAMA* 1980; 224:506–507.
4. *Guidelines for Hospitals and Physicians on "Do Not Resuscitate."* Medical Society of the State of New York, Sept. 20, 1982.
5. *Deciding to Forego Life-Sustaining Treatment.* Report of the President's Commission for the Study of Ethical Problems in Medicine and Biomedical and Behavioral Research, 1983.
6. Bedell SE, Delbanco TL, Cook EF, *et al.:* Survival after cardiopulmonary resuscitation in the hospital. *N Engl J Med* 1983; 309:569–576.
7. Messart B, Quaglieri CE: Cardiopulmonary Resuscitation: Perspectives and problems. *Lancet* 1976; 2:410–411.
8. Hershey CO, Fisher L: Why outcome of cardiopulmonary resuscitation in general wards is poor. *Lancet* 1982; 1:31–34.
9. Nagel EL, Fine EG, Krischer JP, *et al.:* Complications of CPR. *Crit Care Med* 1981; 9:421.
10. Petty TL: Mechanical last rights. *Arch Intern Med* 1983; 142:1442–1443.
11. McPhail A, Moore S, O'Connor J, *et al.:* One hospital's experience with a "do not resuscitate" policy. *CMA Journal* 1981; 125:830–836.
12. Bedell SE, Delbanco TL: Choices about cardiopulmonary resuscitation in the hospital: When do physicians talk with patients. *N Engl J Med* 1983; 310:1089–1093.
13. Zimmerman JE, Knaus WA, Sharpe SM, *et al.:* The use and implications of do not resuscitate orders in intensive care units. *JAMA* 1986; 255:351–356.
14. Perry SE, Schwartz HI, Amchin J: Determining resuscitation status: A survey of medical professionals. *Gen Hosp Psychiatry* 1986; 8:198–202.

7

Withholding Life Support from the Elderly, or Learning to Live with High-Tech Death

JOHN J. REGAN

Until a few years ago, death was like religion and politics—an unmentionable topic in polite society. This taboo has disappeared rapidly, however, as the environment for the process of dying has shifted from informal settings to medical institutions. Death today frequently occurs in a hospital or nursing home in an atmosphere dominated by medical expertise, institutional policies and procedures, and biomedical technology. In some cases, death can even be a controlled event. Its method and timing can be arranged to serve the desires of the patient, his or her family, a potential transplant beneficiary, a fetus within the patient, or even the diagnosis-related group (DRG) reimbursement system.

Robert Veatch has illustrated the dilemmas generated by high-tech death by the case of the 33-year-old man who was deposited unconscious at the door of the emergency room of a metropolitan hospital.[1] Laboratory analysis confirmed the resident's diagnosis that the patient had experienced a heroin overdose. Lack of oxygen to the brain had left the patient in a deep coma. After appropriate resuscitative measures were administered, however, the heart was functioned normally while respiration was being maintained by the ventilator.

After further examination of the patient, three camps emerged within the medical staff. One group agreed with the neurologist that the

---

JOHN J. REGAN • School of Law, Hofstra University, Hempstead, NY 11550.

patient was dead because the latter showed no signs of brain activity. They urged that the patient be declared legally dead and his kidneys retained for another patient in the hospital. A second group, however, argued that because the patient's heart was still beating he was still alive. This group conceded, nonetheless, that he was in an irreversible coma and had reached the point where the respirator should be turned off so that he might die with dignity. The third group agreed with the second in saying that the patient was still alive because his heart was beating, but argued that the physician's duty was to preseve life and that therefore treatment should be continued. The remarkable part of this debate was the fact that all parties agreed on the medical facts of the patient's condition.

This patient is a symbol of a larger group of patients of all ages, but especially the elderly, for whom similar life–death decisions must be made. The decisions involve a complex mix of ethical, legal, and medical issues. This is but a sampling of these questions:

- When is a person to be considered dead?
- When is it permissible not to begin life-sustaining treatment for a living patient, or to withhold such treatment after it has begun?
- Is artificial feeding fundamentally in the same category of medical treatment as a respirator?
- Who should make the ultimate decision to withhold treatment— the physician, the patient, the patient's family, or some combination of these parties?
- Who should decide that a patient is competent or not to make his or her own decision about treatment?
- What is competency? How can it be assessed?
- What weight should be given to a living will?
- What course of action is appropriate when a physician disagrees with the patient's or family's wishes?
- Should termination of treatment decisions be subject to intra-institutional review, or should all such cases be referred to the courts?
- To what extent should economic considerations, the costs of con-tinued treatment, be allowed to influence the decision to withhold or withdraw life-sustaining treatment?

Until the mid-1960s, the question of defining death raised little controversy. Death occurred, from both a medical and legal perspective, when circulation and respiration ceased. But the development of new resuscitative techniques and the ability to maintain circulatory and respi-ratory activity by use of a mechanical respirator, despite loss of brain activity, challenged these traditional standards. The respirator, in effect, became a mechanical substitute for the brain's integrating functions, at least for a short period of time. The problem for physicians and families thus became the need to identify the point at which a patient had ceased to be an individual in need of medical care and had instead become a corpse.

New criteria based on the cessation of brain activity were first proposed in 1968[2] and, in modified form, have since gained almost universal acceptance. The standard adopted in the uniform determination of death act is:

> An individual who has sustained either (1) irreversible cessation of circulatory and respiratory functions, or (2) irreversible cessation of all functions of the entire brain, including the brain stem, is dead. A determination of death must be made in accordance with accepted medical standards.[3]

This formulation, which has been endorsed by the American Medical Association and the American Bar Association, has been adopted in over 40 jurisdictions. New York, through court decision[4] and department regulation, has accepted essentially the same standard, thus permitting physicians to rely on either the traditional heart–lung criteria or the total and irreversible cessation of brain function in determining that death has occurred.

Certain points about this definition should be noted.

1. The brain-based criteria are used only for determining that the death of an individual has already occurred. They are not an attempt to define the ultimate questions of what, in fact, makes a human being a person, or what is the dividing line between life and death. The criteria are thus really only norms for action, not a metaphysical statement about the meaning of personhood.

2. The proposed brain death standard requires the cessation of activity in the entire brain, including the brain stem. In contrast, some philosophers define human existence to include a capacity for social interaction and, therefore, they advocate a higher brain or cerebral definition of death that would permit a patient who has suffered irreversible loss of consciousness or cerebral activity to be declared dead. Thus a person in a permanent coma would be dead under this standard. However, neither New York nor any other state has adopted the higher brain standard because of the troubling questions it raises in the cases of certain patients, such as the retarded or severely demented.

3. The legal definition of brain death does not incorporate specific clinical criteria for making the determination that death has occurred. It requires instead that the determination be made in the light of accepted medical standards, thus permitting further development in the clinical techniques for ascertaining that an individual has died.

4. Some members of the Orthodox Jewish community believe that only the heart–lung criteria may be used in the determination of death. They advocate a conscientious objector exemption from the brain death standard for persons of their beliefs. Such an exemption raises significant questions about the nature of the state's interest in imposing a uniform standard for the determination of death; the procedures hospitals and physicians would follow in implementing the exemption; third-party reimbursement for treatment of a patient who claimed the exemption; and the allocation of scarce intensive care resources to such patients in preference to others still alive.

But what about the living? When may life-sustaining treatment be withheld or withdrawn from a person who is alive even under current criteria?

The starting point for answering this question is the much maligned legal doctrine of informed consent. According to this principle, the competent patient has the right to control medical treatment of his or her body. Under ordinary circumstances in most states, no one else— spouse, family, or physician—has the right to authorize such treatment. Before treating the patient, the physician is required to explain a proposed procedure to the patient and to warn of any material risks or dangers in or collateral to the therapy, so that the patient may make an intelligent and informed decision whether or not to undergo such treatment. Thus, the purpose of the informed consent doctrine is to enable the patient to act as an autonomous and rational decision maker.

The flip side of this principle means that the right to give consent implies the right to refuse treatment. Decision making includes a right to say no as well as to say yes. The logic of this statement, however, escaped many courts until the last 10 years or so. Before that time, courts often found a variety of reasons for overriding a patient's refusal of life-sustaining treatment. Some cited the patient's responsibilities to his or her family, particularly if the patient was the parent of a young child; or to the medical profession whose ethic was thought to mandate an unremitting effort to heal; or to society at large, which had an interest in preserving all human life, regardless of an individual's wishes; or, finally, to the patient himself, who was only a custodian of his own life and not its ultimate master. Some courts also argued that a refusal of life-sustaining treatment was equivalent to suicide.

Judicial and legislative attitudes toward refusal of treatment began to change in the mid-1970s as a result of several landmark court decisions. In the case of Karen Quinlan[5] the New Jersey Supreme Court held that Karen's father, who was also her court-appointed guardian, could authorize the removal of a respirator that was maintaining her in a permanent vegetative state, even though she might then die because of her inability to breathe spontaneously. Significantly, the court noted that, in so acting, her father was really exercising Karen's own constitutional right of privacy, that is, her right to make a major decision concerning continued medical treatment of her body. Around the same time, the highest court in Massachusetts ruled in the case of Joseph Saikewicz that the court could authorize the termination of treatment for a 57-year-old retarded man suffering from leukemia, basing its decision on considerations as close as possible to those which the patient himself would have weighed if he had been competent.[6] Although both court decisions involved incompetent patients, both courts recognized that the patients themselves, were they competent, would have the right to refuse treatment as an exercise of their constitutional right to privacy. In each case, the court rejected arguments that would have allowed the state to override the patient's decision or to prosecute health care personnel who honored the patient's wishes.

New York followed this trend, but only to a certain extent. In the first of the two cases decided in *In re Storar,* the court held that the patient's guardian and the attending physicians could terminate the treatment of Brother Fox, an irreversibly comatose patient, on the basis of clear and convincing evidence of his previously expressed wish not to be kept alive in such circumstances.[7] The decision did not require any particular procedure for the individual to express this desire, and thus opened the door to the use of the living will in New York.

Several recent decisions have further extended the right of refusal established by these early cases. In *In re Conroy*[8] the New Jersey Supreme Court held that "medical treatment" that the patient could refuse included not only a respirator but also artificial feeding and hydration through a nasogastric tube. The highest court of Massachusetts agreed, even in a case where the patient was permanently unconscious but not otherwise suffering from illness likely to cause imminent death.[9] Finally, a California court has upheld the right of Elizabeth Bouvia, a competent patient who is a quadriplegic but not otherwise fatally ill, to refuse involuntary forced feeding through a nasogastric tube, even though death by starvation or dehydration may be the result.[10]

On the legislative front, California was the first state to authorize by statute a document executed by a patient detailing the patient's wishes with respect to the life-sustaining treatment.[11] Under certain circumstances, the document would bind the attending physician to implement its instructions. By July, 1985, 35 states and the District of Columbia had enacted one form or another of living will legislation.[12]

What about the case of the incompetent patient who has never expressed any opinion for or against life-sustaining treatment? Most courts believe all such patients have a right to have someone else make a decision in their behalf. New York is the major exception. In the *Storar* case, the New York Court of Appeals refused to acknowledge the right of a guardian to order discontinuance of life-prolonging transfusions in the case of a profoundly retarded and terminally ill adult, because he himself had never been competent. The court also suggested that the state legislature is the appropriate body to empower guardians to make these decisions for such patients.

What are some of the factors which need to be weighed in terminating treatment of an incompetent patient? This is a list culled from recent court opinions:

1. Is the treatment medically futile (only prolonging the dying process) or will it improve the patient's condition?
2. How intrusive is the treatment?
3. Is the patient suffering?
4. What are possible side-effects of the treatment?
5. What is the relation between the patient's age and the consequences of continued treatment?
6. Can the patient comprehend and cooperate with the treatment?
7. What is the patient's prognosis? Is there terminal illness, permanent chronic or irreversible vegetative coma, or lack of cognitive brain function?

In the *Quinlan* case, the court proposed this formula for weighing these factors: The state's interest in continuing treatment weakens and the individual's right to terminate treatment grows as the degree of bodily invasion increases and the prognosis dims. In Karen Quinlan's case, the guardian's decision to terminate treatment outweighed the state's opposition because her prognosis indicated that she would never resume cognitive life and the degree of bodily invasion was great, in that she required 24-hour intensive nursing care, antibiotics, the assistance of a respirator, a catheter, and a feeding tube. Other courts have generally concurred in this formula, but with one notable qualification: they reject so-called quality of life considerations. These courts consider it impermissible to weigh nonmedical factors, such as mental retardation, old age, or senility, in deciding whether to continue treatment or not. They fear that life-sustaining treatment may be denied to certain types of people, such as the retarded or the senile, on the ground that their lives are not worth preserving.

What I have described thus far about treatment refusal cases are the *substantive* considerations—the nature of the patient's or surrogate's right to refuse treatment, and the relationship of that right to the state's interests in preserving life. There is another set of issues that pervades these decisions, the *procedural* problems. The first of these procedural issues is the question of who should make decisions for an incompetent person. In most states, only a court or a court-appointed guardian has the legal power to authorize medical treatment or its withholding or withdrawal. Despite this principle, two extralegal (even illegal) practices have flourished. The first is where the physician does not consult the patient or the family, but instead makes the decision himself. The second is the custom of seeking consent from the spouse or next of kin. This, too, has no legal sanction in most states, despite the fact that this practice is widespread. The *Quinlan* case exemplifies the guardianship model of substitute decision making, in that Karen's father was appointed her guardian by a court and authorized to use his discretion to order termination of treatment, but subject to ratification of the prognosis by a hospital ethics committee. In the *Saikewicz* case, on the other hand, the court reserved to itself the ultimate power of decision, at least for cases ordering termination of the treatment for a never-competent patient. In both types of cases, the heavy dependence on the courts is time-consuming and expensive. As a result, the search is now on for alternative ways for determining competence and authorizing withdrawal of treatment for those found to be incompetent.

One such alternative is for the state legislature to authorize family consent for incompetent patients. This system would permit the physician and family together to make these decisions without initial recourse to the courts. This proposal has some attractive features, in that it permits those closest to the patient and best informed about the medical facts to make the treatment or no-treatment decision. On closer examination, however, it contains two major difficulties. First, it tolerates an unknown degree of error by the physician and the family in assessing the patient's competence, making the diagnosis and prognosis, or evaluating the best interests of the patient. Second, it fails to incorporate a totally impartial party into the decision-making process. The way out of these difficulties may be to require health care providers to establish institutional committees whose functions would include review of these physician–family decisions and/or a dispute settlement role where there are disagreements among these decision makers. The courts would still be available as an ultimate dispute settler of last resort if the committee was unable to resolve the dispute, but there would be no need to petition them for action as a routine matter. The use of some intermediate review mechanism, such as an institutional committee, still leaves problems but it appears to have promise.

Note that I have referred to three decisions that must be made in the course of this process of deciding what to do about treatment. The first is essentially a medical determination: the making of a diagnosis and a prognosis by the physician responsible for the patient's care.

These opinions, of course, are medical determinations and do not involve the family or other nonphysicians. But that does not mean that they should be entirely unreviewable. There is substantial feeling that these medical determinations should be confirmed by at least one other physician and noted in detail in the patient's record whenever life-sustaining treatment is involved. Such a procedure, in the long run, protects the physician not only from his or her own errors of judgment, but also from the pressures to shorten patient stays generated by hospital policy or the DRG system.

The second decision in this process is the determination that the patient is incompetent. Incompetency is essentially a legal judgment, not a medical or psychiatric determination. In its classic sense, it involved a judge's finding that a person, because of mental illness, disorder, mental retardation, substance addiction, or even old age, no longer had the capacity to manage his property effectively or to make personal decisions affecting life, health, or safety. At one time, incompetency was thought to be an either/or proposition. Either you were totally competent or you were incompetent for all purposes. There was no middle ground. Recent thought, however, views competency as a relative concept, at least for some persons. A person may be competent to make certain kinds of decisions, but not others. It is clear that the comatose or severely retarded patient is incompetent to make any meaningful decision for himself, but many other patients may retain the capacity to evaluate their situation with regard to continuation or not of treatment, even though they can no longer make business or other personal care decisions. A competency determination, therefore, involves a fairly sophisticated estimate of a patient's capacity to appreciate the nature and consequences of his situation and to make a rational decision about it. In this context, competency might be better termed decisional capacity.

Here are some questions that the clinician might put to the patient as a means of assessing the person's decisional capacity.[13] These questions assume that the patient has already received the information needed to give an informed consent, but before he is asked to make a decision.

1. What is your present physical condition?
2. What is the treatment that is being recommended for you?
3. What do you and your doctor think might happen to you if you decide to accept the treatment?
4. What do you and your doctor think might happen to you if you decide not to accept the recommended treatment?
5. What are the alternative treatments available (including no treat-

ment) and what are the probable consequences of accepting each?

Great caution is necessary in drawing conclusions about competency from the assessment process.[13(pp 578–583)] Some psychiatrists recommend that the process involve at least two contacts with the patient on at least two different days. The clinician should try to develop a sense of consistency from case to case in making these determinations. Mere indecisiveness or vacillation by the patient is not incompetence, nor is mental illness alone, although certain delusions may cause incompetence when they prevent a patient from understanding the nature and consequences of his decision. Finally, in making a competency assessment, the psychiatrist needs to be alert to the particular circumstances of each case, especially as to who is challenging the patient's capacity, and why. Whom is the psychiatrist advising? Is it the hospital? The physician? The patient? The family? The answer may raise ethical issues as to who is the real client, and it may also reveal factors that unconsciously affect the ultimate competency determination itself.

The third decision in the decision-making process is the ultimate substantive determination to continue treatment or to terminate it. Once the diagnosis and prognosis are validated, the patient determined to be incompetent, and the appropriate surrogate identified, the decision to continue or terminate treatment should ordinarily not be subject to further review. This statement assumes, of course, that the surrogate and the physician are in agreement as to the appropriate course of action and that no third party challenges their decision.

But suppose the surrogate and the physician are in sharp disagreement. For the physician to insist on treating a patient whom the surrogate does not want treated opens the physician to a possible suit for battery. Similarly, the physician who ceased treatment would risk a suit for malpractice (or worse). One solution to this dilemma is to say that the surrogate's decision is always to be followed. This approach, however, ignores possible medical objections of the physician to the surrogate's decision. A better solution might be to submit the dispute to an institutional review committee for the purpose of mediation, but not adjudication. An airing of each side's position in the presence of the other, together with comment from the committee, might well bring the parties to a meeting of minds. But suppose the mediation effort fails. What next? At this point, the dissenting physician has two options: (a) if he or she believes continued treatment is necessary despite the surrogate's refusal, the physician should withdraw from the case or seek a court order authorizing the continuance; or (b) if the physician believes termination is appropriate but the surrogate insists on continuance, the physician should withdraw but should *not* withhold treatment.

Another type of dispute can arise when a third party (e.g., another member of the family) challenges the decision of the surrogate to terminate treatment. This challenge may be ignored when the termination decision is legally valid, but such an approach may simply invite trouble for both the surrogate and the physician. Again, the better tactic would be to utilize the institutional committee's services as a tool of mediation. The goal would be to persuade the challenger to acquiesce in the surrogate–physician decision. If the effort fails, the challenger can still seek a restraining order from a court, but in the absence of such an order, the surrogate's decision is controlling.

Apart from these procedural considerations, a troubling issue that has emerged in several recent cases is the question of whether artificial feeding and hydration is medical treatment. Should the termination of intravenous feeding and hydration be treated legally as the equivalent of shutting off a respirator? Some argue that these procedures are essentially different. The termination of artificial feeding and hydration, they argue, can lead to a slow and painful death from starvation. The patient would die, they believe, from the physician's act of deprivation rather than from natural disease causes, unlike a case involving discontinuance of a respirator where the body's own inability to function is the cause of death.

This position, however, has not been accepted by the New Jersey,[8] California,[10] and Massachusetts[9,14] courts. Last year, for example, the New Jersey Supreme Court held in the case of Claire Conroy, an 88-year-old nursing home patient, that feeding and hydration tubes could be withdrawn as long as there was no net increase in the patient's level of pain. The court could find no meaningful legal distinction between the termination of a respirator that sustained breathing and the withdrawal of tubes maintaining nutrition and hydration. Both were basically forms of medical treatment and, therefore, their continued use was subject to the consent of the patient or the surrogate.

Note, however, that we are talking about *artificial* feeding and hydration. No court has approved (or is likely to approve) a decision to withhold normal meals or liquids from a patient in order to allow the person to die, regardless of whether the patient or anyone else authorizes it. The patient who wishes to commit suicide by starvation will have to find a nonmedical setting for this course of action.

A survey of the legal issues involved in decisions to terminate treatment would not be complete without a word about three practice issues involving the behavior of physicians. First, the law is placing increased demands on the treating physician to provide a detailed documentation in the patient's record about the physician's conclusions and decisions concerning the patient's condition and recommended courses of action. The day of the "purple dot" for a DNR order or a verbal order to the nurse is gone, at least as far as the law is concerned. The physician should note in the record the basis for a decision to continue treatment or not, as well as a summary of any discussions held with the patient and/or the family with regard to major treatment decisions. A well-documented record is the best safeguard for the physician against the hindsight of the patient, family, hospital administration, or public authorities.

The physician should also encourage patients and their families to plan ahead for decisions involving life-sustaining procedures. One vehicle for formalizing these plans is a living will, a document recording a person's desires about continuation or not of treatment in the event that the individual is being sustained by mechanical devices but is no longer competent to make his or her own decisions. This document does not yet have formal legal status in the tri-state area, but in all theses states a living will would probably be clear and convincing evidence of a patient's wishes and, therefore, would be honored by the courts.

Another legal instrument gaining wider use is a durable power of attorney. A power of attorney is a document appointing another to act as one's agent or attorney-in-fact. This authorization normally ceases if the appointing party becomes incompetent, but many states now permit the agent's authority to continue after incompetency. Hence the power's "durability." In some states, including Connecticut, it is now possible for an individual to execute a durable power of attorney for health care decisions whereby the person appoints an agent empowered to make termination of treatment decisions for the patient in the event the patient becomes incapacitated. Although the legal status of the durable power for health care matters is uncertain in New York and New Jersey, at the present time it appears likely that these states will authorize the use of such documents in the near future.

Someone recently penned a version of the 23rd Psalm with a dose of irony. It read:

Medical Science Is My Shepherd, I Shall Not Want

> It Leadeth me beside the marvels of Technology;
> It Restoreth my brainwaves;
> It Maintaineth me in a persistent vegetative state for its namesake;
> Yea, though I walk through the valley of the shadow of death.
> I shall find no end to Life;
> For Thou art with me;
> Thy Respirator and Heart Machine, they sustain me;
> Thou Preparest intravenous feeding for me;

In the presence of irreversible disability;
Thou Annointest my head with oil;
My Cup runneth on and on and on and on and on.
Surely Coma and unconsciousness shall follow me all the days of my
    continued breathing;
And I will dwell in the intensive-care unit forever.

Both public opinion and the medical profession are rejecting this dismal portrait of the high-tech terminally ill patient. The next step will be for the legal system to restore control over decisions affecting such patients to those to whom it rightfully belongs, the patients themselves and their physicians.

REFERENCES

1. R. Veatch, *Death, Dying and the Biological Revolution,* New Haven, Yale, 1976, p 1.
2. Ad Hoc Committee of the Harvard Medical School to Examine the Definition of Brain Death: A definition of irreversible coma. *JAMA* 1968; 206:377.
3. Report of the Medical Consultants on the Diagnosis of Death to the President's Commission for the Study of Ethical Problems in Medicine and Biomedical and Behavioral Research: Guidelines for the determination of death. *JAMA* 1981; 246:2184.
4. *People v Eulo,* 63 N.Y. 2d 341 (1984).
5. *In re Quinlan,* 70 N.J. 10, 355 A.2d 647 (1976).
6. *Superintendent of Belchertown State School v Saikewicz,* 373 Mass. 728, 370 N.E. 2d 417 (1977).
7. *In re Storar,* 52 N.Y. 2d 363, 420 N.E. 2d 64, 438 N.Y.S. 2d 255 (1981).
8. *In re Conroy,* 98 N.J. 321, 486 A.2d 1209 (1985).
9. *In re Brophy,* 497 N.E.2d 626 (Mass. 1986).
10. *Bouvia v Superior Court,* 225 Cal. Rptr. 296 (Ct. App. 1986).
11. California Natural Death Act, Cal. Health & Safety Code §§7185–7195 (1976).
12. Society for the Right to Die, *The Physician and the Hopelessly Ill Patient* New York, Society for the Right to Die, 1985, p 19.
13. G. Annas and J. Densberger, *Competence to Refuse Medical Treatment: Autonomy vs. Paternalism,* 15 Toledo L. Rev 561, 577 (1984).
14. *In re Hier,* 18 Mass. App. 200, 464 N.E. 2d 959 (1984).

8

Euthanasia and Living Wills

The Rights of Patients to Participate in Their Deaths

EDWIN MARGER AND SUSAN HAMEL

> *On the day when death will knock at thy door*
> *What wilt thou offer him?*
> *I will set before my guest the full vessel of my life.*
> *I will never let him go with empty hands.*

Rabindranath Tagore

INTRODUCTION

Many books have been written about death. Much philosophy dating as far back as Socrates and Aristotle has been expressed on the mortality of our human existence. More recently, excellent works on the "stages of dying" (and the acceptance of death) have provided answers to our society in dealing with it.[1]

"Human mortality may not be denied, for death is the common end to all life."[1(p38)] Every day someone copes with that natural order of things. It is in the newspaper, on the television, over the telephone, even next door. When death occurs swiftly, as in the loss of our seven young astronauts, it is a traumatic blow with little or no time for transition. The word *transition* is pivotal here. Transition is the process or an instance of

---

EDWIN MARGER • Attorney in Private Practice, 6666 Powers Ferry Road, Atlanta, GA 30339; Member: Georgia, Florida, Washington, DC Bars; Member: Board of Governors, Florida Bar; Trustee: Forensic Science Foundation.   **SUSAN HAMEL** • Law Offices of Edwin Marger, 6666 Powers Ferry Road, Atlanta, GA 30339.

113

passing from one form, or state, or stage to another. It is not specifically the transition itself that is pertinent here, (i.e., when death actually occurs) but who, if anyone may make the decision concerning that passage or transition from life to death.

The technological advances of the last 30 years have prolonged the lifespan of many but have also saddled society with the legal, ethical, and medical dilemma of who has the right to die and perhaps even the more menacing undertones of the question Who dies? In this chapter, euthanasia, the right to die, and living wills will be discussed. The positive and negative aspects of euthanasia, voluntary and involuntary, will be noted. Cases involving competent and incompetent patients will be discussed. And lastly, the living will itself—Is it a viable alternative to the often death-delaying judicial decision? Some of the existing legislation on living wills will also be reviewed.

Definition and History of Euthanasia

The word *euthanasia* comes from the Greek *eu*, meaning good, and *thanatos*, meaning death. It is defined as "the intentional causing of a painless and easy death to a patient suffering from an incurable or painful disease."[2] The definition speaks of an action being performed upon someone. Another word, *antidysthanasia*, used less frequently, has been defined as meaning passive euthanasia—"hastening one's death by refusing medical attention."[3]

The subject of euthanasia, or mercy killing, has been debated long before the judiciary was called upon to intervene.[4] In 1935, the British Euthanasia Society was founded and in 1938 a parallel American organization was created. Subsequently, a bill before the House of Lords in 1936 and a bill in Nebraska in 1938 permitting euthanasia both failed.

The horrible crimes of the Nazi regime left the issue inactive until 1959 when Glanville Williams's book, *The Sanctity of Life and the Criminal Law* brought it into the light again.[5] Williams supported euthanasia for any competent adult wishing it, with the certification of two physicians that the patient had a terminal illness or would endure insufferable pain. Yale Kamisar, one of Williams's adversaries, felt that a person in such a position could be incompetent under the influence of sedative drugs. Without continuing into the arguments for or against, let us proceed with the historical background.

The last 25 years have produced a few supporters who have penned their opinions.[6] Within the last 10 years, most of the exchange has centered around the rights of patients who are terminally ill (whether or not they may refuse life-prolonging treatment). Exploring the reasons for this, one has to realize that the causes of death have changed. In the early part of this century, bacterial disease was the main cause of death. But with the advent of antibiotics, the antecedents have moved to cancer and cardiac disease. Though there are as yet no specific cures, there are treatments and methodologies to lengthen life. What is at issue now is whether the quality of life is also prolonged, and most importantly, whether an individual has the right to determine his or her quality of life.

In 1976 the first major judgment in a case of a terminally ill patient was handed down.[4] With the *Quinlan* case, the area of judicial decision regarding life-saving treatment was opened further. In six states decisions have been handed down from the highest courts. Issues have also been brought before appellate and trial courts. Whether or not the judiciary should be involved at all in these life and death decisions is a question in itself. And four of those six highest state courts have indicated that the courts are not the place for such decisions.[7] Who should have the power over death or life—physicians, judges, relatives, or an independent entity?

THE LAW WITH REGARD TO EUTHANASIA

Euthanasia, in the present state of the law, is classified as homicide or suicide. Founded in common law and widely accepted in Western thought is the fact that the state has a basic right to prohibit either. Inherently, this is not only a moral but practical idea, that is, a society must be concerned with the protection and preservation of human life. In an earlier day the common law regarded suicide as a criminal offense and forfeiture of lands and estates was the penalty.[8] But over the last 200 years virtually all of the criminal sanctions against suicide have been dropped. This is certainly not because of a changing belief that the state has no right to interfere but merely that the sanctions are either not effective or injurious to family or friends.[9] Yet suicide is still not legal.[10] Under both English and American criminal law, then, voluntary euthanasia is murder for the person who carries out the act and suicide for the person who accepts.

Advocates of euthanasia or mercy killing would argue as to the motive for the death. But the law does not recognize motive in either case. The law is explained coldly in a short paper in the Law Quarterly of 1889, "Murder From The Best of Motives." Herbert Stephen, the author, states conclusively "that doctors are not to be morally commended when they seek to substitute their individual feelings and judgements for the plain and universal rule supplied by the criminal law."[11] Though this was written nearly 100 years ago, views have not yet changed so radically as to permit new legislation in the area. Gregory Gelfand points out that the possibility of error and the possibility of abuse are disadvantages to the legalizing of euthanasia.[12]

What are the possibilities with legislation? "The proper goal of legal intervention in life-or-death decision making is to broaden the range of individual choice and to make such choices effective."[13] Perhaps the structures of the criminal law with regard to homicide or suicide need to be reviewed and changed, either giving euthanasia its own niche in the law, and sanctions governing it, or retaining the restriction that euthanasia is murder but lifting or lessening the criminal sanctions. Certainly this has already been done in a number of circumstances in the courts. Indeed in these cases it would appear that "the court's action . . . was an attempt to align the law with what is believed to be the public sentiment."[14]

As society looks into a new century, legislation must be dealt with. The fact that life-prolonging measures allow critically ill patients to make decisions with regard to their lives means that the law must therefore allow them to make decisions concerning their deaths. The United States Constitution recognizes the right to personal autonomy in certain instances. But "it seems correct to say there is no constitutional right to choose to die."[3]

If the lack of legislation regarding the personal autonomy of a critically ill patient is appalling, even more frightening are the civil and criminal liability attached to the physician. Doctors, more now than ever before, are directly involved with the death decision-making process. Though the medical profession revolves around the preservation of life and the alleviation of suffering, the question of whether they should prolong life under certain circumstances or terminate life remains unanswered. If the medical profession is to be relied on for their educated judgment and experience, there must be sufficient legislation to protect them from civil and criminal liability. In the same light there should be sufficient sanctions to govern the abuse of such legislation. The following discussion of case decisions involving the death decision-making process will give some insight as to how and why our law must change to accommodate the needs of a society concerned with death.

EUTHANASIA-DEATH DECISION MAKING—A LOOK AT THE JUDICIARY

In 1959, Yale Kamisar did a review of cases involving euthanasia and discovered that in most of the cases the jury did not convict the accused. It appeared that the law when interpreted by the jury and applied conceptually is quite different than when interpreted literally. In *People v Wemer*,[15] a 69-year-old defendant suffocated his wife, a crippled bedridden arthritic. Though originally the defendant entered a plea of guilty to manslaughter, the court after hearing testimony concerning his care and devotion to his wife and reviewing a letter from her doctor regarding her terrible pain and despair, allowed the defendant to change his plea to not guilty. The court then found the defendant not guilty on grounds that the jury "would not be inclined to convict." (Note the use of *would* rather than *could*.) Here is a case where the court did base its decision on the defendant's motives in the face of opposing statutes. Yet in *People v Roberts,* the euthanatizor was convicted of first degree murder and sentenced to life imprisonment.[16] But the examples of cases like the *People v Werner* are much more abundant than *People v Roberts*.[17(pp 2 13)]

On March 31, 1976, a decision was handed down by the Supreme Court of New Jersey that not only opened the door for discussion on the right to die but allowed the patient, through the right of privacy (under the 4th Amendment), the right to refuse life-sustaining medical treatment.[4] It was a case fraught with controversy and the publicity surrounding it made it all the more exceptional. In *In re Quinlan* the court gave the power to decide whether life prolonging treatment should be continued to the guardian of the comatose, an incompetent woman. In doing so the court granted a privacy right to the incompetent patient, but the decision regarding her life/death dilemma was given to her guardian (who in this case was her father).[4]

Over the next 10 years, other state courts adopted this unique right of privacy that arose from *In re Quinlan*. In *Severns Wilmington Medical Center, Inc.*[18] and *Matter of Colyer*,[19] two state supreme courts dealt with the situation wherein the incompetent patient was incapable of asserting the right to terminate the treatment. Guardians in both these instances were allowed to decide if life-sustaining equipment could be withdrawn. Significantly, in *In re Quackenbush*[20] and in *Satz v Perlmutter*,[21] where both patients were competent, the courts relied on *In re Quinlan* in asserting a constitutional right of privacy, that is, a right to die. In these two cases, the right needs of the critically ill patient outweighed the state's interest in the preservation of life.

In 1980, in the case of *Leach v Akron Medical Center*,[22] the legal guardian brought action in a civil court to have a respirator discontinued. The court relied on *Matter of Eichner*[23]:

> Under the circumstances present herein, by no stretch of the imagination can the State be deemed to be "taking life" in a manner analogous to the imposition of a death penalty in a criminal action. This judicial proceeding is not directed towards the imposition of a penalty for criminal activity but, rather,

towards the furtherance of the best interest of the comatose and terminally ill
patient.

The court granted the motion of the guardian to have the respirator removed, and again relied on the Constitution of the United States guaranteeing as part of the right of privacy the right to a terminably ill person the right to decide future medical treatment.[22]

In *John F. Kennedy Memorial Hospital, Inc. v Bludworth,*[24] a decision in favor of the patient was made based on his previously expressed intention. But once again even though the patient evidenced the desire not to be maintained with life-supporting apparatus, the court felt that the "death with dignity" process needed judicial resolution.

The judicial process is a lengthy one and the question of whether courts should be involved in the right to die has been in fact raised by several of the state supreme courts. In *Permutter:*

> It is asserted that the question of "death with dignity" is a matter so complex
> that it should be left to the legislature, and the judiciary should abdicate to
> that process. We are told that the many variations of the problem and multi-
> ple issues subsumed under the general issue make it inadvisable at best for
> the Court to address the issue.[21]

The courts have subsequently made it clear that legislative inaction cannot prevent them from protecting the constitutional right to die with dignity.

THE LIVING WILL

The living will or an advance directive as it has been called is a recent development[25] in right to die legislation. Within a year of the Karen Ann Quinlan case, California passed the first natural death law declaring that adults in a terminal condition, under certain circumstances, could refuse life-sustaining measures.[26] The term *living will* was coined in 1967 by Dr. Louis Kutner. A living will is designed to be a directive written by a competent adult to be used in the event of a terminal illness or severe injury that renders him incompetent, thus making him incapable of making decisions. It is one possible solution to some of the right to die questions. Thirty-five states and the District of Columbia have made the living will legally effective.[27] The use of the living will is a method for the patient to retain the antonomy of his body, and exercise the most important decision of his life by his right to death. The positive aspects of living will legislation include that it (a) must be executed by a competent adult, (b) must be witnessed by competent adults who are not affiliated or related to the declarant, (c) provides immunity for health care individuals, and (d) can be revoked at any time.

The living will is a good concept, setting the standard for the disposition of life under certain anticipated conditions, but it is unfortunate that all the problems have not been worked out. A fairly good living will is set out in the State of Georgia[28] (see Appendix B).

The problems that exist with living wills include the fact that many of the states have procedural prerequisites and revalidation requirements (i.e., a time limit on the will), and that many health care providers are confused as to their liability in following or not following the provisions of the living will.

CONCLUSION

The subject and debate surrounding euthanasia grows in intensity on a daily basis. Those who defend a patient's right to die believe in each individual's right to deal accordingly with his own body. Dietrich Bonhoeffer in his book *Ethics* says

> the body in each individual case is always "my body." It can never, even in marriage, belong to another in the same sense as it belongs to me. It is my body that separates me in space from other men and that presents me as a man to other men.[29]

There are those who feel that when a person's body has reached a given deteriorative state he is indeed not presented as a man to other men, thereby lessening not only the dignity of his life but the dignity of his death. In the Alaskan Indian culture when an elderly member of the community decides to die, he or she gathers his or her friends and family for prayer and farewells. And within hours after that, the person will be dead.[1] There is much to be said for choosing the day on which you will die, especially if you have anticipated that you have reached it.

Those who believe that mercy killing voluntarily or involuntarily should remain classified on the law books respectfully as suicide and homicide have valid arguments for their case. Gregory Gelfand in his article "Euthanasia and the Terminally Ill Patient" says that "the prospect of mercy-killing is certainly, at best, a mixed blessing for the terminally ill patient."[12] He holds that medical error, the possibility of finding a cure, and the threat that future euthanasia will become something discriminatory, "warrants outright rejection of any such practice."[12] Indeed, it is a very real and frightening possibility that any or all of those possibilities could become reality.

What are the answers, the solutions? Should we be able to legislate our own demise? Should others be able to have power over our lives? If so, what are the guidelines and how should we extend them? What role should the judiciary have? As we progress into a technological future that our ancestors could only dream of and few of us can comprehend, our definitions of life and death and of human existence must change. We must approach it with a courageous mind and humane heart.

REFERENCES

1. Kubler-Ross E: *Death, the final stage of growth*. Englewood Cliffs, NJ, Prentice-Hall Inc, 1975.
2. *Webster's II new riverside university dictionary*. Springfield, Mass, Meriam-Webster Inc, 1984, p 447.
3. Ytreberg DE: Right to die, wrongful life. *American Jurisprudence 2nd*, New Topic Service, 1979, p 3.
4. *In re Quinlan*, 70 NJ 10, 355 A.2d 647, 79 ALR.3d 205 (1976).
5. Williams G: *The Society of Life and the Criminal Law*. New York, Alfred A Knopf Inc, 1957.
6. MacGuire D: *Death by choice*. Garden City, NY, Doubleday, 1974; Delgada R: Euthanasia reconsidered. *Arizona Law Review* 1975; 17:474.
7. *Severns v Wilmington Medical Center, Inc.*, 421 A.2d 1334, 1346, (DEL. 1980); *Satz v Perlmutter*, 379 So.2d 359 (Fla. 1980); *In re Quinlan*, 70 NJ 10, 50, 355 A.2d 647, 669 (1976) cert. denied; *Garger v New Jersey*, 429 U.S. 922 (1976); *Matter of Colyer*, 99 Wash.2d 114, 139, 660 P.2d 738, 746 (1983).
8. Kutner L: Due process of euthanasia: The living will, a proposal. *Indiana Law Journal* Summer 1969; 44:539.
9. Sherlock R: For everything there is a season: The right to die in the United States. *Brigham Young University Law Review* 1982; 545–616.
10. Model Penal Code §3.07(5).
11. Stephen H: Murder from the best of motives. *Law Quarterly of 1889* 1889; 189.
12. Gelfand G: Euthanasia and the terminally ill patient. *Nebraska Law Review* 1984; 63:777.
13. Chapman, Joan Du C., Fateful Treatment Choices for Critically Ill Adults Part I: The Judicial Model. *Arkansas Law Review* 1984; 37:908.
14. *Holmstedt v Holmstedt*, 383 Ill. 290, 49 N.E.2d 25, 28 (1943); *People v Weimer*, Criminal Law No. 58-3636 Cook County, Illinois (Dec. 30, 1958); Hirschfield JC: Criminal law—Euthanasia—Defendant allowed to withdraw guilty plea of manslaughter to accommodate finding of not guilty on arraignment. *Notre Dame Lawyer, Recent Decisions* 1959; 34:460–464.
15. *People v Werner*, Criminal Law No. 58-3636, Cook County, Illinois (Dec. 30, 1958).
16. Reported in *Notre Dame Lawyer, Recent Decisions* 1959, 34:461.
17. *People v Roberts*, 211 Mich. 187, 178 N.W. 690 (1920).
18. *Severns v Wilmington Medical Center, Inc.*, 421 A.2d 1334, 1346 (Del. 1980).
19. *Matter of Colyer*, 99 Wash. 2d 114, 139, 660 P.2d 738, 746 (1983).
20. *In re Quackenbush*, (1978) 156 NJ Super 282, 383 A.2d 785.
21. *Satz v Perlmutter*, (1978, Fla. App. D4) 362 So.2d 160.

22. *Leach v Akron Medical Center,* (1980) 68 Ohio Misc. 1, 22 Ohio Ops.3d 48, 426 NE.2d 809.
23. *Matter of Eichner,* (1980), 73 A.D.2d 431, 426 N.Y.S.2d 517.
24. *John F. Kennedy Memorial Hospital, Inc. v Bludworth,* (1983), 432 So.2d 611 (Fla. App. 4 Dist. 1983).
25. Martyn SR, Jacobs LB: Legislating advance directives for the terminally ill: The living will and durable power of attorney. *Nebraska Law Review* 1984; 63:741.
26. California (Sept. 30, 1976).
27. Alabama (1981), Arizona (1985), Arkansas (1977), California (1976), Colorado (1985), Connecticut (1985), Delaware (1982), District of Columbia (1982), Florida (1984), Georgia (1984), Idaho (1977), Illinois (1984), Indiana (1985), Iowa (1985), Kansas (1979), Louisiana (1984), Maine (1985), Maryland (1985), Mississippi (1977), Missouri (1985), Montana (1985), Nevada (1977), New Hampshire (1985), New Mexico (1977), North Carolina (1977, amended 1979, 1981, 1983), Oklahoma (1985), Oregon (1977), Tennessee (1985), Texas (1977), Utah (1985), Vermont (1982), Virginia (1983), Washington (1979), West Virginia (1984), Wisconsin (1984), and Wyoming (1984).
28. Georgia Code 31-32-1. See Appendix B.
29. Bonhoeffer D: *Ethics,* Bethge E (ed). New York, Macmillan Publishing Co Inc, 1955, pp 158–159.

APPENDIX A

Cases Involving the Right to Die

In re Quackenbush, 156 NJ Super 282, 383 A2d 785 (1978). Upon finding that the patient in question was competent, the court upheld the patient's refusal to consent to an operation to amputate his gangrenous legs. The court heard conflicting medical testimony regarding the patient's competency to make a decision respecting his medical treatment, and the judge himself visited with the patient. The court accepted the patient's reliance on *In re Quinlan,* 70 NJ 10, 355 A2d 647, 79 ALR3d 205, cert den 429 US 922, 50 L Ed 2d 298, 97 S Ct 319 (1976), in asserting a constitutional right of privacy, and the court noted that it had been pointed out in that case that the state's interest weakens and the individual's right of privacy grows as the degree of bodily invasion increases and the prognosis dims, until the ultimate point when the individual's rights overcome the state's interest in preserving life. Under the circumstances of the present case, the court held, the extensive bodily invasion involved—the amputation of both legs above the knee and possibly the amputation of both legs entirely—was sufficient to make the state's interest in the preservation of life give way to the patient's right of privacy to decide his own future regardless of the absence of a dim prognosis.

Satz v Perlmutter, Fla. App. D4, 362 So. 2d 160 (1978). The court held that a competent, 73-year-old man who was suffering from amyotrophic lateral sclerosis (Lou Gehrig's disease) had a right to have the mechanical respirator that was keeping him alive discontinued, and that this right was based upon the constitutional right of privacy. The court noted that the patient's condition was terminal, his situation wretched, and the continuation of his life was temporary and totally articial, and that he sought, with the full approval of his adult family, to have the respirator removed. The court said that because the patient would have the right to refuse treatment in the first instance because he was a competent adult, he also had the concomitant right to discontinue it. Because the court treated the right to refuse and the right to discontinue life-saving medical treatment as concomitant rights, its comments on the considerations involved in determining whether a patient has a right to have such medical treatment discontinued would also appear to be significant in determining whether a patient has the right to refuse life-saving medical treatment in the first instance. In the instant case, the court said that (a) there was a substantial distinction between the state's interest in preserving the life of one suffering from a curable affliction and the artificial preservation of the life of one who is suffering from a terminal illness; (b) the fact that the adult members of the patient's family were in agreement with the patient's wishes made the state's interest in protecting innocent third parties inapplicable; (c) the state's duty to prevent suicide was to be considered, but as the patient had expressed the wish to live, but under his own power, and he did not induce his affliction, this was not a suicidal wish; and (d) the state's interest in helping to maintain the integrity of the medical profession was not lessened by the prevailing medical standards that do not require that all efforts to prolong life be made in all circumstances.

Leach v Akron General Medical Center, 68 Ohio Misc 1, 22 Ohio Ops 3d 48, 426 NE2d 809 (1980). In action to remove 70-year-old housewife from respirator, wherein by clear and convincing evidence, it was established that the housewife: (a) was suffering from amyotrophic lateral sclerosis, terminal illness of nervous system, and had sustained irreversible brain damage, (b) was in permanent, chronic, vegetative state, (c) was not cognitive nor, within reasonable medical certainty, likely to have cognitive powers restored, (d) was incurably ill, and (e) would have elected not to be placed on or continued on life support system, and that sole motive of husband and two children in bringing present action was to end present vegetative condition of family member, constitutional right to privacy was held to encompass freedom of terminally ill to choose for herself, or by mechanism of parens patriae concept of substituted judgment, whether or not to decline medical treatment; inasmuch as compelling state interest was insufficient to outweigh housewife's right to refuse medical treatment, life support system was ordered to be discontinued.

Re Severns, 425 A2d 156 (1980, Del Ch). Request of husband and prospective guardian of injury victim who had remained in coma for over a year since her injury that authorization be given for discontinuance of all medical supporting measures designed to keep victim alive in her existing state would be granted where victim's physician testified that when victim's neck was broken apparently that part of her brain had been destroyed that has to do with normal human capability to be aware of one's environment, to think, recall, and speak, as well as to possess personality and to engage in intellectual activities, and where victim had expressed wish prior to her accident that in event she were ever to become unable to reason and care for herself as a result of accident or illness that she did not want to be kept alive in vegetative state but would like to be allowed to die with dignity.

Re Welfare of Colyer, 99 Wash.2d 114, 660 P.2d 738 (1983). Court properly granted request by husband of woman in chronic vegetative state to have life-sustaining systems discontinued, where there was no compelling state interest opposing removal of life-sustaining mechanisms that outweigh patient's privacy right to refuse such treatment, where two physicians agreed patient was incurable and probably would never recover to sapient state, where each physician had appropriate qualification to understand her condition, neither physician had treated her previously, and therefore neither appeared to have any interest in decision outside medical considerations, and where, under governing statute requiring guardian, in effort to assert "rights and best interest" of incompetent person, guardian is permitted to use his best judgment and exercise, when appropriate, an incompetent's personal right to refuse life-sustaining treatment. In cases where physicians agree on prognosis and close family member uses his best judgment as guardian to exercise rights of incompetent, intervention by courts would be little more than formality.

John F. Kennedy Memorial Hospital, Inc. v Bludworth, Fla. App. D4, 432 So.2d. 611 (1983). Application of court-appointed guardian of a comatose and terminally ill individual requires review by a court of competent jurisdiction before life-sustaining procedures may be suspended. Upon application for termination, court must determine from known facts and circumstances what the individual would want done were he conscious and competent. Such holding is limited to case of comatose patient in hospital, whose condition has been certified by two physicians as being terminal, whose interest are protected by a court-appointed guardian, and where the patient at some earlier point has executed a living will or has expressed or evidenced a like intention. *Terminal condition* means an inaudible physical state caused by injury, disease, or illness that, regardless of application of life-sustaining procedures would produce death within a reasonable degree of medical probability, and where application of life-sustaining procedures serves only or primarily to postpone the moment of patient's medicolegal death. *Certification* means that hospital patient has been diagnosed and certified in writing to be in terminal condition and where prospects of regaining cognitive brain function are extremely remote.

APPENDIX B

The Georgia Living Will

  Living will made this ________ day of ____________ (month, year).

  I, ___________________, being of sound mind, willfully and voluntarily make known my desire that my life shall not be prolonged under the circumstances set forth below and do declare:

1. If at any time I should have a terminal condition as defined in and established in accordance with the procedures set forth in paragraph (10) of Code Section 31-32-2 of the Official Code of Georgia Annotated, I direct that the application of life-sustaining procedures to my body be withheld or withdrawn and that I be permitted to die;

2. In the absence of my ability to give directions regarding the use of such life-sustaining procedures, it is my intention that this living will shall be honored by my family and physician(s) as the final expression of my legal right to refuse medical or surgical treatment and accept the consequences from such refusal;

3. This will shall have no force or effect seven years from the date I signed this document as stated above; however, I understand that, if at the end of said seven years I am incapable of communicating with the attending physician, this will shall remain in effect until such time as I am able to communicate with the physician;

4. I understand that I may revoke this living will at any time;

5. I understand the full import of this living will, and I am at least 18 years of age and am emotionally and mentally competent to make this living will; and

6. If I have been diagnosed as pregnant, this living will shall have no force and effect during the course of my pregnancy.

Signed _______________________________

_____________________________ (City), _____________________________ (County), and

_____________________________ (State of Residence).

I hereby witness this living will and attest that:

(1) The declarant is personally known to me and I believe the declarant to be at least 18 years of age and of sound mind;

(2) I am at least 18 years of age;

(3) To the best of my knowledge, at the time of the execution of this living will, I:

(A) Am not related to the declarant by blood or marriage;

(B) Would not be entitled to any portion of the declarant's estate by any will or by operation of law under the rules of descent and distribution of this state;

(C) Am not the attending physician of declarant or an employee of the attending physician or an employee of the hospital or skilled nursing facility in which declarant is a patient;

(D) Am not directly financially responsible for the declarant's medical care; and

(E) Have no present claim against any portion of the estate of the declarant;

(4) Declarant has signed this document in my presence as above-instructed, on the date above first shown.

Witness _______________________________
Address _______________________________
Witness _______________________________
Address _______________________________

Additional witness required when living will is signed in a hospital or skilled nursing facility.

I hereby witness this living will and attest that I believe the declarant to be of sound mind and to have made this living will willingly and voluntarily.

Witness: ________________________

Medical director of skilled nursing facility or chief of the hospital medical staff

(Code 1981, § 31-32-3, enacted by Ga. L. 1984, p. 1477, § 1.)

9

Determining When or If a New York Patient May Be Allowed to Die

Navigating between the Perils of Scylla and Charybdis

JON M. BEVILACQUA

The recent focus being given to do-not-resuscitate (DNR) orders[1] by the New York State Task Force of Life and the Law and other groups was due in part to the 1984 Queens Grand Jury Report[2] concerning DNR orders. This report resulted from the events surrounding the deaths of two elderly patients at a Queens hospital and the uncertainty with which the legality of withholding extraordinary means of life support concerning the terminally ill is viewed by both physicians and lawyers.

This perception in this area of the law has apparently caused health care professionals to embark on an uncertain course between civil and criminal liability in treating some terminally ill patients who either undergo cardiac arrest or enter a vegetative state.

In order to understand fully the need for legislative action concerning these problems and their potential civil and criminal consequences, a review of the essential findings of the Queens Grand Jury report will be undertaken, followed by a discussion of the law in New York concerning withholding or withdrawing extraordinary means of life support, the

JON M. BEVILACQUA • Deputy Attorney General's Office for Medicaid Fraud Control, New York State Office Building, Room 4A10, Veterans Memorial Highway, Hauppauge, NY 11787.

right to decline medical treatment and who, if anyone, may exercise this right.

QUEENS GRAND JURY REPORT

In March of 1984, a Queens Grand Jury that was empanelled by the New York Special Prosecutor for Medicaid Fraud Control issued a report after a lengthy investigation[3] concerning DNR orders at a certain hospital[4] in Queens County and the issuance of DNR orders in the general medical community in New York State. This report was released to the public after a New York State Justice of the Supreme Court reviewed both the report and the grand jury testimony and evidence upon which the report was based pursuant to the state's criminal procedure law. The report noted the confusion in the medical and legal community in New York concerning DNR orders and "found that some of those doubts are shared by highly responsible members of the medical profession despite virtually universal agreement that such decisions are medically appropriate in certain limited cases."

The grand jury's concern over abuses in the DNR decision-making process was precipitated by the death of a 78-year-old patient "Maria M" in Hospital X's medical intensive care unit (MICU) where she had been a patient for approximately 2 months prior to her death.

Because Mrs. M.'s family physician was away, she was brought to the emergency room of a local hospital on January 11, 1981, suffering from what appeared to her daughter to be the flu. The tentative admitting diagnosis by the emergency room physician was myeloplithisis, that is, atrophy of the spinal cord.

Within 2 weeks, Mrs. M. began showing difficulty breathing and was transferred to the intensive care unit, where she remained for approximately 2 months until she died on March 27, 1981.

Early in the morning hours of March 27, 1981, when the telemetry monitor showed that Maria M. was undergoing bradycardia, the nursing staff immediately responded to the patient's room. The nurses failed to summon a resuscitation team because as they testified, a medical student on duty had instructed them that Maria M. was a "no code," that is, not to be resuscitated.

The Special Prosecutor's Office obtained an exhumation order from the court and the autopsy was performed by the medical examiner's office. The cause of death, according to the medical examiner, was cardiac arrest following disconnection from the mechanical respirator. Moreover, the alarm tab switch for this patient's respirator was found by

MICU nurses in the off position and the respirator tubing was found tucked up and under the patient's pillow.

The grand jury found that based on Mrs. M.'s physical size and frail condition, the distance involved, the barriers posed by her bed rails, there was no evidence that she sought to take her own life.

In particular, the grand jury noted that:

> Mrs. M.'s medical chart revealed no definitive diagnosis as to the cause of her breathing difficulty, and none of the physicians who testified was able to attribute it to any terminal illness. No one ever suggested to Mrs. M. or any of the members of her family who visited her regularly that cardiopulmonary resuscitation might be withheld.

After her death, the family was told that everything possible had been done to prolong her life.

During the course of this inquiry, the grand jury learned that DNR orders were not to be written in the patient's chart according to official hospital X policy. Instead, according to medical personnel, a small ad hesive-backed paper purple dot was to be attached to the nurse's Kardex file, which was to be routinely destroyed when the patient died or was otherwise discharged. The grand jury found as a result that the "no code" order could thusly never be attributed to any particular physician; nor could the grand jury determine who ultimately issued the DNR order, which of course was the reason behind the ambiguity utilized in this hospital's practice.

The grand jury found, as a result, that

> the failure to uniformly involve patients and their families in the DNR decision-making process was the logical consequence of the standing policy not to leave evidence which might be used against the hospital in a law suit.

The hospital administration's view set forth in an internal memorandum was that "no code orders [were] of doubtful legality and are, therefore, not permitted."

After hearing the evidence and experts, the grand jury strongly recommended

> that the state expressly acknowledge that DNR decisions are regularly made as a matter of medical necessity and recognize the prevailing legal view that carrying out such a decision does not, in and of itself, constitute a violation of the civil or criminal law.

The grand jury recommended that the state legislature enact legislation that should establish, at a minimum, the following basic standards:

1. Such decisions shall be reached jointly by a responsible, licensed physician and the patient or, if the patient is not competent, a family member or other appropriate surrogate.

2. Such decisions shall be made only after the physician has, with a reasonable degree of medical certainty, diagnosed the underlying illness as incurable or untreatable and expected to result in imminent death.

3. The patient or his surrogate shall be meaningfully apprised of the nature of that illness and its consequences.

4. If an affirmative decision is reached to withhold cardiopulmonary resuscitation, an order to that effect shall be entered permanently in writing on the medical chart of the patient by a licensed physician who participated in the decision-making process.

5. Such a decision shall be readily revocable by the physician or the patient or surrogate in the event of a remission or for any other reason.

The grand jury also recommended that the State Commissioner of Health, pursuant to his already existing statutory authority and pending the enactment of the legislation proposed in the above recommendation, promulgate regulations embodying the minimum basic standards enumerated in that recommendation.

WITHHOLDING OR WITHDRAWING EXTRAORDINARY MEASURES OF LIFE SUPPORT

Common-Law Right to Accept or Decline Medical Treatment in New York

The modern and almost daily technological advances in medicine have unduly distorted our perception of the law in some cases whereas in others it has exposed a clear need for the law to catch up with medicine.

Justice Mollen in *Eichner v Dillon* 73 A.D.2d 431 (2nd Dept. 1980) *modified sub nom. In re Storar* 52 N.Y.2d 363 (1981) noted that modern technological advances in medicine "generally outpaced the ability of the judicial system to deal comprehensively with them in a manner consistent with the fulfillment of social policy objectives" (p. 531). Justice Mollen further commented that certain medical procedures discussed over 15 years ago by visionaries such as Ray Bradbury, Arthur C. Clarke, and Isaac Asimov are now reality and pushing the traditional limits of the law.

At present, the law in New York concerning the withholding or withdrawing of extraordinary measures depends on the patient's competency or lack thereof. There is a distinction to be made for (a) the competent patient, (b) the presently incompetent patient who has previously when competent solemnly expressed an opinion concerning the withholding of extraordinary life sustaining measures, and (c) the lifelong incompetent patient.

Competent Adult Patient

The prevailing legal view in New York is that a competent adult has a common law right to accept or decline medical treatment and as a result the issuance of a DNR order or an order to terminate life support within certain parameters does not violate the criminal or civil laws of New York State.

The keystone decision in New York concerning the common law right to accept or decline medical treatment was set forth long ago in *Schloendorff v Society of N.Y. Hosp.*, 211 N.Y. 125 (1914). In *Schloendorff*, Justice Cardozo stated that

> every human being of adult years and sound mind has a right to determine what shall be done with his own body; and a surgeon who performs an operation without his patient's consent, commits an assault, for which he is liable in damages. . . . This is true except in cases of emergency where the patient is unconscious and where it is necessary to operate before consent can be obtained. (p. 130).

This common law right to accept or decline medical treatment, which is also known as the bodily right of self-determination, yields only to "compelling state interests" of sufficient magnitude.

Compelling state interests have been defined as (a) protection of the patient's minor children dependents, (b) prevention of suicide, (c) preservation and sanctity of life in absence of no hope of recovery, and (d) maintenance of the ethical obligation of the medical profession. (*Matter of Storar*, 106 Misc.2d 880 [Sup. Ct. Monroe Co., 1980], *Matter of Fox*, 102 Misc. 2d 184, 203 [Sup. Ct., Nass. Co. 1979], and Application of *Lydia E. Hall Hosp.*, 116 Misc. 2d 477 [Sup. Ct., Nass. Ct., 1982].) As the court of appeals stated,

> the State has a legitimate interest in protecting the lives of its citizens. It may require that they submit to medical procedures in order to eliminate a health threat to the community (see, E. G. *Jacobson v Massachusetts*, 197 U.S. 11). It may, by statute, prohibit them from engaging in specified activities, including medical procedures which are inherently hazardous to their lives (*Roe v Wade*, *supra*, pp. 105, 154) *Matter of Storar*, 52 N.Y.2d 363, 377 (1981), *cert. den.* 454 U.S. 858.

The physicians respect for a competent adult patient's right to decline medical treatment is now regarded as a defense to criminal liability by virtue of the court of appeals decision in *Matter of Storar, supra,* wherein the court stated that,

> In this State, however, there is no statute which prohibits a patient from declining necessary medical treatment or a doctor from honoring the patient's decision. To the extent that existing statutory and decisional law manifests the State's interest on this subject, they consistently support the right of the competent adult to make his own decision by imposing civil liability on

those who perform medical treatment without consent, although the treatment may be beneficial or even necessary to preserve the patient's life.

. . . .

... A state which imposes civil liability on a doctor if he violates the patient's right cannot also hold him criminally responsible if he respects that right. Thus a doctor cannot be held to have violated his legal or professional responsibilities when he honors the right of a competent adult patient to decline medical treatment. (p. 377)

The patient's right to decline medical treatment in the context of a criminal prosecution concerned the court in *People v Robbins,* 83 A.D.2d 271 (4th Dept., 1981). In *Robbins,* a competent adult, who had suffered with diabetes and epilepsy, decided to stop taking all medication. The diabetic had discussed this decision with her husband and a self-professed minister and they all agreed that she should not continue taking her medication. As a result the wife became ill and lapsed into a coma from which she died. The husband had failed to summon medical aid for his wife and was indicted for criminally negligent homicide (Penal Law, §125.10). The minister was indicted as an accessory (Penal Law §20.00), primarily for having counseled Mrs. Robbins to refrain from taking medication or seeking medical assistance. The defendants moved to dismiss the indictment. The lower court granted the motion to dismiss the indictment, and the People appealed. The Appellate Division, Fourth Department, affirmed the lower court.

The district attorney had argued that, despite the wife's decision to stop taking all medication, the husband had a common law marital duty to provide medical care for his spouse. The court disagreed, holding that the husband and the minister could not be prosecuted because the competent adult spouse has a right to decline medical care relying upon the holdings in *Schloendorff* and *Storar.*

The court (in *People v Robbins, supra* 83 A.D.2d at 275) stated that

it would be an unwarranted extension of the spousal duty of care to impose criminal liability for failure to summon medical aid for a competent adult spouse who has made a rational decision to eschew medical assistance. In New York such a rationale would be in direct conflict with the related rule that a competent adult has a right to determine whether or not to undergo medical treatment (*Matter of Storar,* 52 N.Y.2d 363; *Schloendorff v Society of N.Y. Hosp.,* 211 N.Y. 125). There is no basis under New York law for denying an adult the right to refuse medical care where such refusal does not pose a threat to the life or health of others. "In this State, however, there is no statute which prohibits a patient from declining necessary medical treatment or a doctor from honoring the patient's decision. To the extent that existing statutory and decisional law manifests the State's interest on this subject, they consistently support the right of the competent adult to make his own decision by imposing civil liability on those who perform medical treatment without consent, although the treatment may be beneficial or even necessary to

> preserve the patient's life. (see, e.g., *Schloendorff v Society of N.Y. Hosp.*, 211
> NY 125, *supra; Matter of Erickson v Dilgard*, 44 Misc. 2d 27; *Matter of Melideo*,
> 88 Misc. 2d 974; Public Health Law, §§ 2504, 2805-d; CPLR 4401-a). The
> current law identifies the patient's right to determine the course of his medi-
> cal treatment as paramount to what might otherwise be the doctor's obliga-
> tion to provide needed medical care. A State which imposes civil liability on a
> doctor if he violates the patient's right cannot also hold him criminally re-
> sponsible if he respects that right. (*Matter of Storar*, 52 N.Y.2d 363, 377,
> supra.)

In order to exercise the right to accept or decline medical treatment
in a meaningful fashion, competent patients must be informed of all
necessary information concerning their condition.

The court in *Matter of Fox,* 102 Misc.2d 184 (Sup. Ct., Nass. Co.
1979), discussed the requirements for informed consent and stated that

> the law clearly requires that the consent be informed, based upon all infor-
> mation necessary under the circumstances. (*Abril v Syntex Labs.*, 81 Misc.2d
> 112, 113; *Barnette v Potenza*, 79 misc. 2d 51, 54–55). The statutory and ad-
> ministrative law of the State embodies the same concern for self-determina-
> tion on the part of the patient. (See Public Health Law, §2803-c). It is re-
> quired by 10 NYCRR 405.25(a)(5) that hospitals establish written policies
> which afford each patient, among other things, the right to "receive from his
> physician information necessary to give informed consent" and the right to
> "refuse treatment to the extent permitted by law." (10 NYCRR 405.25 [a]
> [6].)

Thus, the competent adult patient's right to decline medical care is
inviolate save for compelling state interests.

The Incompetent Adult Patient

The incompetent adult patient and the right to withhold or withdraw extraordinary means of life support were discussed by the court of
appeals in *Matter of Storar, supra.* The *Storar* case involved two incompetent adult patients, in two separate cases, *Eichner v Dillon* and *In re Storar.*
They were consolidated by the court of appeals in view of their com-
monality of subject matter, that is, the withholding or withdrawing of life
prolonging medical treatment.

Despite the fact that both patients had died, the court of appeals in a
majority decision by now Chief Judge Wachtler found the issue to be of
sufficient public importance to warrant resolution. Both cases involved
patients who had been (a) terminally ill; (b) in a chronic vegetative coma;
and (c) deemed incompetent in the hospital to give consent to the withdrawal or withholding of extraordinary life-support measures.[5] Moreover, both patients had legal guardians who applied to the court for an
order to terminate their medical care.

The court held that Mr. Storar's legal guardian could not make a determination to decline medical treatment for Mr. Storar. However, the court held that the authorization by the lower court to Brother Fox's guardian to discontinue the use of the respirator was proper because it was based on a stated intention to decline medical treatment asserted while he was still competent. The lower court had found that by clear and convincing evidence that Brother Fox had, while competent, expressed a desire not to have extraordinary means of life support used if he entered a vegetative state.

The court of appeals in sum found that the operative distinction between the two cases, *Eichner* and *Storar,* was that in *Eichner,*[6] the patient, Brother Fox had previously and solemnly declared while competent that he did not want to be maintained by extraordinary means if he entered a vegetative state; whereas John Storar in contrast had been a lifelong incompetent and thus could not make an intelligent choice.

The lower court in *Eichner* had held a hearing[7] and found that Brother Fox made his position known by a discussion with the fellow members of his religious society following the Karen Ann Quinlan case. This conversation further involved a discussion of Pope Pius XII's address to respiratory therapists concerning and confirming the patient's right to refuse "extraordinary" life sustaining measures.[8]

In the *Storar* case, the patient, John Storar, was incompetent since birth and therefore unable to make an intelligent and articulate choice concerning the withholding of "extraordinary life-support measures." The court held that the patient's guardian, his mother, did not have the power to speak on his behalf, as the patient was incompetent to make a decision because of his lifelong cognitive inability.

The court of appeals in *People v Eulo,* 63 N.Y.2d 341 (1984) summed it up this way:

> in the absence of such evidence of personal intent (there, due to the patient's incompetence), a third party has no recognized right to decide that the patient's quality of life has declined to a point where treatment should be withheld and the patient should be allowed to die. (p. 357)

It is therefore clear in New York that only the competent adult has a right to determine the course of his or her medical treatment. Neither the incompetent patient nor his legal guardian may legally decline medical treatment.[9] Accordingly, the major problem in New York concerns protecting and advancing the rights of terminally ill incompetent patients[10] who are either lifelong incompetents or who while competent have not expressed a position concerning the withholding or withdrawing of extraordinary means of life support.

EFFECTUATING THE COMPETENT ADULT'S POSITION CONCERNING MEDICAL TREATMENT

Recently, the attorney general in New York issued Formal Opinion No. 84-F16 concerning the use of a durable power of attorney as a means of communicating a competent adult's position concerning withholding extraordinary means of life support.[11]

The attorney general maintained that, based on a reading of the statutes and case law, a competent person may use the durable power of attorney to make known his or her position concerning the withholding of extraordinary life support measures should they become incompetent or enter a vegetative coma similar to Brother Fox.

Although the "durable power of attorney" has traditionally been used in financial or business contexts, the Attorney General believes it can be extended to cover health care decisions by competent adults.

The opinion notes that a principal (competent person) under General Obligations Law §5 1601 is authorized "to invest an attorney-in-fact with a power of attorney that will survive even if the principal should become incompetent or otherwise disabled." This is known as a durable power of attorney.

The formal opinion discusses the authority to make a decision on behalf of the incompetent patient and concludes that

> a durable power of attorney may not be used to delegate to an agent *generally* the authority to make health care decisions on behalf of an incompetent principal. However, a durable power of attorney may be used by a competent adult to delegate *specifically* to an agent the responsibility to *communicate* the principal's decisions to decline medical treatment under defined circumstances.

The opinion went on to point out with respect to a durable power of attorney that there is a lack of legislative intent "either to grant or to deny an agent the power to make health care decisions for an incompetent principal." Thus this opinion provides only the competent principal with one method to communicate his health care decision, that is, through a durable and properly limited power of attorney expressing his specific decision to decline medical treatment under defined circumstances.[12] This decision by a competent adult would survive his or her incompetency.

SUMMARY

If there is an informed consent given by a competent patient concerning his right to decline medical treatment, there should be no problem for physicians concerning the withholding or declination of medical care, absent any of the previously mentioned compelling state interests.

With respect to incompetent patients, the court of appeals in *Storar, supra,* disapproved of the *mandatory* complex procedures set up by the appellate division to handle the matter of judicial authorization of the declination of medical treatment. The *Storar* court did, however, allow for an *optional* judicial procedure, so that parties would not have to act at their peril where the legal consequences of their contemplated action were uncertain.

In this regard, the *Storar* court stated that

> However, responsible parties who wish to comply with the law, in cases where the legal consequences of the contemplated action is uncertain, need not act at their peril (*New York Public Interest Research Group* v. *Carey,* 42 N.Y.2d 527, 530). Nor is it inappropriate for those charged with the care of incompetent persons to apply to the courts for a ruling on the propriety of conduct which might seriously affect their charges. We emphasize, however, that any such procedure is optional. Neither the common law nor existing statutes require persons generally to seek prior court assessment of conduct which may subject them to civil and criminal liability. If it is desirable to enlarge the role of the courts in cases involving discontinuance of life sustaining treatment for incompetents by establishing, as the Appellate Division suggested in the Eichner case, a mandatory procedure of successive approvals by physicians, hospital personnel, relatives and the courts, the change should come from the Legislature. (p. 383)

Although this judicial procedure is undoubtedly cumbersome and time-consuming, it does provide a safe vehicle for resolving complex issues facing physicians and lawyers, such as (a) whether the adult patient is or was competent, (b) what the previously competent adult patient's position was, if any, concerning medical treatment, (c) whether compelling state interests exists sufficient to override the patient's decision to decline medical treatment, (d) whether the patient has a terminal illness and is hopelessly ill, (e) the veracity of the witness or witnesses to the patient's decision to decline medical care, and (f) whether the patient's decision to decline medical treatment was solemnly made. Additionally, if there is a fear that a competent patient may become incompetent, there should be serious thought given to memorializing the patient's position with the use of the aforementioned durable power of attorney.

The incompetent patient who has never expressed while competent an opinion concerning the withholding or withdrawing of life prolonging medical treatment and the lifelong incompetent patient[13] like John Storar cannot have their medical care terminated by virtue of the court of appeals decision in *Storar*. It is especially urgent with regard to the incompetent patient that the Legislature act to adopt the substance of the aforementioned Queens Grand Jury's recommendations in order to protect and advance their rights by allowing an appropriate surrogate or family member to make a decision on terminating health care for the incompetent patient. The incompetent patient should not have less rights than the competent patient.

Accordingly, such devices as the judicial resolution machinery in *Storar,* and the durable power of attorney, should be helpful in avoiding the perils of Scylla and Charybdis at least until the legislature provides guidance through legislation[14] in order to clarify once and for all the rights of all patients whether competent or incompetent.

REFERENCES

1. A DNR order has been defined as an order to withhold resuscitation from a patient who undergoes cardiac arrest.
2. The author is a Special Assistant Attorney General in the Office of Deputy Attorney General Edward J. Kuriansky. Mr. Kuriansky is the Special State Prosecutor for Medicaid Fraud Control. Mr. Bevilacqua presented the investigation concerning do-not-resuscitate orders to the Queens Grand Jury, which resulted in a Grand Jury Report.
3. During the course of this investigation, the grand jury subpoenaed the hospital records of the two elderly patients from Hospital "X," which had unsuccessfully sought to block the grand jury from obtaining these records. In *Grand Jury Proceedings (Doe),* 56 N.Y.2d 348 (1982), the Court discussed the investigation and stated that:

   > We hold today that a hospital being investigated by a Grand Jury in connection with possible crimes committed against its patients by the hospital staff may not successfully assert the physician-patient (CPLR 4504) or social worker-client (CPLR 4508) privileges, or the patient's right to privacy, in opposition to a Grand Jury subpoena calling for the hospital's records. . . .

   > In July, 1981, the Deputy Attorney-General for Medicaid Fraud Control issued two Grand Jury subpoenas, one to appellant hospital and the other to its executive vice-president, calling for the production of hospital records relating to a patient, Maria M., who had died in the hospital. The hospital moved to quash the subpoenas based on the assertions that the physician-patient privilege applied, that the subpoenas violated the patient's right to privacy, and that portions of the requested matter were material prepared for litigation. In response to the motion to quash, the Deputy Attorney-General demonstrated that the subpoenas were issued as part of an investigation into possible "no coding" at the hospital, a procedure under which various lifesaving and support measures are selectively denied to certain seriously ill patients (p. 351).

4. The hospital, patients and medical staff were not identified pursuant to the restraints of the New York Criminal Procedure Law §190.85[2]b, as it bears on those as to whom the Grand Jury Report may be deemed critical.
5. Extraordinary treatments are those that involve excessive expense, pain, or other inconvenience or, in the alternative, are those that do not offer a reasonable hope of benefit (*Matter of Eichner,* 73 A.D.2d 431, 441, n. 5 [2nd Dept., 1981]).
6. Father Eichner was the committee for the incompetent patient Brother Joseph Fox.
7. The lower court in *Storar* in holding a hearing applied the "clear and convincing standard" of proof to the factual issue to be determined (426 N.Y.S.2d. 545–546). The Court of Appeals in *Storar* agreed that this is the appropriate standard of proof (*Storar, supra* at 378–379). The Court reasoned that the highest standard in civil cases should be used (p. 274). Such a standard serves "to impress the factfinder with the impor-

tance of the decision" (*Addington v Texas,* 441 U.S. 418, 427), and "forbids relief whenever the evidence is loose, equivocal or contradictory" (*Backer Mgt. Corp. v Acme Quilting Corp.* 46 N.Y.2d 211, 220).

8. In this regard, it should be noted that, according to newspaper reports, when Terrence Cardinal Cooke was advised he had a terminal illness, he stated that he did not want any "extraordinary" measures used to keep him alive (October 7, 1983 edition of *The Washington Post*).

9. The only exception to this rule was set forth in *Storar* where the incompetent patient Brother Fox had previously been competent and made a solemn declaration concerning his desire to refuse extradinary measures if he entered a vegetative state. Accordingly, a competent patient's position concerning "extraordinary measures" will be carried out should the patient later become incompetent, provided of course that the patient's position was solemnly made in front of witnesses.

10. Of course based upon established "law" a court may exercise its powers, as parens patriae to authorize medical treatment on behalf of an individual where the interests of the individual so warrant and there is clear and convincing proof that the individual is not competent to make a rational decision (*Matter of Harris* 91A.D. 1142 [3rd Dept., 1983] and *Storar, supra*).

11. The opinion does recognize that there has not yet been any judicial interpretation concerning the use of a durable power of attorney in a health care context and thus this conclusion is not free from doubt.

12. The opinion points out that in Pennsylvania and California a durable power of attorney may be used for making health care decisions on behalf of an incompetent principal (Shearer GM, Levy RB: Decision making for incompetent patients by designated proxy. *N Engl J Med* 1984; 310:1598–1601 [citing 20 Pa. Cons. Stat., §§5601–5606; Chapter 1204 of the 1983 California Statutes]).

13. There is some hope on the horizon for the mentally disabled patient in a mental hygiene facility. Recently, the New York State Commission on Quality of Care for the Mentally Disabled has circulated a document, Surrogate Decision-Making Committees. According to the Commission's counsel, a demonstration project to test the feasibility of a nonjudicial alternative for medical decision making on behalf of medically incompetent residents of mental hygiene facilities was enacted into law by c. 354, L 1985 and MHL ¶45.07(j). This statement was set forth in a letter dated January 27, 1986 to the N.Y.S. Bar Association, Committee on Health Law by Paul F. Stavis, Esq.

14. The New York State Task Force on Life and the Law has formulated proposed legislation concerning guidelines and procedures for withholding and terminating care.

IV

Diagnostic and Clinical Factors

10

Biological and Clinical Considerations in Geriatric Medicine

MICHAEL L. FREEDMAN

DEFINITION OF AGING

Aging is a basic biological process that occurs in virtually all animal species. Only animals that continue to grow in size after reaching adulthood do not age. There are several fish and reptile species that fall into this category. They are not immortal, however, as they ultimately die of disease. Certainly all mammalian species age and as human beings we all must face this inevitability.[1]

Aging may be defined as a slow gradual decline in all body functions that ultimately will lead to the death of the organism. Inherent in this definition is the concept of a maximal life span or biological limit to each species longevity that is brought about by this gradual decline in function.[2] The basic problem of aging to us, then, is at what point, if ever, does aging severely incapacitate the human so that he can no longer function.

The Bible, a source of much of western civilization's thinking, gives us two diametrically opposed views of aging. In the description of the

Portions of this chapter were supported by a grant from the Metropolitan Jewish Geriatric Center, Brooklyn, New York.

MICHAEL L. FREEDMAN • Division of Geriatrics, New York University and Bellevue Medical Centers; Department of Medicine, New York University School of Medicine, 550 First Avenue, New York, NY 10016.

141

death of Moses, we read of an optimistic view. "And Moses was a hundred and twenty years old when he died; his eye was not dim nor his vigor abated."[3]

The pessimistic view is told in the story of the death of King David at age 70. "Now King David was old and stricken in years; and they covered him with clothes but he got not heat."[4] In spite of his servants attempts to rejuvenate him his wife Bath-sheba found "the King was very old."[5] After David gave his final charge to his son Solomon, "he slept with his fathers and was buried in the city of David."[6]

Thousands of years have passed since these two descriptions of aging were first written down and yet we still have not resolved which of these are "normal aging." Certainly we know that there are a few fortunate individuals who go through life with relatively little loss of "vigor" whereas others are stricken with multiple problems at a relatively early age and lose their "heat." One obvious factor in early aging is the influence of disease which will certainly lead to incapacity or death before one reaches one's biological potential. If we look at the history of mankind, we recognize that people in the past were considered old at ages we now consider young because they had accumulated several illnesses. Thus, we must carefully define what is aging and what is illness.

PHYSIOLOGICAL AGING

Aging results mainly in a loss of reserve at activity with a preservation of function at the resting state. Although there are some aging changes that affect function at rest, such as presbyopia and presbycusis, these do not threaten survival even though they may have important psychological impact. The loss of reserve available for coping with activities above the resting state beings at about age 30 and by age 70 is usually very obvious. Particularly if the individual is stressed by an illness, this loss of reserve can lead rapidly to incapacitation and death.[2]

Categories of Age-Related Changes

When categorizing changes with age it should be understood that not everything goes down at the same rate, and rarely some functions seem to increase. Five categories of physiologic aging may be seen in people[2]:

1. Function is totally lost: the most obvious is the incapacity of the female to reproduce.
2. Units of function are lost but the remaining units function the

same as in a young person. We see this in the structural loss of nephrons and muscle mass but the remaining nephrons and muscle fibers work perfectly normally.

3. The reduced efficiency of each unit even though there is no anatomic loss; this is seen in the reduced conduction velocity of nerve fibers in older individuals.

4. An interruption of control function resulting in a secondary aging change; this is seen in a large rise in gonadotropins in the female as the feedback control exerted by the ovarian hormones is reduced.

5. Increased function may also be seen rarely; such as the secretion of antidiuretic hormone in response to osmotic challenge.

An example of the losses seen in a 70-year-old man as compared to a young man is shown in Table 1.

Changes in Appearance

A person usually reaches the maximal height in the early 40s and then becomes shorter at a rate of 1 cm per decade and probably even more than that after age 70.[7]

The loss of height is due to a flattening of the foot arches, increase in spinal curvature and a collapse of the vertebral column as the intervertebral disc connective tissues dries out. The long bones, however, do not shorten with age. The chest increases in both diameters due to loss of the elasticity of the lungs. If osteoporosis is present, the lateral diameter of the chest may fall. The skull diameter increases and the continued growth of nose and ear length results in what we can all recognize as the older person's face.[8]

We also tend to have a greater percentage of our body weight as fat as we age. Muscle mass and bone density diminishes.[8-10] Fat tends to distribute in central areas in the omentum, the perinephric area, and within organs. Subcutaneous fat is lost from the limbs but increases on the trunk, usually around the waist in men and on the hips and buttocks in women. Total body water falls and the greatest part of this loss is from the intracellular compartment. This reduction represents a decline in cellular mass rather than tissue dehydration. As a result total body potassium, which is mainly intracellular, also falls with age.[10-12]

Body weight often begins to fall in the seventies due to this loss in lean body mass. Thus the elderly person's weight is much more due to fat rather than muscle or bone. All tissues and organs (except for the prostate gland) are losing functional mass. The muscles lose the most mass while in the viscera, the kidneys and liver are most affected.[13]

Table I. Anatomic and Functional Decline in the 70-Year-Old Human Man Expressed as a Percentage of That Seen in the Young Adult

Percent decline	Anatomic or functional decline
0–5	Body temperature
	Resting heart rate
	Body fluid pH
	Total body weight
	Body height
	Blood volume
	Extracellular fluid
5–10	Brain weight
	PaO_2
10–20	Total body water
	Cerebral blood flow
	Kidney weight
20–30	Maximal heart rate
	Resting O_2 usage
	Exchangeable potassium
	Intracellular fluid volume
	Muscle mass
	Cell "solids"
30–50	$\mathring{V}_E$ (maximal)
	$\mathring{V}_{O_2}$ (maximal)
	Resting cardiac output
	Maximal urinary concentration
	Kidney blood flow
	$\mathring{V}_{O_2}$
	Cardiac reserve

Much of our visual recognition of how old people are comes from our perception of the skin. Because the skin is our major interface with the environment, it reflects both the accumulation of external damage and the aging process itself. The most commonly seen cutaneous manifestation of aging is a rough, scaly epidermal surface that gives people a dry appearance.[14] Upon closer inspection, one sees fine superficial epidermal crevices and wrinkles as well as uneven pigmentation and the presence of benign seborrheic keratoses. This aspect of the aging skin seems to result from the decreased ability of the epidermis to replace and to repair itself.[15–17] With a diminished epidermis the barrier function of the skin is also diminished, which leads to the development of benign, premalignant and malignant epidermal tumors. Chronic sun exposure plays a crucial role in the development of basal cell carcinomas, certain forms of Bowen's disease, and squamous cell carcinomas.[18] In contrast, the development of seborrheic keratoses, the most common form of epidermal growth in the elderly, is not related to sun exposure.

The dermis is also undergoing changes with age that result in laxity and wrinkling of the skin.[14] Wrinkling results in a large measure from sun exposure but eventually will also occur in sun-protected areas.[19] The laxity of the skin due to alteration of the elastic fiber network is also exacerbated by sun exposure but will also ultimately occur all over.[19] Aged skin also has a fragile parchment-like quality with a thin translucent appearance and a propensity for injury from minor trauma. This quality results in "senile" purpura. Hair follicles also diminish with age, which will result in some degree of balding in all individuals.

Hair color is determined by the presence and activity of melanocytes within the hair bulb.[20] Many melanocytes are still present in their normal locations in gray hair bulbs but contain large cytoplasmic vacuoles or incompletely melanized melanosomes. Melanocytes are infrequently seen or totally absent in white hairs.[21] The matrix of the hair follicle is one of the most rapidly growing tissues in the body, which is thought to explain why there is an eventual loss of melanocytes.[22] It is estimated that 50% of the scalp hair of 50% of people over age 50 will be gray.[23]

The Aging Brain

It is obvious that we look different as we age and that our physical capacities have diminished. Most people accept these changes relatively well particularly because they are gradual and give us time to become accustomed to them. The function that worries most of us, however, is what will happen to our minds. A deterioration of our central nervous system with its resultant alteration in normal function and behavior is the major threat to quality of life of the elderly and imposes a tremendous emotional and economic burden on families and society.

In aging there are selective losses in the sensory system.[24] Older people show a loss of vibratory sensation, two-point discrimination, and tactual recognition. Other sensory modalities, such as light touch, pressure, and joint position sense appear to remain fairly much intact.[25]

Sleep is altered with aging. Older people have a prolongation of sleep latency and the total time awake over the sleep period is increased. There are frequent awakenings particularly in the second half of the night.[26] Rapid eye movement (REM) latency is reduced with aging and REM sleep time decreases.[27] These findings certainly are consistent with the complaints of many older people that they cannot sleep well and seem to be exaggerated in patients with dementia.

Many older people complain of a decrease in memory functions. Both human and animal research has found that given a complex task, the elderly require more practice than the young to perform equivalently.[28-30] Given the same number of trials, the elderly do not perform as reliably.[31-33] It is now clear that the loss of neurons (estimated at 50,000 to 100,000 cells per day) does not alone explain the decrease in memory in older people. One must consider the changes in the glial cells, connective tissue and endothelial cells, neurotransmitter mechanisms, and synapse function.

Complaints about declining memory are so common in the elderly that if one cannot diagnose dementia the patient's problem is labeled benign forgetfulness of senescence. Objectively, in a normal elderly population the ability to memorize arbitrary collections of verbal information and slowness in retrieval for familiar and unfamiliar material are found. In psychiatric illnesses, such as depression, there are considerable complaints of memory dysfunction that are not borne out by formal testing. However, most patients with subjective complaints about memory are not depressed nor do they respond to anti-depressants. At the present time, there is still considerable overlap between "normal" forgetfulness, confusion, and dementia. Recent work here at NYU by Reisberg *et al.*[34] has done much to clarify this overlap and will be discussed in the following.

MAJOR CHRONIC PROBLEMS WITH ADVANCING AGE

As a result of the declines in body function, there usually is a point in an elderly person's life where there are limitations on the activities of daily living. The only way most of us can escape these is to die. There are five chronic problems that may be identified, the so-called 5 "I"s (all begin with the letter I). There are: (a) impaired homeostasis, (b) immobility, (c) incompetence, (d) incontinence and (e) iatrogenic disease.

Impaired Homeostasis

Homeostasis is the maintenance of an optimized internal environment in spite of what is happening in the external environment. Maintenance of homeostasis requires a complex interplay between receptors of a central control system and effectors. The messages being carried between these points go via nerves or the endocrine system. Aging effects all of these elements.[2]

Internal Receptors.   In general, there is a loss of sensitivity of these receptors. For example, the baroreceptors that monitor the arterial blood pressure have a blunted response, which may result in postural hypotension and a sensation of dizziness in older people.[35]

External Receptors.   Aging changes in the skin result in a loss of cutaneous receptors and a diminished response in the remaining ones.[36] Older people have a much poorer ability to recognize temperature changes than young people. This explains, in part, the reason that older people are less liable to take corrective actions either to conserve heat if it is too cold, or to produce more heat if it is too warm. As a general rule, older people seem to prefer environmental temperatures warmer than young people.[2] When exposed to cold, the elderly begin to shiver later than do the young.[37] Conversely, when it is hot, the elderly have a delayed sweating response and the heat load persists longer.[38] Therefore, the elderly are at much greater risk for both hypothermia and hyperthermia even though the basal body temperature is the same.[39]

Regulatory Center.   These centers are within the hypothalamus or the brain stem and lose precision.[2] Therefore, greater deviations from the basal state are necessary to initiate corrective action. This loss also plays a role in the delay of the elderly to avoid temperature extremes.

Behavioral Output.   After a change in the environment is perceived, the elderly have a slower output of action via the voluntary motor system. The loss of memory, regulatory, and motor neurons are all involved here and discomfort needs to be extreme before corrective action is taken.[2]

Endocrine Regulation.   There is considerable variation in the endocrine gland output with age.[40] For example, the elderly have a decrease in aldosterone secretion that makes sodium conservation more difficult.[41] On the other hand, antidiuretic hormone secretion tends to increase.[42] These two changes make the older person more likely to have the syndromes of hyponatremia.

Autonomic Outflow.   There are few studies of this in humans but animal studies show that all stages of autonomic outflow change dramatically with aging.[43] Certainly the elderly seem to have altered receptors and a blunted response to even high circulating neurotransmitters.[44,45] This can be seen in the decrease of vasodilation with isoproterenol in the elderly[44] and the difficulty in achieving beta-blockade with propranalol due to a possible decrease in affinity of the beta-receptors.[46]

Behavioral Effectors.   The decreased mobility with age due to the decline in the musculoskeletal system, the loss of proprioception and balance makes it difficult for the elderly to perform voluntary corrective acts. For example, the elderly often cannot get to food, water, or cannot move to a more comfortable temperature. Geriatric care must take this into account.[47]

Regulatory Effectors.   These involve involuntary muscle, myocardium, respiratory muscles, glands, etc. As a general rule, there is a deterioration of function with age. For example, in the heart, as one loses nodal pacemaker tissue, the heart rate goes down.[48]

Immobility

The elderly often become much less mobile due to a loss of bone, muscle, homeostatic mechanisms, and the development of illness such as osteoarthritis.[49] Some of this decline in mobility is accentuated by the physical deterioration due to a lack of exercise. Exercise declines as one gets into one's mid-thirties and again upon retirement.[50] It is now well established that aging changes in the musculoskeletal system can be slowed and even somewhat reversed by exercise (even though there is a real decline in capability.)[50,51] Physical deconditioning, however, is a major problem in the elderly.[52] Three weeks of bed rest results in the loss of conditioning equal to that of 25 years of aging.[53,54]

Musculature. Muscle mass declines with age[55] and together with a loss of respiratory[56] and cardiac reserve,[48] as well as connective and supporting tissues,[2] makes it difficult for older people to carry out the same degree of exercise as the young.[55] Hemoglobin concentration and the number of circulating red cells do not change with normal aging[57–59] but 2,3 diphosphoglycerate (DPG) does go down, making unloading of oxygen to the tissues more difficult.[2] Nevertheless, physical conditioning is still possible taking into account the physical work capacity of the average 70-year-old is approximately one half that of the 20-year old.[60]

Bone. Bone loss is also an inevitable consequence of aging.[49] It is more pronounced in women at the time of menopause. Men lose bone more slowly and lose cortical bone as they lose tissue in general and not preferentially, as do women. Because women live longer than men and start with less bone, they have a more fragile skeleton than men.

The problem of osteoporosis, or loss of bone, is a major problem particularly for elderly women. Table II lists the risk factors that have been epidemiologically linked to this illness.[61] It is also important to differentiate osteoporosis from osteomalacia, or vitamin D deficient bone mineral loss. Table III lists the causes of vitamin D deficiency in the elderly.[61]

For women who are currently undergoing menopause, the deformities caused by fractures of the vertebrae, wrists, and hip will make up the single largest cause of hospitalization. Death from hip fractures is now the 12th leading cause of death in the United States.[49] The best established prophylaxis is now low-dose estrogen-progesterone therapy. The very low dose estrogen and high dose calcium treatment seems very promising.[49]

**Table II. Risk Factors Associated
with Osteoporosis**

Female sex
Early menopause
Family history
Nonblack race
Small stature
Low muscle mass
Nonobesity
Lack of sunlight exposure
Low calcium diet
Lack of exercise
Cigarette smoking
Use of corticosteroids
High protein diet
?Coffee drinking

At the present time, it is recommended that all perimenopausal women should ingest 800 to 2000 mg of calcium per day, least 400 IU of vitamin D per day, expose themselves to sunlight at least 15 minutes twice a week, and engage in regular exercise.[61]

Incompetence

The dementias are among the most tragic illnesses that affect the elderly. Certainly, the loss of brain function takes away that which is so characteristic of a person and that distinguishes the human from other animal species. A demented patient places an awesome burden on the

**Table III. Causes of Osteomalacia of Vitamin-D
Deficiency**

Lack of sunlight exposure
Poor dietary intake
Steotorrhea
Cholestasis
Gastrectomy
Chronic renal failure
Chronic alcoholism
Cirrhosis of liver
Medications
    anticonvulsants (increased degradation of vitamin D)
    cholestyramine (malabsorption)
Age-related impairment of renal conversion of 25-hydroxy
    vitamin D to 1,25 dihydroxy vitamin D

family, physically, emotionally, and financially. It is incumbent upon any physician when faced with a patient with a dementia that any correctable condition be vigorously searched for and treated.

Types of Dementia.   Dementia is defined as a global deterioration in memory, affect, judgment, orientation, and intelligence.[62,63] Dementia is a chronic condition with a slow, insidious onset that is usually first noted by the patient's family and friends.[63] Dementia must always be differentiated from delirium, which has an acute onset accompanied by restlessness, sleep disturbances, fluctuating alertness, diminished attention span, and often hallucinations. Delirium is always secondary to a definable cause and is, therefore, potentially reversible (if the cause is reversible).[62,63] Dementia is sometimes reversible—estimates range as high as 20% of cases[64] but may actually be less.[64,65] Table IV lists the reversible causes of dementia, delirium, and cognitive decline in the elderly.

At least 80% of the dementias are irreversible with senile dementia of the Alzheimer's type (SDAT) representing 60% to 70% of these.[63] Multi-infarct dementia accounts for 15% to 25% of cases of irreversible dementia and most of the remainder are caused by a combination of these. Rare causes of irreversible dementia include Creutzfeld-Jakob

Table IV. Examples of Reversible Causes of Dementia

Therapeutic drug intoxication
Depression
Metabolic abnormalities (includes azotemia, hyponatremia, volume
    depletion, hypoglycemia)
Endocrine disorders (includes hyper- and hypothyroidism)
Systemic infection
CNS infection
Congestive heart failure
Cardiac arrhythmias
Cerebrovascular accidents
Subdural hematoma
Brain tumors
Normal pressure hydrocephalus
Sensory deprivation (blindness, deafness)
Environmental change, isolation
Anemia
Malignancy
Chronic pulmonary disease (hypoxia and/or hypercapnea)
Deficiency of essential nutrients (vitamin B_{12}, folate, ? iron)
Chemical intoxication
Alcoholism

disease, dementia pugilestica, Parkinson's disease, Huntington's chorea, hepatolenticular degeneration (Wilson's disease) and Pick's disease.[63]

Dementia is not part of the normal aging process; impairment of cognitive function is a pathologic condition that does not occur in the majority of elderly people. However, significant cognitive impairment affects approximately 10% of all persons over age 65 and 30% of those over 80. Over the more than one million elderly in United States nursing homes, 50% to 75% have dementia. Of those living at home, an estimated 2 million or more also suffer from significant cognitive impairment.[66]

Senile Dementia of the Alzheimer's Type (SDAT). The etiology of Alzheimer's disease is still unknown even though various promising leads are being investigated. Included in the possible causes are infectious agents; genetic causes; heavy metal poisoning; decreased neurotransmitters (mainly acetylcholine); and decreased cerebral blood flow. In spite of the progress in research in this area, we still have no effective therapy.

Alzheimer's disease is a clinical entity that in recent years has been well characterized by Reisberg *et al.* here at NYU.[34,67] In their "global deterioration scale," people are classified into seven stages ranging from normal (no cognitive decline) to very late dementia (very severe cognitive decline).[34] This is summarized in Table V. In many older people, we see a progression from Stage 1 through Stage 4 but no actual progression to Stage 5 where dementia may definitely be diagnosed. Even though these people suffer cognitive decline they do not necessarily progress to what we can currently diagnose as SDAT. Once a person is in Stage 5, the decline is inevitable. This then appears to be a range in older people from Stage 1 (normal) to Stage 2 (benign forgetfulness). Some become slightly confused (Stage 3 and 4) but do not progress beyond that point. However, a patient with SDAT will inevitably progress from Stage 1 through Stage 7 in a variable number of years.[34,67] Particularly in the early stages (2, 3, and 4), we must look very carefully for the reversible causes of cognitive decline[63] (see Table IV).

Multi-infarct Dementia. Multi-infarct dementia occurs after repeated strokes and affects multiple areas of the brain. This form of irreversible dementia is characterized by a relatively rapid onset, stepwise fluctuating mental decline, focal or lateralizing neurological signs, pseudobulbar palsy, and marked emotional lability. It is more common in men and is often accompanied by diabetes, hypertension, and evidence of atherosclerosis elsewhere in the body.[68] Imaging techniques of the brain shows evidence of previous infarcts, which is not seen in Alzheimer's disease.[69] Although this disorder cannot be reversed, progression of dementia may be halted by preventing further brain infarction.

                                                                    MICHAEL L. FREEDMAN

Table V. Global Deterioration Scale (GDS) for Age-Associated Cognitive Decline and SDAT

GDS stage	Clinical phase	Clinical characteristics
1 No cognitive decline	Normal	No complaints No defects
2 Very mild cognitive decline	Benign forgetfulness	Subjective complaints of memory deficits such as misplacing object, forgetting names. No objective deficits noted in employment or social situations.
3 Mild cognitive decline	Early confusion	Often deficits noted at work and in social situations. Patient may become lost; word and name finding deficiencies; difficulty retaining reading material; loses objects. Patient becomes anxious but uses denial.
4 Moderate cognitive decline	Late confusion	Unable to perform complex tasks. Decreased knowledge of current and recent events including parts of one's personal history. Difficulty concentrating, handling finances, or travelling. Denial is prominent and flattened affect and withdrawal begins.
5 Moderately severe cognitive decline	Early dementia	Patient can no longer survive alone. Forgets major aspects of personal life. Disorientation sets in and patient has difficulty choosing proper clothes and dressing oneself. More withdrawal and flattened affect.

Table V (*Continued*)

GDS stage	Clinical phase	Clinical characteristics
6 Severe cognitive decline	Middle dementia	May forget name of spouse upon whom they are totally dependent. Unaware of all recent events, their surroundings and is disoriented to time and place, but usually retains knowledge of their own name. Often becomes incontinent. Personality changes and has delusions, obsessions, agitation and may become violent. Eventually becomes quieter and loses ability to carry a thought long enough to direct a purposeful course of action.
7 Very severe cognitive decline	Late dementia	All verbal abilities are lost—may retain one word for long period of time. Incontinent; cannot feed self. Loses all psychomotor skills and finally cannot walk or perform any purposeful action.

Incontinence

Conservative estimates are that at least 10 million Americans suffer from some form of incontinence. Even though exact figures are not available, approximately 10% of the community dwelling people over 65, 50% of these in institutions, and 30% of those over age 80 suffer from acute or chronic incontinence. The yearly cost of caring for the incontinent elderly in United States nursing homes is estimated to be $500 million to $1.5 billion.[70,71] In recent years, studies in geriatric patients have shown that this condition is often correctable and almost always manageable.[70,71] However, until recently, most patients and health professionals were reluctant to discuss the problem and incontinence often led to social isolation, limitation of normal life, and institutionalization.

Pathophysiology.   To maintain continence intraurethral pressure must remain higher than intravesical pressure. Incontinence, therefore, results from a decrease in intraurethral pressure, an increase in intravesical pressure, or a combination of the two. Intravesical pressure is dependent on the muscle tone of the detrusor, bladder volume, and intra-abdominal pressure. Intraurethral pressure is dependent on intra-abdominal pressure, the tone of the striated pelvic muscle, urethral and bladder smooth muscle, and thickness of the urethral mucosa. In women, estrogen helps in maintaining urethral mucosal thickness. Normally, urethral pressure rises more than intravesical pressure when abdominal pressure is increased. The maintenance of continence, of course, will therefore depend on the intact neurological loops controlling these forces.[72,73]

Urinary incontinence may be classified clinically into five types: urge, stress, overflow, reflex, and functional. Urge incontinence is associated with a strong desire to urinate either just before or during the incontinent episode. Stress incontinence occurs with a sudden increase in intra-abdominal pressure, such as with coughing, laughing, or sneezing. Overflow incontinence occurs when the bladder is overdistended. Reflex incontinence is involuntary urination without any sensation of a full bladder. This may be caused by spinal cord injury, herniated disc, multiple sclerosis, or a peripheral neuropathy. The bladder empties completely with large volume incontinence. Functional incontinence refers to the inability of the patient to get to the toilet in time. In the elderly patient this may be a problem, particularly in an institutional setting.

Acute Incontinence.   Acute onset of incontinence is often curable.[70,71] The causes of reversible acute incontinence are shown in Table VI. Fecal impaction is common in the bedridden elderly and decreases bladder storage capacity, which may lead to urge incontinence. This course is curable if the impaction is relieved. Atrophic vaginitis/ureth-

**Table VI. Common Causes of
Reversible Acute Incontinence**

Fecal impactions
Atrophic vaginitis and urethritis
Acute inflammation
Metabolic abnormalities
Acute illness
Confusion
Immobility
Psychological factors

ritis can lead to stress incontinence and is often cured by local application of estrogen cream. Medication can cause various types of incontinence; examples include diuretics, anticholinergics, muscle relaxants, sedatives, and some antihypertensives. A patient with urge incontinence of acute onset must always have acute inflammation from an infection, stone or tumor excluded. Metabolic abnormalities including hyperglycemia, hypercalcemia, and hypokalemia must always be considered in a patient with a brisk diuresis.[74]

Obviously immobility, restraints, or poor access to toilets can all lead to an acute incontinence as can any confusional state due to any acute illness. Psychological factors, such as regressive behavior or hostility towards caregivers, can be manifested by incontinence. Approximately 75% of patients who first become incontinent during hospitalization can be cured.[75]

Chronic Incontinence. Chronic incontinence is less often curable but is still a manageable condition. This can be viewed as caused by four pathophysiologic mechanisms: the detrusor is overactive or underactive and the urethra is too tight or loose.

The most common cause of chronic incontinence in the elderly is an overactive detrusor (detrusor instability, detrusor hyporeflexia, spastic bladder, uninhibited bladder).[70,71] The cause is either stimulation of the bladder (inflammation, radiation, chemotherapy, tumors, stones, etc.) or escape of the bladder from CNS inhibition. This interference in CNS inhibitions is seen in cerebrovascular disease, tumors, multiple sclerosis, Parkinsonism. cervical spondylosis, and normal pressure hydrocephalus. It is also seen in Alzheimer's patients but as a late (Stage 6) finding. Detrusor instability can also occur as the bladder initially responds to an obstructed outlet (e.g., an enlarged prostate). Detrusor instability may be caused by deconditioned urinating reflexes. A person who has had one episode of incontinence may voluntrily urinate frequently to avoid another one, a practice which will decondition the detrusor and lead to a small bladder capacity.

Symptoms of detrusor instability include urgency, frequency, nocturia, and small-volume voiding. Bladder capacity is small but there may be a post-voiding residual. Leakage may occur in any position but is more common in the upright position.[70,71]

Incontinence also results from detrusor hyporeflexia or insufficiency.[70,71] The bladder is hypotonic or atonic and not enough pressure can be generated to overcome urethral resistance. Clinically, the picture is that of overflow incontinence with a distended bladder and large residual volume. The patient complains of a decreased urinary stream or a need to strain to urinate.

Detrusor hyporeflexia can be myogenic secondary to severe chronic bladder outlet obstruction with secondary bladder distention, or neurogenic from a herniated disc or neuropathy (diabetes, tabes dorsalis or alcoholism). If the problem is neurogenic, perineal sensation, sacral reflexes, and control of the anal sphincter are commonly impaired also.[70] Damage to pelvic nerves from surgery or trauma can also lead to a hypotonic bladder. Medications, such as muscle relaxants or anticholinergic drugs, are common precipitants. If sensory input from the bladder is decreased but the other innervations and muscles remain intact, the patient can maintain continence by voiding at periodic intervals.

Overflow incontinence can also be secondary to urethral obstruction. The sphincter may not open when the bladder contracts (detrusor-sphincter dyssynergia). The external sphincter also may not properly relax, which is called functional obstruction. These causes of incontinence are uncommon in the neurologically normal older person.[70] There may also be an anatomic abnormality such as prostatic hypertaphy, fecal impaction, urethral stricture, or pelvic tumors. These are by far the most common cause of urethral obstruction and are often curable.[71]

Urethral insufficiency leading to stress incontinence is very common in women. Men also can develop this post-prostatectomy if the sphincter is damaged. In women childbirth, obesity, estrogen lack, weak pelvic muscles, and an abnormal position of the bladder can all contribute to this condition. In both sexes pelvic nerve injury, sympatholytic drugs, or trauma to the sphincter can produce urethral insufficiency.[70,71]

Clinically, the patient complains of loss of urine during times of increased intra-abdominal pressure. The diagnosis is confirmed by reproducing the condition in the office. Usually only small amounts of urine are lost and there is no post-voiding residual. Up to one fourth of women with stress incontinence also have elements of detrusor instability.

Evaluation and Treatment of Incontinence.   The evaluation of a patient with urinary incontinence is relatively simple and has been summarized in several recent reviews.[70,71,76,77] Once the type of incontinence has been determined the decision as to treatment is usually clear. In the elderly, one must weigh the dangers of surgical and drug regimens against the benefit.[78] Certainly in all people schedules in voiding routines and the use of disposable incontinence undergarments are useful adjuncts to avoiding embarrassing situations. The patient should also avoid diuretic beverages containing such things as caffeine and alcohol.

There are new drugs now being tried to treat incontinence. In detrusor instability nifedipine and ibuprofen are currently being evaluated in addition to the measures shown in Table VII. Alpha-adrenergic stimulants to increase urethral resistance to bladder contractions may be a useful adjunct.[70,71]

In the elderly with detrusor hyporeflexia the pharmacologic treatments are often unsuccessful and have a high rate of side effects. However, frequent voidings and double-voiding with Crede and Valsalva maneuvers are very helpful in the elderly because outlet resistance decreases with age. If residual urine is not reduced to acceptable amounts with these measures, intermittent catheterization should be employed.[70,71]

In uretheral obstruction, any anatomic obstruction should be removed surgically if possible. Drugs may be useful in those cases where obstruction cannot be eliminated. Recently, prazosin, an alpha antagonist, has been used with some success even in patients with benign prostatic hypertrophy.[79]

With urethral insufficiency, initial approaches include weight loss

Table VII. Treatment of Urinary Incontinence

Detrusor Instability
    Imipramine 25–150 mg once daily
    Oxybutynin 5 mg 2–4 times daily
    Propantheline 7.5–30 mg 3 times daily
    Bladder retraining
    Disposable incontinence undergarments

Detrusor Hyporeflexia
    Bethanechol 10–30 mg 3–4 times daily
    Phenoxybenzamine 10–60 mg once daily
    Intermittant catheterization
    Frequent voidings
    Disposable incontinence undergarments

Urethral Obstruction
    Relieve obstruction
    Phenoxybenzamine 10–60 mg once daily
    Sphincterotomy
    Disposable incontinence undergarments

Urethral Insufficiency
    Weight loss
    Pelvic exercises (Kegel exercises)
    Estrogens
    Imipramine 25–150 mg once daily
    Phenylpropanolamine 25–50 mg twice daily
    Surgery
    Pessary
    Disposable incontinence undergarments

for the obese patient and Kegel's exercises to strengthen pelvic musculature.[80] If these measures alone are not sufficient, they can be combined with estrogens[81] to restore the integrity of the urethral mucosa and thus resistance to outflow. Alpha-adrenergic agents can also be added.[82] If loss of the urethrovesical angle is the primary cause, surgery will be useful. If surgery cannot be done a pessary can be fitted.[70,71]

With all of these treatments there will still be many elderly people who either fail therapy or who cannot tolerate surgery or drugs. In these patients the use of disposable incontinence undergarments are immensely important. It is the rare patient who needs indwelling bladder catheterization. When using the incontinence undergarments good skin care is essential. In recent years the introduction of disposable discreet, well-fitting, very absorbent undergarments have made urinary incontinence much easier to deal with. The physician should take a positive attitude in caring for the incontinent patient, as this condition is certainly manageable and need not be the cause for social isolation.

Iatrogenic Disease

The final major chronic problem is iatrogenic disease. This includes problems with medications, surgical or other diagnostic and therapeutic procedures, and finally the environment of our health care facilities.[83]

Drug Problems. The elderly are at special risk for adverse drugs reactions.[84,85] As shown in Table VIII there are multiple factors increasing the risk with advanced age.[83] The elderly take approximately twice the number of prescribed medications. With decreasing dexterity and the increase of illnesses that cause memory loss, compliance is often poor. With advancing age there is a change in pharmacokinetics.[86,87] In brief, there are changes in distribution volume that make fat-soluble drugs last longer in the biology. As renal excretion diminishes, drugs excreted by this route will last much longer in the body. Hepatic metabolism remains fairly intact and the only consistent impairment is oxidative metabolism. As a general rule, the hepatic metabolism decrease with age is relatively of little importance. Clinically, smoking induction and inhibition by other drugs of the hepatic enzymes are much more important. Receptor sensitivity tends to decrease with age as has been shown with beta-sympathetic agonists and antagonists. On the other hand, increases in end-organ sensitivity may occur as has been shown with opiates and warfarin.

The incidence of adverse reactions rises significantly with both increasing age and increasing numbers of medications. Two groups of drugs seem to cause nearly two thirds of all drug reactions. They are cardiovascular drugs (digoxin, antihypertensives, and diuretics) and

Table VIII. Factors Increasing Risk of Adverse Drug Reactions in the Elderly

Increased usage of medications
Compliance
Altered pharmacokinetics
   (particularly distribution volumes and diminished
   renal excretion)
Altered receptor sensitivity (pharmacodynamics)
Increased end-organ sensitivity

central nervous system drugs (psychotropics, antidepressants, hypnotics, tranquilizers, or anti-Parkinsonism drugs).[88]

Among the most serious side effects of drugs are postural hypotension anticholinergic effects, confusion, neurologic, renal, and cardiac complications. Table IX shows common examples of drug classes causing these complications.

Postural hypotension is a common symptom even in the healthy elderly.[33] Certainly one must be careful in prescribing vasodilatory antihypertensive and diuretic medications. In general, one must begin with low doses and raise them extremely cautiously. In treating hypertension, for example, one should be far less aggressive than in younger people and should not try to lower the blood pressure as rapidly. Many times, to avoid the symptomatic orthostatic hypotension, the blood pressure cannot be lowered to "normal" levels. Psychotropic medications also have this side affect and for that reason small doses are usually given at bedtime.

The anticholinergic drugs are very widely prescribed and can cause confusion, urinary retention, constipation, and can precipitate acute narrow angle glaucoma. Pupillary dilation, tachycardia, arrhythmias, hypotension, and absence of sweating can also occur.[83]

Virtually any drug can cause confusion in the elderly even when the blood level is in the "therapeutic range."[63] When faced with a patient whose confusion started after a drug was instituted, the first step has to be withdrawal that drug.

Drugs also can cause convulsions, ataxia, peripheral neuropathies, and extra pyramidal symptoms. The elderly seem to be at particular risk for these side effects.[89] When prescribing major tranquilizers, the physician should be aware that the elderly develop tardive dyskinesia much more commonly than the young.[90] In addition, akathisis (motor restlessness) may be caused by these drugs. This may present as pacing and raising the dose will worsen the symptom. Haloperidol is more likely to produce these extrapyramidal side effects then is thioridazine.

**Table IX. Example of Common
Drugs Classes That Cause Major
Side Effects**

Postural Hypotension
    Antihypertensives
    Psychotropics
    Calcium channel blockers
    Opiates
    Anti-Parkinson drugs

Anticholinergic
    Antidepressants
    Antispasmodics
    Antihistamines
    Psychotropics

Confusion
    Analgesics
    Anticholinergics
    Antidepressants
    Anticonvulsants
    Antiinflammatories
    Anti-Parkinson's
    Barbituates
    Beta-blockers
    Benzodiazepines
    Calcium channel blockers
    Cimetidine
    Digitalis preparations
    Hypoglycemics
    Lidocaine
    Vincristine

Neurologic
  Convulsions
    Lidocaine
    Antidepressants
  Ataxia
    Anticonvulsants
  Peripheral neuropathy
    Vinca alkaloids
    Nitrofurantoin
  Extrapyramidal symptoms
    Phenothiazines
    Butyrophenones
    Metoclopramide
    Antidepressants

Table IX. (*Continued*)

Renal
      Antibiotics
      Diuretics
      Antiinflammatories

Cardiac
      Digitalis preparations
      Beta-blockers
      Calcium channel blockers
      Antiarrhythmics
      Antidepressants
      Psychotropic drugs

Rival complications are seen with aminoglycoside antibiotics, diuretics, and anti-inflammatory drugs. This is more common in the elderly because of their already compromised renal function.[91]

Cardiac complications can occur in the elderly when treated with drugs that have a negative inotropic effect. Thus, beta-blockers, calcium channel blockers, and various anti-arrhythmics can precipitate congestive heart failure in the elderly. Digoxin is one of the more common drugs to cause toxicity in the elderly and often can be withdrawn.[92]

In addition to all of these side effects, there are allergic, hematologic, skin and gastrointestinal side effects which are apparently equal in occurrence between older and younger patients.

The physician must also be aware of the dangers of drug interactions in the elderly who are often taking multiple medications. These include both prescription and nonprescription drugs.[93,94]

Other Iatrogenic Diseases. Besides medications, older people do not do as well in hospitals.[95] Delirium, depression, incontinence, immobility, malnutrition, thromboemboli, and complications of all procedures occur in higher frequency in the elderly. The use of restraining devices seems to increase the risk of harm and these should be discouraged.[96,97] Overprotection of the geriatric patient impairs potential for functional independence.[98] Incontinence may be induced by the patient's inability to reach the bathroom or by staff failure to respond to the patients call.[99] Falls may result from patients trying to get over side rails, by poor room lighting, and lack of contrast between floor and walls. Patients with fragile skin may have breakdown as a result of lack of proper skin care.[100] When dealing with the elderly patient, innovative environments such as the NYU Cooperative Care Unit, where the patient is admitted with a care partner who lives-in with the patient and participates in the total hospital experience may be the answer to avoid at least a portion of this type of iatrogenic disease.[101]

Finally, in treating geriatric patients, physicians must also be careful that acts of omission are not carried out. The elderly do respond to therapeutic maneuvers and attitudes of ageism must not persist. Because someone is elderly, they should not be denied the right of good medical care, the cure of disease, and the relief of pain and discomfort. Dying with dignity does not mean the withholding of therapies that can reverse a disease or that can relieve pain and suffering. The role of the physician must always be to cure what is curable and to make the incurable patient as comfortable as possible.[102]

SUMMARY

The Biblical description of the death of Moses at age 120 without the infirmities of old age seems to be a realistic goal of medicine. Geriatric medicine has made important contributions in the past decade and the health care professions are responding to the "geriatric imperative" of the nation. Better health habits and health care in early life are allowing most people to live into their seventies, eighties, and above. We are now attacking the chronic infirmities of aging and are finding that they do respond to treatment. We still have a long way to go, but we have made a good beginning.

REFERENCES

1. Hayflick L: The cell biology of aging, in Geokas MC (ed): *The Aging Process, Clinics in Geriatric Medicine*, Philadelphia, WB Saunders, 1985, vol 1, pp 15–27.
2. Kenney RA: Physiology of aging, in Geokas MC (ed): *The Aging Process, Clinics in Geriatric Medicine*, Philadelphia, WB Saunders 1985, vol 1, pp 37–59.
3. The Holy Scriptures, Deuteronomy. New York, Hebrew Publishing Co, 1951, Ch 34:v.7.
4. The Holy Scriptures, First Kings. New York, Hebrew Publishing Co, 1951, Ch 1:v.1.
5. The Holy Scriptures, First Kings. New York, Hebrew Publishing Co, 1951, Ch 1:v.15.
6. The Holy Scriptures, First Kings. New York, Hebrew Publishing Co, 1951, Ch 2:v.10.
7. Miall WE, Ashcroft MT, Lovell HG *et al.:* A longitudinal study of the decline of adult height with age in two Welsh communities. *Hum Biol* 1967; 39:445–454.
8. Rossman I: Anatomic and body composition changes with aging, in Finch CE, Hayflick L (eds): *Handbook of the Biology of Aging*, New York, Van Nostrand-Reinhold Co., 1977.
9. Forbes, LP: The adult decline in lean body mass. *Hum Biol* 1976; 48:161–173.
10. Novak LP: Aging total body potassium, fat-free mass and cell mass in males and females ages 18 and 85 years. Gerontology 1972; 27:438–443.

11. Boddy K, King PC, Womersley J *et al.:* Body potassium and fat-free mass. *Clin Sci* 1973; 44:622–625.

12. Lye M. Distribution of body potassium in healthy elderly subjects. *Gerontology* 1981; 27:286–292.

13. Larrson L: Morphological and functional characteristics of aging skeletal muscle in man: A cross-sectional study. *Acta Physiol Scand* 1978 (Suppl 457).

14. Gomez EC, Berman B: The aging skin, in Geokas MC (ed): *The Aging Process, Clinics in Geriatric Medicine,* Philadelphia, WB Saunders, vol 1 1985, pp 285–305.

15. Grove GL, Kligman AM: Age-associated changes in human epidermal cell renewal. *J Gerontol* 1983; 38:137–142.

16. Leyden, JJ, McGinley, KJ, Grove GL: Age-related differences in the rate of desquamation of skin surface cells, in Adelman RD, Robert J, Cristofalo VJ (eds): *Pharmacological Intervention in the Aging Process,* New York, Plenum Press, 1978, pp 297–298.

17. Grove GL: Age-related differences in healing of superficial skin wounds in humans. *Arch Dermatol Res* 1982; 272:381–385.

18. Gilchrist BA: *Skin and the Aging Process.* Boca Raton, Fl, CRC Press Inc, 1984.

19. Sams WM, Smith JC: Alterations in human dermal fibrous connective tissue with age and chronic skin damage. *Adv Biol Skin* 1965; 6:199–210.

20. Fitzpatrick TB, Brunet P, Kukita A: The nature of hair pigment, in Montagne W, Ellis RR (eds): *The Biology of Hair Growth,* New York, Academic Press, 1958, pp 255–303.

21. Orfanos C, Ruska H, Mahle G: White hair of older people. *Arch Klinische Experimentele Dermatologie* 1970; 236:395–405.

22. Bullough US, Laurence EG: The mitotic activity of the follicle. *J Anat* 1964; 98:171–187.

23. Keogh EV, Walsh RJ: Rate of greying of human hair. *Nature* 1965; 207:877–878.

24. Vernadakis A: The aging brain, in Geokas MC (ed): *The Aging Process, Clinics in Geriatric Medicine,* Philadelphia, WB Saunders 1985, vol 1, pp 61–94.

25. Simpson DM, Erwin CW: Evoked potential latency change with age suggests differential aging of primary somatosensory cortex. *Neurobiol Aging* 1983; 4:59–63.

26. Loewenstein RJ, Weingartner H, Gillin JD *et al.:* Disturbances of sleep and cognitive functioning in patients with dementia. *Neurobiol Aging* 1982; 3:371–377.

27. Ulrich RF, Shaw DH, Kupter DJ: Effects of aging on EEG sleep in depression. Sleep 1980; 3:31–40.

28. Goodrick CL: Behavioral rigidity as a mechanism for facilitation of problem solving for aged rats. *J. Gerontol* 1985; 30:181–184.

29. Peak DT: A replication study of changes in short-term memory in a group of aging community residents. *J Gerontol* 1970; 25:316–319.

30. Taub HA: Memory span. practice and aging. *J Gerontol* 1973; 28:335–338.

31. Erber JT: Age differences in recognition memory. *J Gerontol* 1974; 29:177–181.

32. Gordon SK, Clark WC: Application of signal detection theory to prose recall and recognition in elderly and young adults. *J Gerontol* 1974; 29:64–72.

33. Medin DL, O'Neil P, Smelz E *et al.:* Age differences in retention of concurrent discrimination problems in monkeys. *J Gerontol* 1973; 28:63–67.

34. Reisberg B, Ferris SH, deLeon MJ *et al.:* The global deterioration scale for assessment of primary degenerative dementia. *Am J Psychiatry* 1982; 139:1136–1139.

35. Gribbin B, Pickering TG, Steight, P *et al.:* Effect of age and high blood pressure on baroreflex sensitivity in man. *Circ Res* 1971; 29:424–431.

36. Konshalo, DR: Age changes in touch, vibration, temperature, kinesthesis, pain sen-

sitivity, in Bioren JE, Schaie KA (eds): *Handbook of the Biology of Aging*, New York, Van Nostrand-Reinhold, 1977.

37. Wagner, JA, Robinson S, Marino RP: Age and temperature regulation of humans in neutral and cold environments. *J Appl Physiol* 1949; 2:233–239.

38. Hellon RF, Lind AR, Weiner JS: The physiological reactions of men of two age groups to a hot environment. *J Physiol* 1956; 133:118–132.

39. Fox AH, MacGibbon R, Davies L *et al.:* Problems of the old and the cold. *Br Med J* 1973; 1:21–24.

40. Noth RH, Mazzaferri MD: Age and the endocrine system, in Geokas MC (ed): *The Aging Process, Clinics in Geriatric Medicine*, Philadelphia, WB Saunders, 1985, vol 1 pp 223–250.

41. Flood C, Gherondache C, Pincus G *et al.:* The metabolism and secretion of aldosterone in elderly subjects. *J Clin Invest* 1967; 46:960–996.

42. Hellerman JJ, Vestal RE, Rowe JW *et al.:* The response of arginine-vasopressin to intravenous ethanol and hypertonic saline in man: The impact of aging. *J Gerontol* 1978; 33:39–47.

43. Collins KJ, Exton-Smith AN, James MH *et al.:* Functional changes in autonomic nervous response with aging. *Age Ageing* 1980; 9:17–24.

44. Rowe JW, Troen BR: Sympathetic nervous system and aging in man. *Endocr Rev* 1980; 1:167–179.

45. Young JR, Rowe JW, Pallota JA *et al.:* Enhanced plasma nor-epinephrine response to upright posture and oral glucose administration in elderly human subjects. *Metabolism* 1980; 29:532–539.

46. Feldman RD, Limbird LE, Nadeau J, *et al.:* Alterations in leukocyte B-adrenergic sensitivity in the elderly. *N Eng J Med* 1984; 310:815–819.

47. Culpepper L, Murphy J, Fretwell M: Biology, primary care, family and community: A basis for rational geriatric care, in Magenheim MJ (ed): *Geriatric Medicine and Social Policy, Clinics in Geriatric Medicine*, Philadelphia, WB Saunders, 1986, vol 2, pp 37–51.

48. Klausner SC, Schwartz AB: The aging heart, in Geokas MC (ed): *The Aging Process, Clinics in Geriatric Medicine*, Philadelphia, WB Saunders, 1985, vol 1, pp 119–141.

49. Gordon GS, Genant HK: The aging skeleton, in Geokas MC (ed): *The Aging Process, Clinics in Geriatric Medicine*, Philadelphia, WB Saunders, 1985, vol 1, pp 95–118.

50. Shephard RJ: *Physical Activity and Aging.* Chicago, Croon Helm, 1978.

51. DeVries HA: Physiological effects of an exercise training regimen upon men ages 52-88. *J Gerontol* 1970; 25:325–336.

52. Bartz WH: Disuse and aging. *JAMA* 1982; 248:1203–1208.

53. Saltin B, Blomquist G, Mitchell J *et al.:* Response to exercise after bedrest and after training. *Circulation* 1968; 37, 38:(suppl) 1–78.

54. Taylor HL, Henschel A, Brozek J *et al.:* Effect of bedrest on cardiovascular function and work performance. *J Appl Physiol* 1949; 2:233–239.

55. Larsson L: Morphological and functional characteristics of aging skeletal muscle in man: A cross sectional study. *Acta Physiol Scand* 1978; (Suppl 457).

56. Krumpe PE, Knudson RJ, Parsons G *et al.:* The aging respiratory system, in Geokas MC (ed): *The Aging Process, Clinics in Geriatrics Medicine*, Philadelphia, WB Saunders, 1985, vol 1, pp 143–175.

57. Htoo, MSH, Kofkoff RL, Freedman, ML: Red cell parameters in the elderly: An argument against new geriatric normal values. *J AM Geriatr Soc* 1979; 27:547–551.

58. Freedman ML, Marcus DL: Anemia and the elderly: Is it physiology or pathology? *AM J Med Sci* 1980; 280:81–85.

59. Marcus DL, Freedman ML: Clinical disorders of iron metabolism in the elderly, in

Freedman ML (ed): *Hematologic Disorders, Clinics in Geriatric Medicine*, Philadelphia, WB Saunders, 1985, vol 1, pp 729–745.

60. Sidney KH, Shephard RJ: Maximum and submaximum exercise tests in men and women in the seventh, eighth and ninth decades of life. *J Appl Physiol* 1977; 43:280–287.

61. Freedman ML, Ahronheim JC: Nutritional needs of the elderly: Debate and recommendations. *Geriatrics* 1985; 40:45–62.

62. McEvoy JP: Organic brain syndromes. *Ann Intern Med* 1981; 95:212–220.

63. Freedman, ML Orgain brain syndromes in the elderly: When are they reversible. *Postgraduate Med* 1983; 74:165–176.

64. Task Force sponsored by National Institute on Aging. Senility reconsidered: treatment possibilities for mental impairment in the elderly. *JAMA* 1980; 244:259–263.

65. Wolff ML: Reversible intellectual impairment. *J Am Geriatr Soc* 1982; 30:647–650.

66. Gruenberg EM: Epidemiology of senile dementia, in Haynes SG, Feinleib M (eds): *Epidemiology or Aging*, Bethesda, MD, National Institute of Health 29:104 NIH publication No 80-969, 1980.

67. Reisberg B (ed): *Alzheimer's Disease*. New York, Free Press, 1983.

68. Hackinski VC, Iliff LD, Zihka E *et al.*: Cerebral blood flow in dementia. *Arch Neurol* 1975; 32:632–637.

69. Jarvik LF: Diagnosis of dementia in the elderly: A 1980 perspective. Ann *Rev. Gerontol Geriatr* 1981; 1:180–203.

70. Snustad DG, Rosenthal JT: Urinary incontinence in the elderly. *Am Family Practitioner* 1985; 32:182–196.

71. Resnick NM, Subbarao VY: Management of urinary incontinence in the elderly. *N Eng J Med* 1985; 313:800–804.

72. deGroat WC, Booth AM: Autonomic systems to the urinary bladder and sexual organs, in Dyck PJ, Thomas PK, Lambert EH, Beinge R (eds): Peripheral Neuropathy, ed 2, Philadelphia, WB Saunders, 1984, pp 285–299.

73. Gosling JA, Chilton CP: The anatomy of the bladder, urethra and pelvic floor, in Mundy AR, Stephenson TP, Wein AJ (eds): *Urodynamics: Principles, Practice and Application*, New York, Churchill Livingstone, 1984, pp 3–13.

74. Resnick NM: Urinary incontinence in the elderly. *Med Grand Rounds* 1984; 3:281–290.

75. Resnick NM, Paillord M: National history of nosocomial urinary incontinence. International Continence Society Proceeding 1984; 471–472.

76. Blaivas JG, Zayed AAH, Lubeb KB: The bulbocavernosus reflex in urology: a prospective study of 299 patients. *J Urol* 1981; 126:197–199.

77. Williams, ME, Pannill FC: Urinary incontinence in the elderly: Physiology, pathophysiology, diagnosis and treatment. *Ann Intern Med* 1982; 97:895–907.

78. Hilton P, Stanton SL: Algorithmic method for assessing urinary incontinence in elderly women. *Br Med J* 1981; 282:940–942.

79. Hedlund H, Anderssen KE, Ek A: Effects of prazosin in patients with benign prostatic obstruction. *J Urol* 1983; 130:275–278.

80. Castleden CM, Duffen HM, Mitchell EP: The effect of physiotherapy on stress incontinence. *Age Ageing* 1984; 13:235–237.

81. Mohr JA, Roger J, Jr, Brown TN *et al.*: Stress urinary incontinence: a single and practical approach to diagnosis and treatment. *J Am Geriatr Soc* 1983; 31:476–478.

82. Beisland HO, Fossberg E, Moer A *et al.*: Urethral sphincteric insufficiency in postmenopausal females: treatment with phenypropanolamine and estriol separately and in combination: A urodynamic and clinical evaluation. *Urol Int* 1984; 39:211–216.

83. Patterson C: Iatrogenic disease in later life, in Magenhiem MJ (ed): *Geriatric Medicine and Social Policy, Clinics in Geriatric Medicine*, Philadelphia, WB Saunders, 1986, vol 2, pp 121–136.

84. Hurwitz N: Predisposing factors in adverse reactions to drugs. *Br Med J* 1969; 1:536–539.

85. Seidl LG, Thornton GF, Smith JW *et al.:* Studies in the epidemiology of adverse drug reactions. *Bull Johns Hopkins Hosp* 1966; 119:299–305.

86. Crooks J, O'Malley K, Stevenson IH: Pharmacokinetics in the elderly. *Clin Pharmacokinet* 1976; 1:280–296.

87. Greenblatt DJ, Sellers EM: Drug disposition in old age. *N Eng J Med* 1982; 306:1081–1088.

88. Williamson J, Chopin JM: Adverse reactions to prescribed drugs in the elderly: a multicenter investigation. *Age Ageing* 1980; 9:73–80.

89. Critchley EMR: Drug-induced neurological disease. *Br Med J* 1979; 1:862–865.

90. Lang, AE: Tardive dyskinesia. *Mod Med Can* 1984; 39:549–554.

91. Cline DM, Stoff JS: Renal syndromes associated with nonsteroidal anti-inflammatory drugs. *N Eng J Med* 1984; 310:563–573.

92. Dall JLC: Maintenance digoxin in the elderly. *Br Med J* 1970; 2:705.

93. Gasney M, Tallis R: Prescription of contraindicated and interacting drugs in elderly patients admitted to hospital. *Lancet* 1984; 2:564–567.

94. Ouslander JG: Drug therapy in the elderly. *Ann Int Med* 1981; 95:711–722.

95. Warshaw GA, Moore JT, Friedman SW *et al.:* Functional disability in the hospitalized elderly. *JAMA* 1982; 248:847–850.

96. Lancet Editorial (2) Cotsides: Protecting whom against what? *Lancet* 1984; 2:557–558.

97. Powell C, Edmund L, Fingerote E: Freedom from restraint: consequences of reducing physical restraints in the treatment of elderly persons. *Ann RCPS Can* 1983; 16:343.

98. Avorn J, Langer E: Induced disability in nursing home patients: A controlled trial. *J Am Geriatr Soc* 1982; 30:379.

99. Gellick MR, Serrell NA, Gellick CS: Adverse consequences of hospitalization in the elderly. *Soc Sci Med* 1982; 16:1033.

100. Anderson KE, Jensen O, Kvorning SA: Prevention of pressure sores by identifying patients at risk. *Br Med J* 1982; 1:1370.

101. Grieco A, Freedman, ML: Cooperative care: A model for the acute hospital care of the elderly. *The International Association of Gerontology*, New York, NY, July, 1985.

11

Depressive Illness in the Elderly

Legal and Ethical Issues

KAREN BLANK

Legal issues arise frequently in the treatment of the depressed elderly. These issues have received far less attention than those encountered in the care of persons suffering from cognitive decline secondary to dementia. The clinical treatment of elderly depressed patients raises special considerations in the areas of informed consent and competency, civil commitment, involuntary treatment, liability reduction, confidentiality and the right of the elderly to participate in decisions about their own deaths.

Although the incidence of depressive illnesses in the elderly has not been fully determined,[1,2] depression appears to be the most common psychiatric disorder among elderly patients. Making the diagnosis of depression in this age group is complicated and can be so difficult that, at times, the diagnosis is missed entirely. Some elderly patients present with physical complaints or evidence of dementia in lieu of reports of the emotional experience of depression or dysphoria. Further compounding the difficulty, medical illnesses can present as depression and commonly prescribed medications can induce depressive conditions. It is well established that the risk of suicide increases with age; suicide among

KAREN BLANK • Department of Psychiatry, Beth Israel Medical Center, First Avenue at Sixteenth Street, New York, NY 10003; Mount Sinai School of Medicine, New York, NY 10003.

the elderly occurs at rates more than triple those of the general population[3] and it is a safe assumption that a high percentage of these individuals suffered from some form of major depression. The detection and prompt treatment of depression in the elderly is essential.

Many of the clinical features of geriatric depression can impact on the treating physician in such a way as to increase the physician's use of a paternalistic stance. This professional posture runs counter to trends toward increased protection of patients' autonomy and civil rights. Depressed elderly patients often dramatically display their suffering and their suffering can be painful to witness. This prompts in the physician a wish to take over and act without delay. Depressed elderly patients characteristically are needy, helpless. and dependent on their caretakers and can exhibit marked difficulties in concentrating, attending to information presented to them, and in making decisions for themselves. In more seriously ill patients, judgment with regard to treatment decisions may be further compromised by hopelessness, a sense of worthlessness, guilt, apathy, fatigue, self-destructive thought content, and delusions. Such patients often overtly and covertly abrogate responsibility during the height of their suffering, looking to the physician for treatment decisions. In commencing treatment of the moderately to severely ill elderly depressive the physician is faced with thorny questions as to how to alleviate his patient's suffering while protecting that person's autonomy.

HOSPITALIZATION AND CIVIL COMMITMENT

Hospitalization of elderly depressives is a common occurrence. Depressive disorders have been found to account for almost half of new admissions of persons over age 65 to psychiatric hospitals.[4] Usually if a patient voluntarily admits himself to a psychiatric facility few legal questions arise. The patient is behaving in such a way as to promote a reasonable outcome and as such is usually acting in accordance with his physician's wishes. Remembering that patients are entitled to care under the least restrictive setting and that questions of competency are most often raised when the patient and treatment staff disagree about the decision at hand, the patient signing in voluntarily to a psychiatric hospital most often does so without the staff questioning his ability to make this decision. Where questions do exist as to the patient's competency to understand the meanings of a voluntary admission, the physician should evaluate the patient's decisional capacity using the risk/benefit analysis that is described later in this chapter. When a patient is not competent to make the decision about hospitalization, civil commitment and the use of substitute decision makers should be arranged.

Commitment to a psychiatric facility occurs in a significant minority of patients. In one setting 22% of psychiatric inpatients over age 65 were admitted involuntarily.[5] When a depressed patient in need of treatment is overtly refusing hospitalization or lacks the ability to understand the basic nature of the decision to enter the hospital, commitment is usually necessary. Seriously depressed patients refuse hospitalization and often all treatment for a number of reasons. Obviously there exists considerable prejudice against and fear about psychiatric treatment and this can be exascerbated in some older persons compared with younger peers. Past associations of psychiatric facilities with long-term sanitoriums, state hospitals, or "snake pits" can be resurrected. The need for psychiatric care can be equated by such patients with being hopelessly insane. Feelings of disgrace and shame may follow as well as the fear that admission to a psychiatric facility may result in being "locked up for life." Some may fear being labeled as psychiatric patients. Whether based in reality or fantasy some patients may fear that the loss of autonomy and the implications of being diagnosed "insane" may be used against them in future proceedings such as guardianship or conservatorship. As a result some older patients who recognize their need for treatment request to be treated by their internists on a general medical unit. In addition, refusal of hospitalization in general (not specific to psychiatric facilities) can result from the elderly patient's adoption of a position of increased conservatism and inflexibility. Such a person fearfully avoids new surroundings as they may overwhelm his coping abilities and pose him with new, unmanageable emotional and financial stressors. Patients may also associate hospitalization with serious illness and death or view it as the avenue through which they will be coerced into nursing homes. These types of understandable fears are enhanced by the presence of such classic features of depression as hopelessness, helplessness, preoccupations with illness and death and, at times, with paranoid delusions. Delusional thinking can interfere in other ways with a patient's judgment about hospitalization through the presence of nihilistic delusions, somatic delusions, delusions of extreme worthlessness or of poverty. In the latter case such patients, however much one attempts to reassure them, are convinced that they are destitute and that the cost of hospitalization will pose extreme financial hardship for their families. Still some patients who had been opposed to hospitalization, if provided with proper education, explanation, and reality testing, can legitimately make the decision to sign voluntarily into the hospital. This is an argument for the use of vigorous psychoeducation to facilitate the use of voluntary hospitalization.

The most clear-cut situation necessitating commitment is that in which the patient is refusing hospitalization and poses an obvious danger to himself through suicide. A potential legal and ethical pitfall here is the failure of the clinician to recognize the full suicidal potential of an elderly patient. Countertransference and attitudinal issues may interfere with the assessment of suicidality in older patients. An old, frail, exhausted, depressed patient may harbor suicidal ideas that are underappreciated or overlooked by the clinician especially if he fails to do a complete assessment (raising malpractice potential). This underestimation may follow from unconscious ageism[6] with the physician believing that the old person just would not take such a drastic, decisive measure. Alternatively, it could stem from unconscious (or conscious) collusion with the patient's belief that he has lived long enough and that his suffering and his life should end. Though little question arises about the justification to commit the overtly suicidal patient, many of the depressed patients refusing (or unable to comprehend) the need for hospitalization are not overtly suicidal. Rather they are unable to care for themselves because of their illnesses and this disability poses a serious threat to their health and well-being. Such patients are often suffering from anorexia and become nutritionally jeopardized to the point of dehydration and electrolyte disturbances. They may spend so much time in bed that ambulating becomes hazardous, putting them at risk for falls. A common situation is that in which the patient stops taking medications that are critical for health maintenance, such as cardiac medications or insulin.

In most jurisdictions involuntary commitments can be accomplished under the state's inherent *parens patriae* power, which "requires a determination that the person is mentally incompetent to provide for basic life requirements."[7] Many states allow for commitment of this "unable to care for oneself" population. New York for example is a state with such a provision in the mental health law. In those states with more stringent legal requirements for commitment, obtaining hospitalization for patients unable to care for themselves can be more difficult; those patients lacking family or other involved advocates to petition the court are at risk for being undertreated.

LIABILITY REDUCTION

The presentation of depression in an elderly patient can differ significantly from that in a younger patient and this introduces the very real liability risk of failure to make the correct diagnosis. The most common situations where this can occur are when depressive symptoms accompany or are caused by physical ailments; when depression is exacerbated or caused by medications; when the patient presents with a masked depression; and when the patient appears demented rather than depressed. The psychiatrist must, at all times, take a thorough medical as well as psychiatric history, must be aware of the physical conditions and medications that can present clinically as depression, must work collaboratively with the patient's primary physician, and must consider depression as the underlying etiology in new onset dementia cases, especially if the onset is abrupt or where there is a previous history of affective disorder. When doubt exists about the diagnosis it is advisable to obtain a consultation from a geriatric psychiatrist.

Once the decision to treat an older patient's depression has been made, the psychiatrist must be aware of the significant risks such treatment can pose in elderly patients. Cyclic antidepressants are frequently used in the treatment of depressive illnesses in the elderly. The use of these medications in this age group requires caution and modification of dosages and dosing schedules. Common classes of side effects and risks include sedation, cardiovascular toxicity, and anticholinergic toxicity. Sedation by itself usually does not cause more than discomfort and inconvenience for the elderly patient but when combined with hypotension can contribute to falls and other accidents. Cardiovascular toxicity can, at times, pose serious risks to the patient. At therapeutic doses, cyclic antidepressants can cause sinus tachycardia, which can be a risk to a patient who has ischemic heart disease or other serious limitations of cardiac reserve. These medications have quinidine-like effects that can cause intracardiac conduction to be prolonged and produce a variety of conduction defects. Cyclic antidepressants have also been associated with acute coronary insufficiency and acute myocardial infarction. Orthostatic hypotension can lead to falls, fractures, strokes, or heart attacks.[8] Anticholinergic toxicity can produce a myriad of symptoms from dry mouth and constipation to more serious problems, such as urinary retention, precipitation of narrow-angle glaucoma, memory and cognitive deficits, and full-blown central anticholinergic delirium. The newer generation antidepressants, lithium and MAO inhibitors similarly carry with them considerable risks, though a full review is beyond the scope of this chapter.

Neuroleptic medication is often used in conjunction with antidepressant medication especially in severely agitated and delusionally depressed patients. These medications carry well-known risks that are greatly enhanced in the older patient. These include orthostatic hypotension, extrapyramidal syndromes, tardive dyskinesia, anticholinergic side effects, and less frequent problems, such as liver dysfunction, agranulocytosis, seizures, and temperature regulation dysfunction. Tardive dyskinesia deserves special mention in that older patients, especially females, may be at increased risk for its development.[9] "It is estimated that 40% of patients over the age of 60 who are being treated with neuroleptic drugs will develop tardive dyskinesia."[10]

Electroconvulsive therapy (ECT) is a relatively safe and highly effective treatment for the seriously depressed elderly patient. Properly administered it is safe in most patients, even those with serious medical illnesses. Obtaining consent for ECT is often handled differently from consent for the other treatments offered for depression. Whether justified or not, ECT is viewed by many to be a more hazardous and invasive procedure than treatment with antidepressants. ECT is often utilized for the treatment of the most severely ill patients, those most often unable to give consent themselves. It is crucial that the physician anticipate wherever possible, a situation where a depressed patient is becoming so seriously depressed that administering ECT may be needed urgently. Situations in which a patient's competency to consent to ECT may deteriorate, are best handled in advance so as to prevent prolonged patient suffering, to include the patient maximally in the decision and to avoid unwarranted delays in obtaining necessary legal clearance for treatment. In some jurisdictions, obtaining legal guardianship may be too lengthy a process to endure while treatment is being delayed. At times a court order to treat or an internal procedure in the involved hospital may be possible to ensure timely treatment while protecting the patient's rights.

Electroconvulsive therapy is at times a life-saving treatment. There exists a depressed population for whom alternative treatments are too hazardous. ECT carries anesthesia risks and the common side effects of short-term memory deficits and confusion. In an elderly subject these cognitive deficits can create a situation of dangerousness for the patient where one did not exist previously. During a course of ECT patients may become disoriented and forgetful; as outpatients they may fail to handle dangerous appliances properly or may fail to take adequate safeguards while out of their homes (e.g., need to look before crossing streets, turn off their stove, etc.). All patients and their families must be fully informed about these temporary problems so as to ensure the patient's safety. The physician is obliged to be proficient at performing functional assessments of patients receiving ECT and, if supervision for an organically impaired patient is inadequate, to advise and help arrange for satisfactory safeguards.

The informed consent procedure for frail elderly patients should involve the anesthesiologist. He is in a far better position than the psychiatrist to explain fully anesthesia risks that for some patients are the riskiest part of the treatment.

Finally, with the shift to prospective payment of hospital costs it is likely that the use of ECT in elderly patients will increase so as to shorten lengths of hospital stays. ECT may be recommended increasingly to patients who in the past were given an antidepressant trial as their first treatment. Medicolegal and ethical questions are sure to arise specifically around this type of issue. It can be argued that patients should be informed of such financial considerations if these pressures have significantly affected the process by which the physician chose a recommended treatment.

INFORMED CONSENT AND TREATMENT DECISIONS

The legal criteria for competency to consent to treatment are not uniformly defined[11] but generally require that the patient has a reasonable and rational understanding of the treatment's risks, benefits, and alternatives. Requiring a patient to demonstrate factual understanding may be an overly stringent standard for certain decisions. A growing body of studies suggests that significant numbers of patients who consent to treatment do not factually understand them.[12,13] Factual comprehension standards are biased against inarticulate patients, those of lower intelligence, or those with psychotic process. Several other standards for determining competency have been proposed.[14] The stringency of the criteria generally varies with the risk/benefit ratio of the decision to be made. One set of criteria that has been proposed, listed from the least demanding to the strictest criteria, includes (a) the evidencing of a choice, (b) determining that the choice is reasonable, (c) demonstration of the ability to manipulate information, (d) demonstration of the ability to understand, (e) the demonstration of actual factual understanding and, (f) the ability to "appreciate" the facts.[15,16] In making judgments about clinical (de facto) competency the clinician must consider two factors, the patient's clinical condition or mental status, and the nature of the decision the patient is being asked to make. Roth *et al.*[14] have outlined the use of a risk/benefit assessment in informed consent decisions: when the risk of treatment is low and the benefit high, less stringent standards of competency should be applied in assessing a patient's consent to treatment but more stringent standards may be applied to a refusal. When risk of treatment is high and the benefit low a patient's refusal should be judged by a lower standard of competency and his consent by a higher one. In the decision to hospitalize an elderly patient as discussed earlier, in most instances a patient's consent would be held to a lower standard as hospitalization, in and of itself, is a lower risk decision. But when a significantly depressed patient in need of treatment refuses hospitalization, such refusal should be held to a higher standard as the risks of refusal can be high.

There is controversy as to the effects of aging in general on the

ability to give informed consent. Some have attempted[17] to generalize
the abilities of the aged as a group but this is hazardous in that they are a
markedly heterogenous collection of individuals. Taub[18,19] had found
some impaired ability in the elderly to remember consent information
but as Stanley and others have pointed out, poor recall may have little
bearing on the ability to give informed consent. Stanley[20] offers the only
empirical study of the capacity of elderly depressed patients to give
informed consent. In a small sample she found that depressive patients
performed similarly to normal elderly in rendering informed consent
but that compared with younger cohorts both groups of elderly patients
(nondepressed and depressed) demonstrated poorer understanding and
comprehension of consent material. They also showed less reasonable
decision making in that elderly patients agreed significantly more often
to participate in a high-risk/low-benefit research study. Stanley's investigation did not address the problem of those patients with more serious
complications of depression, such as delusions or depression with reversible dementia. More research in this area would greatly help to clarify
the qualities of the elderly depressives' decisional capacity.

Many commonly occurring features of depression may interfere
with a patient's ability to give informed consent. These features occur in
nonpsychotic as well as psychotic and cognitively impaired patients. The
special problems posed by the suicidal depressed older patient will be
discussed later in this chapter. Factors in nonpsychotic depression that
can interfere with informed consent for treatment are (a) decreased
attention and concentration, which can diminish the comprehension and
retention of pertinent information, (b) apathy or feelings of hopelessness or worthlessness, (c) undue fearfulness of all offered interventions, (d) marked agitation, which is often accompanied by the patient's
demand for immediate relief and apparent intolerance for all but the
briefest exchange of verbal information, and (e) masked depression. In
the latter condition, a patient does not appreciate or portray the emotional experience of sadness or depression and usually focuses instead
on physical symptoms. Thoroughly informing such a patient about the
risks and benefits of antidepressant medications can be exceedingly difficult in the face of the patient's insistence that he is not depressed but
rather needs medical intervention, such as x-rays or analgesics. With
sufficient time and explanation, such patients often can be educated
about the nature of their depressive illness and the needed treatment
and can then give informed consent.

Another factor that may interfere with the elderly depressive's ability to give informed consent is the presence of the syndrome called
pseudodementia. This condition will here in be referred to by the more
accurate nomenclature of depression with reversible dementia.[21] This

form of reversible dementia can occur in depressed patients who premorbidly are free of discernible signs of dementia or can occur in patients with an underlying milder dementing illness. In either case these patients require an approach to informed consent decisions similar to that used with patients demented from other causes; the physician needs to assess the patient's ability to understand the treatment information presented and to offer a voluntary decision. Where this ability is lacking the physician may have to ask the family to provide informed substituted judgment on behalf of the clinically (de facto) incompetent patient, or to pursue formal guardianship or power of attorney arrangements.

Delusions are not infrequently present in geriatric depression and these and other psychotic symptoms often interfere with a patient's ability to give informed consent. In a study limited to elderly depressives, Post[22] reported that 40.2% were delusional. Meyers[23] reported 56% of 34 elderly endogenous depressives were delusional in his pilot study. Delusions in depression are often somatic and self-deprecating. They typically include delusions of bodily dysfunction or decay, of being dead, of poverty, or worthlessness. Paranoid delusions may incorporate those trying to give aid to the patient.

Ms. A., a 77-year-old hispanic woman involuntarily committed to a geriatric psychiatry inpatient service suffered from the delusion that she was dying from brain and liver cancer. She refused all food, fluids, and medication, stating that she was waiting for a surgeon to take her to the operating room and that she knew she was to have "nothing by mouth" before such a procedure. She did not believe that the treating physicians were doctors but thought they were persons trying to deprive her of her surgical procedure. She assented to intravenous hydration. As her nutritional status was rapidly worsening this involuntarily hospitalized patient received electroconvulsive therapy (ECT) on the basis of consent provided by her family.

Mr. A., a 69-year-old married man with a 4-month history of progressively worsening depression, shunned all personal contact because he believed that his decaying body emitted an odor so strong and powerful that it repulsed all who came near him. He insisted that no person come closer to him than the door to his room and would apologize profusely in embarrassed tones for his odor. If the treating staff came closer he would become extremely tense and nearly mute. As ECT was repeatedly discussed with him the patient demonstrated an adequate understanding of its risks, benefits, and the alternatives, but continued for another week to refuse to give consent as he feared that the nurses and doctors in the ECT suite opposed his being brought there and that his odor might cause lasting damage to those facilities. With neuroleptic medication the intensity of these delusional fears diminished and he consented to ECT.

Depressed patients such as these require the responsible physician to decide when and how to proceed legally. The clinician's judgment is the threshold mechanism for challenging competency and the threshold should be set high enough so that courts are not overburdened and treatment unduly slowed. The legal requirements for treating a patient unable to give informed consent vary in different jurisdictions. It behooves each clinician to know the requirements in his or her jurisdiction. In many states, treatment of such persons can proceed only with a court appointed guardian or committee giving consent through the exercise of substituted judgment. In others a person who is de facto incompetent to consent to or refuse treatment can be treated with the hospital's medical director or his designee's approval or with family providing consent. Formal legal procedures are called into operation more often when the risks of treatment or its refusal are high. Recently in Massachusetts, the judgment in *Rogers v Commissioner*[24] required that a judge decide when it is appropriate to give neuroleptic medication to incompetent patients. The impact of this judgment on the practice of psychiatry is potentially enormous for the geriatric psychiatrist who deals regularly and frequently with large numbers of patients who are incompetent whether from dementia or severe depression.

In order to expedite treatment it often happens that clinicians proceed with low-risk treatments in patients who are not competent by strict standards but who assent to offered treatment. At times the clinician may be relying on unconsciously communicated or nonverbal indications of assent from patients. More careful attention should be given to studying factors that indicate unconscious motivation or nonverbal assent, as overreliance on the patient's understanding as a test for competency to consent to low-risk treatments in problematic and overly restrictive.

A central question in informed consent procedures is how much information to give. There are several standards detailing the amount and type of information that should be conveyed to the patient by the clinician. The most cautious standard is that of full disclosure, the imparting of "all known or knowable significant information concerning the proposed intervention."[25] Certainly, accompanied by sufficient documentation this is the best defense a health care professional can have against a claim of lack of informed consent. But is this reasonable, practical, or fair to patients? Many patients, and especially depressed ones, can become overwhelmed by full disclosure especially as they usually lack the sufficient background to appreciate fully and weigh the information presented. Full disclosure would require, for example, describing serious and infrequent cardiac complications to patients beginning tricyclics. In the final analysis, full disclosure is not possible. Other standards of disclosure are the medical standard (reasonable practitioner) or the "reasonable man" standard first enunciated in *Canterbury v Spence*.[26] Clinicians should be aware of the applicable standards in their jurisdiction. The following case history illustrates some of the problems raised by the issue of disclosure.

> A 74-year-old successful businessman was suffering from symptoms of an endogenous depression with agitation. The patient had received several courses of cyclic and newer generation antidepressants and had experienced serious side effects including urinary retention. His physician decided to discuss electroconvulsive therapy with the patient and his wife. The patient had been increasingly unable to work, was extremely agitated each morning, and was painfully and obsessively ruminating. He desperately wanted relief from his symptoms and during descriptions of ECT by his physician he would wave off such disclosures with "whatever you say doctor, let's start immediately, where do I sign?" A rather passive and dependent man by nature, his wife was controlling, distrustful, and blaming. The physician perservered in disclosing to both the patient and wife all of the commonly encountered risks of ECT including cognitive dysfunction, complications of blood pressure elevations, temporomandibular pain, etc. In other words she engaged them in an unusually thorough discussion. The relevant information was understood by them both. The patient received 5 treatments without complication. On the 6th treatment the anesthesiologist apparently failed to properly insert a rubber mouth guard and the patient severely bit his tongue causing blood loss and requiring 11 sutures. The wife was enraged and briefly threatened legal action against the psychiatrist who, despite extensive experience with ECT, had never witnessed such an occurrence. The patient's wife briefly persisted in her view that she "should have been told" that this was a possibility, however remote.

Exceptions to the Informed Consent Doctrine with the Elderly Depressed

There are well-recognized exceptions that exist to informed consent requirements and have applicability to the depressed geriatric patient. The emergency exception applies when an immediate medical treatment is required, no informed consent can be obtained from patient or guardian, and the patient, if able, would not be likely to refuse said treatment. With careful documentation of such an emergency the physician is authorized to perform necessary emergency treatment. There are many such situations that may arise with treatment of elderly depressives, especially with the delusional, demented or actively suicidal patient. The following vignette illustrates one of several instances where the emergency exception was invoked:

A 78-year-old widow residing in a nursing home was transferred to a geropsychiatry unit for treatment of a 4-month-long agitated psychotic depression. The patient had a history of a prior depression 12 years earlier during which she took an overdose of diazepam. She had made a complete recovery with hospitalization and antidepressant medication. Her history also included serious hypertension and rheumatic heart disease requiring a prosthetic mitral valve replacement and chronic anticoagulant treatment. On admission the patient began demonstrating the self-destructive behavior that had so worried the nursing home staff, episodically throwing herself violently to the floor crying out "just let me die!" The behavior necessitated neuroleptic treatment and maximal observation. Several days into her stay the patient broke away from an attendant and threw herself to the floor sustaining a large scalp hematoma with subsequent mental status changes. She refused suturing and the CT scan necessary to rule out a subdural hematoma. She continued to insist that she be allowed to die and also appeared cognitively impaired and unable to understand basic information about the CT scan. She received suturing and the scan over her objections. She did not in fact have a subdural hematoma but an antidepressant blood level drawn at the time of her fall revealed toxic levels and the antidepressant was discontinued. After the patient's cognition and sensorium cleared she was restarted on a lower dose.

Another exception to informed consent is the patient's waiver. "Allowing a patient the opportunity to waive the right to be informed is 'paternalism with permission' and is a necessary part of the full recognition of the patient's claim to self-determination."[25] This has implications in the treatment of the depressed elderly as they often behave in a dependent, frail, and vulnerable manner, are easily overwhelmed, and may imply that they "can't understand" or are "too frightened" to hear about the risks of their treatment. A waiver should ideally be explicit rather than implied. The clinician should still make it clear that the proposed treatment has risks and that he or she is willing to discuss them further with the patient or family.

Therapeutic privilege is a controversial issue that can serve as another exception to adherence with the informed consent doctrine. Under therapeutic privilege a clinician determines that disclosure of medical information would likely seriously complicate treatment, cause severe psychological harm, or so disorganize a patient as to render him unable to come to rational decisions. This may have applicability in the treatment of depressed elderly patients where, for example, disclosing a diagnosis of cancer may cause the already hopeless, suicidal (and perhaps somatically delusional) patient to refuse further antidepressant treatment. However, it is essential to keep in mind that this privilege is strictly limited and that full disclosure should occur once the patient has improved. Therapeutic privelege may be improperly invoked when the clinician so modifies his disclosure of antidepressant risks, in an effort to be persuasive and reassuring, that the patient has not been adequately informed. This of course touches on the very real dilemma of how much a depressed patient needs to be told (or wants to be told) about treatment, especially about low probability but high morbidity risks. Where to draw the informational line is a difficult and wholly unresolved issue. Empirical research would be of great use in this area.

Transference, Countertransference, and Informed Consent

Transference and countertransference reactions can take a myriad of forms, some of which may have potential legal implications; the most common will be listed in the following. Frequently a patient views the clinician as parent/protector; the patient behaves in such a fragile, pained manner as to prompt in the clinician wishes to protect and to take over. Along with this, some depressed, helpless, desperate patients, can provoke in a therapist feelings of omnipotent control. Elderly patients may attempt to infantilize the therapist and treat him or her like a child or grandchild because of the narcissistic injury of feeling frail and helpless in the face of a young, competent clinician. This patient attitude may also enhance the therapist's tendency to behave in an overly controlling manner or may have the opposite effect, leaving the therapist feeling less competent and physicianly. A considerable number of elderly depressives behave in a help-rejecting manner that demands enormous expenditures of staff time and energy in order to effect treatment. This in turn leads to the wish on the part of the staff to distance themselves, to feelings of irritation, and to diminished empathy and respect for the patient. Finally, the depressed patient with delusions can set off numerous, complicated countertransference reactions. When the patient casts the clinician in the role of torturer, for example, the clinician can unconsciously try to protect the patient (often from his own aggressive feelings), he may attempt to prove his benevolence or he may collude with the patient and treat him more like a prisoner than an autonomous individual. If each of these examples of countertransference reactions is considered in turn it is not difficult to see how each might provoke in the clinician a temptation to forgo rigorous adherence to the informed consent doctrine.

EVALUATION OF THE ELDERLY SUICIDAL PATIENT

Another area which raises special questions about treatment of depressed older persons is that of distinguishing depressively based suicidal ideation from a patient's competently decided wish to die. A case illustration will demonstrate some of the difficulties in this area:

> Mr. A., an 85-year-old man, was transferred to an inpatient geropsychiatry unit by his nursing home where he and his wife had resided for four years. Described by his children as a relatively isolated, quiet and dependent man, he was extremely devoted to his wife of 62 years and had been upset for 2 years about her declining health. She suffered from multi-infarct dementia, had recently become blind from diabetic retinopathy, had a below the knee amputation and recently failed to recognize her husband. At that point the patient declared to the nursing home staff that he had no reason to go on living and stopped eating. He had never made a satisfactory adjustment to the nursing home placement that was necessary because of his multiple medical problems, including congestive heart failure, left bundle branch block, prostatic cancer, and renal insufficiency. On admission to psychiatry the patient was cachectic, nearly mute, and negativistic saying little more than "let me die" or "leave me alone." There was considerable disagreement among staff as to whether a depressive illness was causing his passive suicidal ideation or whether the patient had made a legitimate decision that he wanted to die based on his life stressors and quality of life considerations. His children similarly demonstrated indecisiveness. One daughter expressed the belief that this stance was consistent with her father's personality and devotion to his marriage. In contrast his son insisted that, although his father was a quiet man, he had never been nearly mute, that he appeared depressed and despondent, and that he must be treated. The latter opinion prevailed and after the son obtained durable power of attorney and was fully informed about the treatment, ECT was administered. After 8 treatments the patient's mood was somewhat brightened, he was eating well, and when questioned said he would not deny a wish to be dead but denied that he wanted to take any measures to hasten his death.

As seen in this case, discriminating depression with suicidal ideation from an elderly infirm patient's nondepressively based wish to die can be extremely difficult to do with certainty.[27] The clinician must rely on all available sources of data in determining the patient's baseline, past history of affective disorder, and habitual ways of responding under stress. The relatives must be actively involved in treatment decisions where appropriate. When necessary, independent consultation from another psychiatrist should be obtained. As physicians generally feel obliged to prevent a suicide we may occasionally find ourselves in a position of thwarting the patient's autonomy in deciding on their time to die and of perhaps overtreating patients for "depression." There are several reasons why physicians would be biased towards taking an active treatment stance in such a situation. A significant problem is the fact that the law is murky in this area; we lack legal guidelines in the areas of treatment withdrawal and of competency to decide to die. Other pressures towards treatment include the frequent presence of unresolved emotional issues in the physician around death, the limits of his or her own responsibility, and issues of control.

THE ROLE OF THE PHYSICIAN IN PATIENT LIFE DECISIONS

As discussed, depressive illness in the elderly can interfere in many ways with the ability to make competent decisions. The need for temporary guardianship or power of attorney to aid in treatment decisions has already been discussed. Numerous situations arise where the depressed patient may need special assistance in other areas of his life. Seriously depressed people often pay far less attention to their personal and financial affairs or act irresponsibly around them. This is especially true of bipolar patients (whose illnesses classically worsen with age), psychotic patients, or the suicidally depressed who begin to dispose of property and money. Depressed patients may be in need of competency examinations when attempting to execute wills or contracts or when called upon to give testimony. An interesting issue is the competency to work and to make the decision to retire.

Mr. C., a 74-year-old married man, was a successful partner in a consulting firm. He was being treated with antidepressants as an outpatient for a major depression with melancholia. Although he showed no cognitive impairments on routine mental status examination or on a Weschler Memory Scale he was concerned about his ability to function on the job. He had noticed several occasions where he became unusually aggravated and inattentive in business meetings with clients and, in his wish to have the meetings terminate, would prematurely agree to contract terms that were less than satisfactory and far from his usual standards. In psychotherapy, his work duties and that of his partners were explored, his responsibilities temporarily restructured, and his hours reduced. He was much relieved to be protected from financial or contractual decisions. His condition, however, continued to worsen and he arrived at the "solution" to retire from his position. Fortunately he discussed this with his psychiatrist, who made the determination that this decision was based on his depressive hopelessness and nihilism and he was advised to take instead a temporary leave of absence and to hold off with his decision to retire. During his leave he received a course of outpatient ECT to which he had a full response complicated only by a brief (one week) hypomania. During this week he was "raring to go" and nearly insistent that he return to work. In fact he showed extensive organic cognitive impairment and impairment of judgment and was instructed by his psychiatrist to stay out of work. His cognitive impairment cleared and he successfully returned to full employment 4 weeks later.

RECOMMENDATIONS

Special attention must be paid to the medical and medicolegal requirements of treating elderly depressives. The clinical diagnosis and treatment of such disorders vary in some important ways from their presentation and management in younger patients. A clinician who takes responsibility for the treatment of elderly depressed patients must have knowledge and training specific to this area. Without this background serious risks of liability may be incurred through misdiagnosis or failure to choose appropriate treatment. Considerable experience with geriatric patients is required in situations of diagnostic uncertainty. The psychiatrist must formulate a differential diagnosis that includes the possibility of depression in patients who present with certain somatic complaints, with symptoms of dementia, or with the wish to die. It is also essential that the psychiatrist understand those medical conditions and medications that may cause or exaserbate depression. As treatment of depression can carry significant risks, especially in the multiply infirm patient, an empirical trial of antidepressant treatment to help solve the diagnostic dilemma is often hazardous and unacceptable.

Guidelines for the decision to commit elderly patients are the same as those used for younger patients. The older depressed patient often requires commitment because of inability to care for oneself, this situation probably arising more frequently in this age group than in younger patients, who more often are committed for the more clear-cut indications of danger to oneself or others. Psychiatrists must be adept at eliciting a thorough history of patient functioning in order to make these determinations. Also, the psychiatrist must be aware of some families' attempts to utilize inappropriately civil commitment procedures as a first step in obtaining nursing home placement.

Because the presence of a depressive illness in an elderly patient can affect that patient's competency in a number of ways the patient often requires an assessment of competency to participate in medical decisions and aftercare. It is essential that physicians recognize that competency is not a static determination but one that must be reevaluated in light of the decision posed to the patient, that partial competency can exist even in extremely ill and disturbed patients, and that the ability of a patient to give informed consent can shift with changes in his clinical condition and changes in the way information is presented.

Where questionable decisional capacities exist, several modifications of the consent process are recommended. These include:

1. The use of the sliding scale, risk-benefit analysis model[14] for the proper selection of the criteria to be used in determining competency
2. Developing a rapport with the patient that promotes question asking[20] and, for those unable to generate relevant questions, carefully providing patients with necessary structured information
3. Tailoring the consent information to the needs of the patient[20] and delivering it in a way that considers their special limitations (e.g., taking longer time, frequent repetitions, simplified language, louder volume)
4. Making written consent information more readable and leaving this material with a patient for review when possible[20]
5. Using visual, nonverbal means (eg: drawings) to clarify information when applicable
6. Involving relatives in the process[20] and encouraging the use of language most familiar to patients

When questions about the patient's competency arise it is essential that family members be actively informed and engaged in helping to make decisions for the incapacitated elder. In many jurisdictions the physician may proceed with treatment (especially with low-risk procedures) with a competent family member's permission without pursuing formal legal procedures. Many statutes governing informed consent and involvement of family are quite unclear and leave much open to interpretation and the exercise of common sense. In some jurisdictions most significant treatment decisions for the de facto incompetent patient would have to lead to a formal, de jure competency proceeding. To treat without such a hearing and provisions for substituted judgment, even with family permission to proceed, might constitute grounds for battery. Where the recommended treatment carries high risk or where there will be multiple, successive decisions to be made, formal arrangements such as guardianship, committee, or power of attorney are usually necessary to protect best all parties. In such instances, further protection can be afforded the physician by obtaining a consultation from another physician before proceeding with the treatment. As the treatment of depression often vastly alters a patient's competency, formal legal arrangements should be time-limited and a physician should be actively involved in the evaluation to terminate such arrangements.

A special word should be added about confidentiality. Clinicians must pay special attention to this patient right as it may inadvertently be compromised in treating elderly depressives. A competent depressive patient may, through his helpless dependence, leave the impression that his progress and problems should be openly discussed with relatives or friends. Every safeguard should be employed to adhere to the usual rules of privacy unless there is express permission from the patient to do otherwise. Where the existence of incompetence dictates that a family member be involved, priveleged communications about the patient should be limited to those in a decision-making position.

In summary, the presence of a depressive illness in an elderly person can raise many concerns about legal and ethical issues including civil commitment, informed consent, potential liability, and the physician's role at the end of life. Certain features of depression, especially the presence of delusions, self-destructive inclinations, and reversible dementia can serve to increase the tendency of the physician to exercise a paternalistic approach towards the patient. This runs counter to the precept that a patient's autonomy must be protected. If certain safeguards are exercised, the physician can deliver quality care to his patient while diminishing his own risk of liability and the older patient can be treated without significant compromise of his rights and autonomy.

REFERENCES

1. Blazer DG, Williams CD: The epidemiology of dysphoria and depression in an elderly population. *Am J Psychiatry* 1980; 137:439–444.
2. Weissman MM, Myer JK: Affective disorders in a US community: The use of research diagnostic criteria in an epidemiologic survey. *Arch Gen Psych* 1978; 35:1304–1311.
3. Schaie KW, Birren JE: *Handbook of the Psychology of Aging.* Van Nostrand Reinhold, New York, 1977.
4. Myers JN, Sheldon D, Robinson SS: A study of 138 elderly first admissions. *Am J Psychiatry* 1963; 120:244–249.
5. Blank K, Schwartz HI: Civil Commitment of Elderly Psychiatric Patients. Submitted for publication
6. Butler, R: *Why Survive? Being Old in America.* New York, Harper and Row, 1975.
7. Lebegue B, Clark L: Incompetence to refuse treatment: a necessary condition for civil commitment. *Am J Psychiatry* 1981; 138:1075.
8. Salzman C: *Clinical Geriatric Psychopharmacology.* New York, McGraw Hill, 1984, p 88.
9. Jeste DV, Wyatt RJ: *Understanding and Treating Tardive Dyskinesia.* New York, Guilford, 1982, p 96.
10. Smith JM, Baldessarini RJ: Changes in prevalence, severity, and recovery in tardive dyskinesia with age. *Arch Gen Psychiatry* 1908; 37:1368–1373.
11. Appelbaum PS, Gutheil TG: *Clinical Handbook of Psychiatry and the Law.* New York, McGraw Hill, 1982.
12. Olin GB, Olin HS: Informed consent in voluntary mental hospital admissions. *AM J Psychiatry* 1975 132:938–941.
13. Grossman L, Summers F: A study of the capacity of Schizophrenic patients to give informed consent. *Hosp and Comm Psychiatry* 1980; 31:205–206.

14. Roth LH, Meisel A, Lidz CW: Tests of competency to consent to treatment. *Am J Psychiatry* 1977; 134:279–284.
15. Drane JF: The many faces of competency. The Hastings Center Report 1985; 15:17–21.
16. Hoffman BF: Assessing competence to consent to treatment. *Can J Psychiatry* 1980; 25:354–355.
17. Bernstein J, Nelson K: Medical experimentation in the elderly. *J Am Geriatric Soc* 1975; 23:327–329.
18. Taub HA: Informed consent, memory and age. *Gerontologist* 1980; 20:686–690.
19. Taub HA, Kline GE, Baker MT: The elderly and informed consent: effects of vocabulary level and corrected feedbak. *Exp Aging Res* 1981; 7:137–146.
20. Stanley B, Stanley M, Pomara N: Informed Consent and Geriatric Patients, in Stanley B (ed): *Geriatric Psychiatry: Ethical and Legal Issues*, Washington, American Psychiatric Press Inc., 1985.
21. Alexopoulos GS, Young RC, Haycox JA, *et al.*: Biological studies in depression with reversible dementia, in Shamoian CA (ed): *Affective Disorders in the Elderly*, Washington, American Psychiatric Press Inc., 1985.
22. Post F: Affective disorders in old age, in Paykel E (ed): *Handbook of Affective Disorders*, New York, Guilford Press, 1982.
23. Meyers BS, Kalayam B, Mei-Tai V: Late-onset delusional depression: A distinct clinical entity? *J Clin Psychiatry* 1984; 45:347–349.
24. *Rogers v Commissioner of Mental Health*, S-2995, Mass. (Nov 29, 1983)
25. Kapp MB, Bigot A: *Geriatrics and the Law: Patient Rights and Professional Responsibilities.* New York, Springer Publishing Co., 1985, p 25.
26. *Canterbury v Spence*, 464 F .2d 772 (D.C. Cir. 1972)
27. Brown JH, Henteleff D, Barakat S, *et al.*: Is it normal for terminally ill patients to desire death? *Am J Psychiatry*, 1986; 143:208–211.

12

Neuropsychiatric Assessment of Dementia

Inadequacy of Test Protocols

HELEN L. MORRISON

INTRODUCTION

The diagnostic, rehabilitative, and intervention questions posed to the medical forensic evaluator of the geriatric patient are often related to the component questions of organic versus functional disorders that impinge on forensic issues.

Questions as to competence, responsibility, or the adaptive significance of brain injury are frequent. Objectifying the presence, quality, and degree of adaptive deficit is provided by the appropriate use of neuropsychiatric examination. I stress the word *appropriate*. Without comprehensive and systematic evaluation, the forensic medical opinion will likely be erroneous.

What is neuropsychiatry? How does the diagnostic approach of this discipline contribute to the understanding of evaluation in the geriatric patient? This chapter will not be able to provide extensive examination of these questions. The purpose of this chapter is to discuss the several neuropsychiatric instruments most frequently used in assessment of the

---

HELEN L. MORRISON • The Evaluation Center, 919 North Michigan Avenue, Suite 3100, Chicago, IL 60611; Pediatric Psychiatry Center, Kingwood Hospital, Michigan City, IN 46360.

geriatric patient. The vital clinical and forensic question is: Are the current available instruments adequate for use in the geriatric population?

A CAVEAT

Luria states "neuropsychology is merely the most complex and newest chapter of neurology, and without this chapter, clinical neurology will be unable to exist and develop."[1]

The value of neuropsychiatric testing has led to an inappropriate dependence on the results. Some psychiatric and psychological clinicians have disregarded a major tenent of practice. That tenet prescribes that test results, without a comprehensive and competent clinical evaluation and interview, have no value. It is not sufficient that test results be exclusively accepted just on the basis of their lack of intrusiveness, and therefore safety, range, or attributes.

Ideally neuropsychiatric evaluation provides for the assessment of abilities represented in the brain: rationality and recollection and recall, familiarization and adaptation, cognition and knowledge, neurological component functions, and attention, among others. The advantages of using noninvasive types of evaluations are beyond measure. They can provide descriptive and prognostic evaluations when utilized appropriately. These descriptions can encompass all areas of function.[2]

With the noninvasive character of the testing, no risk of mortality or morbidity ensues. A careful specification of deficits is important for cases in which questions are posed concerning responsibility for acts, competence to perform acts, and the effects of brain injury. This testing is especially relevant in contrast to other methods of assessing a patient's adaptive capacity. Other methods of appraisal tend to rely more on spontaneous verbal behavior and nonsystemically elicited perceptual-motor and motor behavioral measures.

Several factors account for the advantageous aspect of neuropsychiatric test protocols. The noninvasive character of the tests provides for safety and lack of morbidity or mortality. If one must balance the risks and gains of invasive testing in provision of positive results, risks often outweigh gains. Another influence is the inaccuracy, lack of depth, and insensitivity of aspects of the neurological evaluation.[3] The test characteristics of nonspecificity can often tell us of the presence or absence of organic damage. The location of brain impairment is often not determined with any degree of accuracy.

DEFINITIONS

What are the component questions asked of the neuropsychiatric consultant? What can one expect of the use of standardized and specialized tests? Most authors would define seven major test functions. These include functional versus organic etiologies of the deficits, mental status, presence and location of structural damage, intellectual strengths and weaknesses, prognosis, rehabilitative potential, and recommendations for treatment.[4]

Assessment of these functions requires that the test instruments be systematic, valid, reliable, and accurate. In the geriatric patient, the task of compliance wih these requirements of neuropsychiatric assessment is very difficult to attain. The complexity of assessment is greatest in the geriatric patient. Changes in cognition in the elderly have been likened to changes in patients affected with organic or drug-induced toxicity, depression, or systematic disease such as hypothyroidism. It is not infrequent that reversible conditions are ignored or misdiagnosed as irreversible.

THE GERIATRIC PATIENT

When one tests the geriatric patient, complications inherent to this patient population need to be kept in mind. Greater versatility and fewer presumptions need to be applied to this population. In these patients, fear and insecurity frequently distinguish the test experience.

Age related factors, such as fear, "may indeed cause memory loss which will then affect his entire testing session."[5] Disorientation, subtle confusion, lack of motivation or cooperation frequently occur. The elderly are commonly disadvantaged by sensory deficits. A significant proportion of the elderly suffer enough loss of visual and auditory acuity to create real difficulties in communication. Auditory sensory deficits have a negative correlation with neuropsychological test functioning.[6]

Peiffer[7] notes that the bias of the examiner can seriously and adversely affect the geriatric patient. The terms *senility* or *mentally ill* are frequent and significant additional barriers of communication.

Performance anxiety can easily interfere with test performance. In the elderly patient, the most certain way to induce this anxiety is to begin with measures of intellectual functioning. Lack of patience on the part of the examiner denies the need of these patients to be given more time to respond to questions.

Technically, the most frequent problems are those that relate to the lack of established norms in this population. Normal aging intellectual changes when compared to younger patients and their intellectual functioning may show inaccurate deficits in intellectual functioning.[8]

The range of scores that are obtained during testing are called norms. This range is obtained when a standardized group is measured with the typical performance of one individual.

TEST INSTRUMENTS

Hooper Visual Organization Test

Originally the Hooper Visual Organization Test was designed to discriminate functional from organic disorders on the basis of objectively measured visual organization tasks. Common objects are divided into more than two parts, 30 objects in all. Visual organization of the object into a cohesive whole must be accompanied by naming of the object. Hooper studied a group of aged living in residential care. Studies of state hospital patients showed a range of 64% accuracy of results in patients diagnosed as chronically organically damaged. Two groups of chronic schizophrenics obtained results of 78% and 67%. Low scores may indicate distortions in thought processes. Therefore, brain damage is not always indicated by low scores.[9] There are no current standardizations done in respect to age, education, or economic or social status.

Memory for Designs

Graham and Kendall designed a test of memory functions.[10] Fifteen designs characterized by geometrical shapes are shown to the subject for a period of 5 seconds. They are then asked to draw these designs from memory. Points are given for preservation of design, distortion of elements, absent elements, rotations, or reversals.

This test has been extremely popular since its inception. There are several problems in using this test in the geriatric patient. First is that no difference is given in the process of scoring between one forgotten element or totally forgotten designs. If the norms are used based on the results obtained in the original samples, older subjects obtain more frequent diagnoses of brain damage than truly exists. In addition, there are no norms for those persons diagnosed with psychiatric illness.

The correlation between the neurological findings and the score obtained in the Memory-for-Designs Test is very minute.

Benton's Revised Visual Retention Test

Assessment of visual organization has been a characteristic of the test devised by Benton. Visual perception and memory are assessed. Four techniques can be used during the test administration of three sets of 10 cards. Standardization samples of brain-injured patients revealed 57% scored three or more standard deviations below expected scores. Forty-three percent scored within the normal or borderline range of results. No ages are noted in the standardization.

Errors are scored on the basis of misplacements, size errors, perseverations, omissions, and rotations. Diffuse injuries or lesions in the right hemisphere are considered most sensitive to the Benton. Validity studies have stated ranges of 60% to 70% in the organic population.

One study, done by French authors, examined average intellect persons ranging from 45 to 84 years of age. The data obtained revealed good agreement with the Benton. Performance levels relate to age. A progressive increase is seen from age 8 to 15 with a plateau until the 30s. A progressive and continuous decrease is seen in the performance of subjects from 40 onward.

Face-Hand Test

A commonly used neurological test is that of the face-hand category. A patient has a light stimuli applied concurrently to the hand and cheek. Brain-damaged patients may report dislocation of the touch, or the presence of only one of the stimuli. Errors occur outside that are expected with parietal lobe damage. Eighty-seven percent of brain damaged patients continue to make errors after 10 trials. These results are compared with the performance of schizophrenic and normal patients, which are 3% and 0.5% respectively.

In the geriatric patient, a study of 1,077 patients revealed 73% of subjects with significant test results were diagnosed as organic brain syndrome.

Wechsler Adult Intelligence Scale

The most frequently used intelligence scale in the geriatric population is the Wechsler Adult Intelligence Scale (WAIS). The newly defined Wechsler Adult Intelligence Scale-Revised appears more difficult for the older adult.

This test was defined in 1939 to serve as an adult intelligence test. Deviation intelligence quotient measured the results obtained by an individual relative to a group of peers. Eleven subtests are grouped into five tests of performance and six tests of verbal competence. Although the beginning role of this test was to assess high intelligence or mental retardation, currently the test is used for chronic alcoholism brain damage[11] or psychiatric disorders. Greater significance is placed on the pattern revealed in the subtest rather than the Full Scale Intelligence Quotient.

Standardization tables exist for ages 65 to 69, 70 to 74, and 75 and over. Cognitive function is assessed by comparing two scores. The first is the conversion of test scores to scaled scores. Secondly, scaled scores are converted to norms obtained through sampling nationwide. In the elderly, little dependable data is available in the age group 64 or older. Age corrections have been too small in performance subscales.[12] The dangers inherent in using the WAIS in the geriatric population have been well demonstrated in the research by Feinberg *et al.*[13]

This study gathered data from members of the California Retired Teachers Association. Population characteristics included minimum of master's degree, good health, high motivation, and high social activity. Despite their generally superior to very superior intelligence quotients, their performance patterns suggested a diagnosis of brain dysfunction if there was strict adherence to norms. Of course, multiple sources of bias are evident in this study, including sampling bias based on the lack of representatives across the sample, and geographical bias.

Bender-Gestalt Perceptual Motor Test

A projective and nonprojective test was designed by Bender in 1936. It, too, measured visual perceptual skills. The Gestalt psychologists provided the theoretical basis for this test. Nine designs are used to demonstrate the organization or parts of visual stimuli into a cohesive whole (*Gestalten*). Although originally designed for children, the test is used extensively in all types and age groups. Persons with organic dysfunction generally have increased difficulty in copying the figures on paper that has a jigsaw pattern. A greater number of cues are needed for perception in organically impaired persons. Wepman suggests the presence of a disorder of visual praxis if Items 1, 2, 4, and 5 have difficulty associated with their reproduction.

Originally, the test was designed to evaluate maturation in visual-motor perception, academic achievement, intellectual potential, and neurological and emotional difficulties. No one would disagree with findings that clear delineations can be made between non-brain-damaged, non-psychiatric patients and brain-damaged persons. Diagnostic significance is considered to relate to substitution of lines for dots, fragmentation, poor integration, and failure to cross lines in figures 6 and 7, distortions, perseverations, rotation or disorientations, and other significant changes. Discrimination occurs at the level of 20 errors.

The establishment of norms in the geriatric population depended on ages of validation groups of 47 years to 55 years. One easily sees how these are hardly geriatric ages. The "normal" geriatric person performs at a lower level with errors more suggestive of dysfunctions of the central nervous system.

Halstead-Reitan Neuropsychological Battery

Halstead described the neuropsychological battery in his volume *Brain and Intelligence: A Quantitative Study of the Frontal Lobes* in 1947. Five current subtests are used including the Finger Tapping Test, Seashore Rhythm Test, Category Test, Speech Sounds Perception Test, and Tactual Performance Test. Two scales that were included in the original battery, the Time Sense and Critical Flicker Fusion Test, have been discontinued due to their inaccuracy in identification of the brain-damaged patient.[14]

The familiar nucleus of the Halstead-Reitan may be augmented with the Wechsler Adult Intelligence Scale, the Minnesota Multiphasic Personality Inventory, the Halstead Category Test, and other tests. The test was used originally to discriminate frontal lobe lesions, lesions in other areas, and non-brain-damaged subjects. Predictions as to location of a lesion are based on the relationships between test scores.

Standardization of original norms were very poor as they were based on men who had served in the military and were under care for psychiatric illnesses or patients awaiting psychiatric surgery for uncontrollable behavior. The average age was 28.3. Erroneous diagnostic-impressions occur frequently as there is a decrease in most subtests with advancing age.[15] Some newer T-score conversions have been defined for ages from 55 to 79. However, no distinctions have been made for these persons on the basis of normal, emotionally impaired, or neurologically impaired geriatric patients.

Older persons (mean—53 years of age) perform very poorly in complex problem-solving areas such as Tactual Performance and Category Tests when compared to a mean age of 28 years. A significant negative correlation ($r = .57$) has been reported between overall performance and age. Invariably, lack of consideration of these factors leads to inflated false positive diagnoses of brain damage.

Luria-Nebraska Neuropsychological Battery

Since 1980 when this author reported on the value of this test, promising results have been seen in the discrimination of neurological and psychiatric populations.[16]

The prominent Russian neuropsychologist A. A. Luria defined the techniques presented in this battery and organized by Christensen.[17-19] Golden and his colleagues have greatly expanded on the battery.[20-25] Luria views cerebral behavioral relationships as complex behavior that requires the joint operation of several independent functions located in various zones of the cerebral cortex. Although these functions appear to be independent, they operate dependently through the integration of cortical association areas. These areas overlap to permit the smooth coordination of efficient functional systems that govern behavior. Single functions that are disturbed will show characteristic signs of disruptions of behavior. The analysis of these basic discrete units of behavior is viewed from the standpoint of their role in complex acts. This is in order to localize as precisely as possible the points of disruption in the functional system. Simultaneously, the area of the cortex that is responsible for the impaired functional unit is localized.

In Luria's conceptualizations, the human brain is made up of three primary blocks that incorporate basic psychological functions.

The first block, as shown in Table I, regulates the energy level of the cerebral cortex, or "cortical tone." It is located in the brain stem reticular formation. This block controls wakefulness and arousal. As such it is reponsible for the stability of basic dynamic processes. Lesions of the first block will produce a loss of the selectivity of cortical actions and the normal discrimination of stimuli, resulting in disordered control of behavior.

Table II summarizes the second block, which is localized in the posterior regions of the cortex. It is essentially involved in the analysis, coding, and storage of information from optic, acoustic, cutaneous, and kinesthetic stimulation. Within each of these cortical areas, Luria postulates a hierarchical organization: (a) a primary zone or analyzer that sorts and records sensory information, (b) a secondary zone responsible for

Table I. Luria's Concept of Cerebral Function: First Block

Function:	Regulates energy level of cortex or "cortical tone"
Location:	Brain stem reticular formation
Controls:	Wakefulness
	Arousal
	Responsible for stability of basic dynamic processes
Dysfunction:	Loss of selectivity of cortical actions
	Loss of normal discrimination of stimuli
Result:	Disordered control of behavior

Table II. Luria's Concept of Cerebral Function: Second Block

Function: Analysis, coding, storage of information from optic, cutaneous, kinesthetic stimulation
Location: Posterior regions of cortex
Hierarchy/functions:
    1° zone—Sorts and records sensory information
    2° zone—Perceptual organization and coding of sensory information
    3° zone—Overlap among adjacent cortex, overall organization and coordination of behavior
Lesions:
    1°—Sensory defect (central blindness or deafness)
    2°—Disorganization of perceptual processes
    3°—Disruption of synthesis of sensory information (visual disorientation in space, incapacity to comprehend language structure)

the perceptual organization and coding of sensory information, and (c) a tertiary zone that is principally concerned with the overlap among adjacent areas of cortex, and thus plays a central role in the overall organization and coordination of behavior. Lesions of the second block or signaling system produce more specific effects than those of the first block. An injury to the primary subzone results in a sensory defect (such as central blindness or deafness). Lesions in the secondary subzones interfere with the analysis of sensory information. These consequently result in disorganization of the behavioral (perceptual) processes that would normally respond to particular stimuli coded or analyzed in that cortical zone. A lesion of the tertiary subzone produces complex disturbances arising from the disruption of the synthesis of various sensory information inputs of different types. For example, visual disorientation in space, or the incapacity to comprehend complex grammatical logic or language structure.

The third block, shown in Table III, is involved in the formation of plans, intentions, and strategies (programs) of behavior. Sensory and motor functions remain unimpaired after lesions to the frontal lobes alone. Lesions in this area will effect goal directed behavior, adaptive reinforcement, and the inhibition of irrelevant stimuli, psychomotor retardation, defects in abstract thinking, and *witzelsucht*. This is defined as foolishness, dullness of comprehension, a mental state marked by frivolity, and an inability to take anything seriously.

Luria's approach may be illustrated through the example of writing, one form of voluntary movement (Table IV). The first component involves afferent (sensory) signals originating in the postcentral sensory cortex, because the position of the limb changes at each movement and

Table III. Luria's Concept of Cerebral Function: Third Block

Function:	Formation of plans, intentions, strategies of behavior
Location:	Frontal lobe
Controls:	Judgment, reasoning, abstract thinking, psychomotor activity, restraint of emotional impulses, volitional movement
Lesions:	Impaired inhibitions, psychomotor retardation, *witzelsucht*, defect in abstract thinking

thus proprioceptive feedback from muscles and joints is required to guide the program of impulses directed to the motor apparatus. Secondly, the movement of the writing instrument has to be precisely oriented toward a particular point in space; this spatial analysis is carried out by the tertiary zones of the parietal-occipital cortex. Lesions of these areas do not impair the sensory-proprioceptive basis of movement, but rather affect the ability to evaluate spatial relations such as left-right or up-down orientation. A third component of the act of writing would appear to involve sequencing of separate links of motor behavior or the plasticity of writing behavior. Lesions to the premotor cortex will therefore produce an inability to arrest one of the steps of the movement and to make a transition from one step to the next, whereas the sensory feedback and spatial orientation components remain unaffected. Finally, the prefrontal areas of the brain are essential for the goal-directed execution of plans or intentions that govern the purposive nature of the act of writing. The essential point is that writing is not simply a motor or sensory function. It rather involves several behavioral components, each

Table IV. Luria's Concept of Behavioral-Cerebral Dependence

Writing	
Component I:	Afferent sensory signal originates in postcentral sensory cortex
	Proprioceptive feedback from muscles guides impulses to motor apparatus
Component II:	Orientation in space of writing instrument—a spatial analysis completed in 3° zone of parieto-occipital cortex
Lesions:	Impaired left/right or up/down orientation
Component III:	Sequencing of motor behavior or plasticity of writing behavior in premotor cortex
Lesions:	Inability to arrest steps of movement
	Inability to transition from one step to next
Component IV:	Goal directed execution of plan or intention in prefrontal areas
Lesions:	Impairment in execution

of which is subserved by different cortical areas. Any dysfunctional area could produce an impairment in the final behavioral activity.

The comprehensive work of Luria has been widely acknowledged as representing a major contribution to the understanding of brain–behavior relationships. Luria's work, spanning more than 40 years, has generally been associated with extensive clinical examination of the higher cortical processes in brain-damaged patients and has been described in several of this writings. The translation of Luria's theoretical approach to clinical applicability is demonstrated in the Luria-Nebraska Neuropsychological Battery[20] as summarized in Table V.

The full battery is a comprehensive inventory composed of 269 items that are divided into 77 sections. These sections evaluate motor skills, rhythmic and pitch perception, tactile perception, expressive language functions, receptive language functions, reading, writing, arithmetic, memory (mnestic functions), visual-spatial functions, and intellectual processes. The *Motor scale* requires (a) the reproduction of simple movements of the hands, mouth, and tongue, (b) copying figures from a model and from verbal instructions (construction apraxia), (c) motor performance based on tactile-kinesthetic feedback, (d) simple motor coordination, and (e) complex motor sequencing. The *Acoustic scale* requires (a) deciding whether simple and complex tones and rhythmic patterns are the same or different, and (b) the reproduction of simple and complex pitch and rhythm relationships. The *Tactile scale* examines (a) cutaneous sensation, (b) muscle and joint (proprioceptive) sensation, (c) stereognostic perception, (d) perception of direction of limb movements and the reproduction of limb positions, and (e) the evaluation of sensory thresholds. The *Visual scale* evaluates the ability to (a) identify objects and pictures, (b) identify missing parts in geometric patterns, (c) construct patterns from blocks, (d) identify time on clocks and map

**Table V. Luria-Nebraska
Neuropsychological Battery**

I.	Motor scale
II.	Acoustic scale
III.	Tactile scale
IV.	Expressive Speech scale
V.	Receptive Speech scale
VI.	Reading scale
VII.	Writing scale
VIII.	Arithmetic scale
IX.	Memory scale
X.	Visual scale
XI.	Intellectual Processes scale

directions, and (e) perform spatial rotations. The *Receptive Speech scale* requires (a) the recognition of letters, phonemes, words, and sentences by oral repetition and written expression, and (b) responding to statements and questions involving the comprehension of simple and complex grammar. The *Expressive Speech scale* requires (a) the articulation of simple speech sounds, (b) saying familiar and unfamiliar words, phrases, and sentences, first spoken by the examiner and then from written text, (c) naming objects from pictures and verbal descriptions, (d) automatic speech, and (e) production of narrative descriptions from pictures of scenes, stories, and when given a topic to discuss. The *Writing scale* assesses (a) copying of individual letters, letter combinations, and words from both written text and from dictation, and (b) writing sentences and phrases. The *Reading scale* evaluates the ability to read simple letters, sounds, words, sentences and paragraphs. The *Arithmetic scale* examines the ability to (a) write Arabic and Roman numerals, (b) write numbers involving spatial orientation, (c) write complex numbers, (d) perform comparisons between numbers, (e) perform simple and complex arithmetic involving logical operations, and (f) perform serial 7s and 13s. The *Memory scale* examines (a) sensory memory for visual, rhythmic, and tactile materials, and (b) verbal memory for simple and complex materials including memory under conditions of interference. The *Intellectual Processes scale* requires (a) interpreting pictures, stories, expressions, and proverbs, (b) appreciation and recognition of humor, (c) concept formation (whole–part relations, class–category relations), (d) understanding similarities, differences, and analogies, (e) definitions of words, (f) understanding logical relationships, and (g) arithmetic calculations.

Further, three scales are provided that have been demonstrated to be maximally sensitive to (a) brain damage in general (*Pathognomic scale*), (b) left hemisphere dysfunction, and (c) right hemisphere dysfunction. The Pathognomic scale is composed of the 33 items reported to be the most discriminating between normal and neurologically impaired patients. Validation studies have reported satisfactory evidence in discriminating between normal and brain-damaged patients,[22] schizophrenic and brain-damaged patients,[23] left and right hemisphere involvement, and among specific quadrants along the anterior-posterior axis of both cerebral hemispheres.[24]

Raw scores are summed for each clinical scale and are then converted to T scores with a mean of 50 and standard deviation of 10. T scores of 60 or greater fall within the brain-damaged ranges, and in almost all cases with three or more scores above 70, it was reported that such patients carried the diagnosis of brain dysfunction 95% of the time.

The Luria-Nebraska Neuropsychological Battery (LNNB) can be administered and objectively scored by an assistant after a brief period, generally one month, of training. The complete battery can be administered in 2.5 hours over several testing sessions, thus minimizing fatigue effects and fluctuations in the patient's level of cooperation and motivation. The Luria is completely portable and thus does not require extensive apparatus or laboratory space. Accordingly, it can be administered with a minimum of difficulty.

Similar to other neuropsychological tests, the norms change with education and age. In general, T scores greater than 60 in a normal person or 70 in a person diagnosed with a psychiatric illness, may indicate brain damage. There is a 90% chance of this being true if three or more scales are elevated above these levels.

The effects of age on the battery have been studied by many researchers: the use of medical history combined with results obtained on the Luria test to correctly identify 12 normal and one out of two impaired persons in 14 subjects over age 65; assessment of the ability of the battery to discriminate brain-damaged from non-brain-damaged elderly with a mean age of 72.2 years. Expressive speech and writing show significant decreases in age-related changes between 60 and 75 or greater aged persons.

CONCLUSIONS

Based on the research done to date, seven of the eight most frequently used neuropsychiatric instruments are not adequate for use in the geriatric patient.

With the Hooper Visual Organization Test, many patients with brain damage easily complete the test. Poor scores also can indicate disruptions in thought processes. The Memory-for-Designs Test shows that normal age decreases are greater than the norms suggested. Too many false positives occur in older subjects if original norms are used to compare results in the geriatric patient. The Benton's Revised Visual Retention Test corrects results obtained only up to 64 years of age. In the face-hand Test, no norms are given and there is a high rate of inaccurate diagnoses given to geriatric patients. The Wechsler Adult Intelligence Scale does not provide normative data for persons over 64 years of age. This is despite the facts that show deterioration in performance with age in a normal population. The Bender-Gestalt was originally defined for children. Validation is lacking for the test and no standard is available with which to compare the geriatric patient.

The Halstead-Reitan battery bases their scores on a young sample. Subtest performance decreases with age. Performance also may depend on education. The scales most damaging to the geriatric patient in regards to misdiagnosis are the Category Test an the Tactual Performance Test.

The Luria-Nebraska has included norms that take into consideration changes in age and education. The Luria is less complex than the Halstead. It is also less tiring with less disturbance in attention and concentration.

SUMMARY STATEMENT

Clinicians must exercise extreme caution in the interpretation of test data obtained in the majority of neuropsychiatric test batteries. The lack of validated and standardized norms often adds to misdiagnosis of neurological impairment. Inconsiderate test giving, bias of the tester, and lack of knowledge of cognition in the normal geriatric individual add to the inaccurate diagnoses.

The role of the clinician is to evaluate damaged and nondamaged functions. Expected ranges of performance are necessary before the word *pathological* is applied to the geriatric patient. If one uses a young population with which to compare the older population, one will *always* find deficits. We need to know how test scores of older persons correlate with behavior in daily life situations.

We cannot add to the shortcomings of the traditional approaches of neurology and psychiatry by ignoring those of neuropsychiatric test instruments. In addition to an urgent requirement for improved procedures for testing in this population, the clinician has major responsibility for correct and cautious use and interpretation of the current procedures used in the geriatric population.

REFERENCES

1. Luria AR: Towards the mechanisms of naming disturbance. *Neuropsychologia* 1973; 11(4):417–421.
2. Luria AR: Brain and conscious experience: A critical notice from the U.S.S.R. of the symposium edited by JC Eccles (1966). *Br J Psychol* 1967; 58(3):467–476.
3. Filskov SB, Goldstein SG: Diagnostic validity of the Halstead-Reitan neuropsychological battery. *J Consult Clin Psychol* 1984; 43(3):382–386.
4. Swiercinsky DP: Prediction of specific brain damage location and process by the neuropsychological factor approach. *J Clin Psychol* 1976; 32(3):651–654.
5. Oberleder M: Adapting current psychological techniques for use in testing the aging. *Gerontologist* 1967; 7(3):188–191.
6. O'Neil PM, Calhoun KS: Sensory deficits and behavioral deterioration in senescence. *J Abnorm Psychol* 1985; 84(5):579–582.
7. Peiffer RA: Langer-Giedion syndrome and additional congenital malformations with

interstitial deletion of the long arm of chromosome 8 46, XY, del B (q 13-22). *Clin Genet* 1980; 18(2):142–146.

8. Botwinick CY, Quartermain D, Freedman LS, *et al.:* Reversal of cycloheximide-induced amnesia by adrenergic receptor stimulation. *Pharmacol Biochem Behav* 1977; 7(3):259–267.

9. Lezak MD: Subtle sequelae of brain damage. Perplexity, distractibility, and fatigue. *Am J Psy Med* 1978; 57(1):9–15.

10. Joslyn D, Grundvig JL, Chamberlain CJ: Predicting confabulation from the Graham-Kendall Memory-for-Designs Test. *J Consult Clin Psychol* 1978; 46(1):181–182.

11. McFie J, Thompson JA: Variation with age of the effects of cerebral lesions in man. *Brain Res* 1981; 31(2):363.

12. Overall JE, Gorham DR: Organicity versus old age in objective and projective test performance. *J Consult Clin Psychol* 1972; 39(1):98–105.

13. Feinberg I, Fein G, Floyd, TC: EEG patterns during and following extended sleep in young adults. *Electroencephalogr Clin Neurophysiol* 1980; 50(5-6):467–476.

14. Ball C: Measurements of time sense and critical flicker fusion test in the brain damaged patient. *J Psychological Measurement* 1981; 45(2):345–362.

15. Bak JS, Greene RL: Changes in neuropsychological functioning in an aging population. *J Consult Clin Psychol* 1980; 48(3):395–399.

16. Morrison, HL: Relevance of modern neuropsychiatry to forensic psychiatry. Paper presented at the annual meeting of the American Academy of Psychiatry and Law, Chicago, October, 1980.

17. Luria AR, Simernitskaya EG, Tubylevich B: The structure of psychological processes in relation to cerebral organization. *Neuropsychologia* 1970; 8(1):13–19.

18. Luria AR: Neuropsychological studies in the USSR. A review. I. *Proc Natl Acad Sci USA* 1970; 70(3):959–964.

19. Christensen AL: *Luria's Neuropsychological Investigation: Manual and Text.* New York, Spectrum, 1975.

20. Golden CJ, Purisch AD, Hammeke TA: *The Luria-Nebraska Neuropsychological Battery: A Manual for Clinical and Experimental Uses.* Lincoln, University of Nebraska Press, 1979.

21. Golden CJ, Hammeke TA, Purisch AD: Diagnostic validity of the Luria neuropsychological battery. *J Consul Clin Psychol* 1978; 46:1258–1265.

22. Hammeke TA, Golden CJ, Purisch AD: A standardized, short and comprehensive neuropsychological test battery based on the Luria neuropsychological evaluation. *Int J Neuroscience* 1980; 8:135–141.

23. Purisch AD, Golden CJ, Hammeke TA: Discrimination of schizophrenic and brain-injured patients by a standardized version of Luria's neuropsychological test. *J Consul Clin Psychol* 1978; 46:1266–1273.

24. Lewis GP, Golden CJ, Moses JA, *et al.:* Localization of cerebral dysfunction with a standardized version of Luria's neuropsychological battery. *J Consul Clin Psychol* 1981; 47:1003–1019.

V

Administrative and Social Policy Issues

13

Administrative and Forensic Considerations in the Operation of a Geriatric Psychiatry Service

HARVEY BLUESTONE, SHELDON TRAVIN, AND MICHAEL KAUFMAN

This chapter will describe how legislation, judicial decisions, regulatory agency rulings, institutional policies, and funding mechanisms influence and control the clinical care of the geriatric patient. It will be demonstrated how these legal, quasi-legal, and administrative forces interact with the clinical judgments and ethical obligations of the physician, sometimes in a complementary way, but frequently in a manner that presents significant clinical and ethical dilemmas.

The chapter is based on a review of the literature and on the authors' experiences in their own geriatric psychiatry service. Presumably, there will be rapid developments in legislation, court decisions, and administrative policies that will affect future developments. Furthermore, there are wide differences among the various states. We have not made a detailed comparison of various state statutes, and to the best of our

HARVEY BLUESTONE • Department of Psychiatry, Bronx-Lebanon Hospital Center, Bronx, NY 10456; Albert Einstein College of Medicine, Bronx, NY 10456.   **SHELDON TRAVIN** • Department of Psychiatry, Bronx-Lebanon Hospital Center, Bronx, NY 10456; Albert Einstein College of Medicine, Bronx NY 10456.   **MICHAEL KAUFMAN** • Crotona Park Community Mental Health Center, Department of Psychiatry, Bronx-Lebanon Hospital Center, Bronx, NY 10456.

knowledge this has not been done by anyone else. Of necessity, frequent reference will be made to New York statutes and legislation. We urge the reader to become familiar with comparable matters in his own jurisdiction.

It should first be made clear that legislation, judicial decisions, administrative regulations, and financial developments take place in a social and historical context. Secondly, we emphasize that consideration of a geriatric psychiatry service cannot be limited to a specialized inpatient unit; because of the likely health-related vicissitudes of the geriatric patient, consideration applies to the utilization of a variety of health care units, as the elderly patient may be transferred to one or more of the following: a geropsychiatric clinic, a community center, a medical-surgical clinic and ward, a convalescent and nursing home, and a hospice program. In all these stages, careful evaluation of the individual's functioning in the medical, psychiatric, and social sense, each with their administrative-legal considerations, is crucial for appropriate case management. Fundamental to any such evaluation of the elderly is the assessment of the patient's decision-making capacity or competency to give informed consent. This basic forensic issue has exceptional relevance to the geriatric patient and is imperative for both treatment and nontreatment decisions as well as for research participation.

ATTITUDINAL CHANGES

Because popular attitudes towards the elderly influence actual performance and shape administrative-legal policies, a brief discussion of attitudinal changes is presented. It is important to realize that these attitudes vary from culture to culture and may also fluctuate over time in the same culture. The Bible enjoins us in the Book of Deuteronomy: "Honor your father and your mother as the Lord, your God, has commanded you, that you may have a long life and prosperity in the land which the Lord, your God, is giving you."[1] Kebric[2] points out how the lot of the elderly in second century A.D. Rome as described in the letters of Pliny The Younger, a Roman aristocrat, in many respects is comparable to that of our own elderly. The elderly of Pliny's Rome experienced the same fears and pains of growing old and the barriers between the generations that today's elderly experience. The major difference, however, was that the Romans venerated old age, which was also the situation in this country until the advent of the present century's youth-oriented society. But it is erroneous to suppose that in general the elderly in less industrially developed countries fared any better than in modern societies.[3] Confronting his own aging, Hall,[4] who was then in his late seventies and a retired professional psychologist and Clark University's founding president, wrote a gerontologic book called *Senescence, the Last Half of Life* in 1922. In it, he predicted that future scientific research in senescence would result in a resurgence of societally useful roles for the elderly. Despite his appeal for progress, forward movement has been slow. The general negative attitude, systematic sterotyping, and discrimination against the elderly has been aptly called "ageism" by Butler.[5] This pervasive prejudice against the elderly has contributed to the general "tragedy of old age in America."[6] The elderly have been deprived of adequate psychiatric treatment, research, and services.[7] The Group for the Advancement of Psychiatry[8] lists six major categories of reasons why therapists are disinclined to treat elderly patients. A significant number of psychiatrists consider older patients less ideal for their practices than younger ones with the same symptoms.[9] Binstock[10] has commented on a new set of axioms about the elderly that has set the stage for the elderly to be used as scapegoats in American society. These axioms portray the elderly as relatively well off, constituting a potent political force, and already costing too much of the federal budget. There has also been recent concern about a new positive mythology created about the aged that "shows no more tolerance or respect for the intractable vicissitudes of aging than the old negative mythology."[11] The new positive mythology depicts older people as healthy, sexually active, productive, engaged, and self-reliant.

What is particularly upsetting about the current attitudes is that they affect society's willingness to care for and treat its older citizens. Siegler[12] notes that the doctor–patient relationship in Western medicine has gone through an age of paternalism and autonomy, and is now in an age of bureaucratic parsimony. In this new system, decision making will not remain exclusively between doctor and patient, but will be based on institutional and societal efficiencies and cost concerns. Inevitability, medical resource allocation and resource rationing will play an increasingly important role in geriatric decision making.[13]

EPIDEMIOLOGY OF GEROPSYCHIATRIC DISORDERS

In order to plan adequately for the elderly mental patient, it is useful to define first the condition and then try to determine its prevalence. Geropsychiatry has been defined by Verwoerdt[14] as the "psychiatry of late life, encompassing the behavioral sciences, psychodynamic concepts, and psychiatric practice with reference to the aging personality and the mental disorders of late life." In Great Britain, the more frequently used term *psychogeriatrics* has been defined by Pitt[15] to mean "that branch of psychiatry which is concerned with the whole range of psychological disorders developing in the senium (i.e., after the age of 65)."

The recent growth rate of the 65-and-over population has been phenomenal. Weinberg,[16] citing the United States Bureau of the Census 1978 statistics, notes that in the 6 years between 1970 and 1976, the 65-plus population increased by 14.8% and the 85-plus group increased by 39.6%. The 1980 Census computes 25.5 million people over 65, constituting about 11.3 percent of the total population. Regrettably, only recently has organized medicine become concerned about these startling figures.

This increased survival rate of elderly people correlates numerically with what is considered to be an epidemic of senile dementia cases. In a cross-cultural survey of the literature, Gurland and Cross[17] write that the prevalence of severe dementia ranges between 2% and 5% of the elderly in the community with an additional 2% to 3% when institutionalized patients are also included. The rates of dementia rise precipitantly to more than 20% in the segment of the population above 80 years of age.

In an updated chapter, Blazer[18] points out that the increased frequency of certain psychiatric symptoms in the elderly population does not mean that there is an increase in frequency of specific psychiatric disorders. Blazer also believes that there is no conclusive evidence suggesting the need for developing diagnostic classifications specific to older people. Furthermore, he wonders if the most common psychiatric disorders found among elderly people in the community are not equally common at other stages in the life cycle.

The alarming rate of elderly suicide continues to be well documented.[19] Those most at risk are believed to be males, whites, unmarried, in lower socioeconomic classes or low-status occupations, and the lonely and isolated in urban areas.

Alcohol abuse is being increasingly recognized as a significant problem among the elderly.[20] Most elderly alcoholics first used alcohol during young adulthood, but the minority of late-onset alcoholics still underscores the stresses associated with old age. One large geriatric mental outreach program reports that 10% of the patients were diagnosed as primary substance abusers (of which 82% were alcohol problems).[21] Another study estimates that 10% to 15% of people aged 55 and over residing in the community have alcohol problems.[22]

GERIATRIC COMMUNITY MENTAL HEALTH SERVICES

Because the vast majority of elderly people reside in the community, a key component in a geriatric psychiatry service involves the provision of ambulatory mental health services. Unfortunately, as most studies have indicated, the elderly seriously underutilize community mental health services.[23,24] Approximately 4% to 5% of the total number of community mental health patients are aged 65 and older, and private psychiatrists spend about 2% of their professional time with the elderly.[25] One explanation given for this underutilization phenomenon is the economic discrimination against the elderly under Medicare, which allows only $250 as a year's benefits for outpatient psychiatric treatment. Less than 5% of Medicare expenditures for mental illness go for outpatient treatment.[26] Other proposed reasons for this underutilization have been that the elderly are often uninformed about community resources, isolated, homebound because of fear of crime in the neighborhood, physical incapacities, or lack of transportation. These varied reasons have inspired such recent innovations as a natural support network,[27] intergenerational neighborhood network,[28] program for treating isolated elderly patients in housing projects,[29] psychiatric consultation to community agencies,[30] and psychiatric outreach programs.[31,32]

Additionally, Waxman *et al.*[33] point out that ageism may contribute less to underutilization of psychiatric services than the elderly's own biases and fears about mental disorders and mental health professionals. The elderly overwhelmingly prefer to be treated for mental problems by their family physicians rather than by mental health professionals. Despite the confidence of the elderly in family physicians. these general practitioners may fail to recognize, diagnose, and treat mental disorders.[34] Interestingly, the 1975 National Ambulatory Medical Survey found that among all age groups, excluding persons between 15 to 24 years of age, the elderly were seen least frequently during the year by office-based physicians, and that these physicians were unable to provide the kind of comprehensive care the elderly required.[35]

A community-oriented geropsychiatric clinic that combines geriatric psychiatry services with a social gerontologic program may be able to provide a comprehensive approach to elderly mental patients. The initial legal impetus to provide such community program was written into Public Law 94-63 in 1975. This law required community mental health centers to implement 12 services, instead of the previous five services; services for the aged were included among the new services. With the more recent tightening of federal funding, and the scarcity of geropsychiatric resources, many of these services for the aged have had to pare down their ambitious plans. Consequently there are many variations of geropsychiatric community centers with no uniformity of types of service.

Fundamentally, the combination of the normal aging process[36] with physical, psychological, and social problems common to the elderly presents a formidable challenge to the staff. One recently established geropsychiatry center was based on three principles: a biopsychosocial approach, structuring of services to coincide with the individual's changing needs and linkages between the informal case system and professional service system.[37] Another community-based program located in Toronto emphasizes giving indirect service through consultation, education, coordination, and collaboration with individuals and agencies dealing directly with the elderly. The idea is to be highly mobile and available to provide support and expertise to those categories already in place.[38] There is a specialized clinic that provides evaluations and recommendations primarily for persons suffering from dementing illnesses and their families.[39] Also in the geriatric literature is one example of a community-based family and patient group program for Alzheimer's disease,[40] and at least two examples of adult day care centers for dementia patients.[41,42]

The Services for the Elderly Program of Crotona Park Community Mental Health Center, which is sponsored by and fully integrated with the Department of Psychiatry of Bronx-Lebanon Hospital Center and is located in an inner-urban area, is a geropsychiatric clinic that also provides a variety of individual, group, activities, socialization, and recreational programs. Recently a dementia group has been started to assist the patient and family members in coping with the illness. Each patient is carefully evaluated in a biopsychosocial framework and assigned a therapist who provides case management services. A complete medical work-up and report is mandatory. The staffing pattern at the center is a full-time attending psychiatrist, full-time psychologist, and rotating resident psychiatrist; three full-time social work staff members; one full-time occupational therapist; a part-time neuropsychologist; a neurologist one day a week; and a full-time masters-level registered nurse.

There are currently 170 elderly mental patients on the active rolls. These 170 patients represent approximately 5.4% of the total number of ambulatory patients receiving psychiatric treatment at Crotona Park Community Mental Health Center of Bronx-Lebanon Hospital Center, which is somewhat less than the 6.8% that is the estimated proportion of the over-65 population in the hospital's service area. Interestingly, this figure of 6.8% is among the lowest percentages of elderly people in the borough and in the city. The community is comprised of predominately Hispanic and black people.

A demographic breakdown of the current patient population of the elderly program reveals the age range to be between 55 and 94 years with a mean age of 67.5 years. Of the total of 170 patients, there are three patients between 50 and 55 years of age, 35 patients 75 years and older, 4 patients 90 years and older, and the remainder between 55 and 75 years. There are 78 (46%) black; 68 (40%) Hispanic; 19 (11%) white; 2 (1%) American Indian; and 3 (2%) other patients of unspecified ethnic backgrounds. Twenty-five of the patients (15%) are male and 145 (85%) are female. All of the patients reside at home with only three exceptions. One of these three is a 60-year-old female patient who has been with the program for 5 years, and is still continuing here though she has recently entered a nursing home. She is brought to the clinic by ambulette. The primary psychiatric diagnoses of the patients in descending order are the following: Dementia and organic brain syndrome—33; anxiety disorders—33; adjustment disorders—30; schizophrenia—29; affective disorder—16; personality disorder—12; paranoid disorder—4; atypical psychosis—2; substance abuse—2; posttraumatic stress disorder—1; uncomplicated bereavement—1; other unspecified—7. An interesting finding are the minimal number of patients diagnosed with substance abuse disorders and somatoform disorders. Alcoholism at the center is acknowledged as underdiagnosed and will be the worthy subject of a more detailed study.

Obtaining the assistance of family members or other caregiving individuals in the management of elderly patients is extremely important. Contrary to earlier assumptions, most elderly are not abandoned by their adult children but live near at least one of them with whom they have frequent contact.[43] Family members need to be educated in carrying out their responsibilities,[44] guided in coping with persons with cognitive impairment[45] or dementing illnesses,[46] and given support and even psychological treatment in dealing with their own feelings and depression.[47]

By contrast, there has also been increasing awareness of the physical and mental abuse or neglect of the elderly mostly in the family.[48] The Senate and House of Representatives Select Committee on Aging of 1980 reported 500,000 to 2,500,000 such cases in the country.[49] Staff members need to become aware of the various forms of abuse or neglect[50] and case detection.[51,52] The problem is difficult to deal with because many elderly victims are unwilling to discuss their predicament. They are fearful and dependent on their caretakers. At least 16 states have therefore passed legislation mandating the reporting of suspected cases. Generally, a physician reporting in good faith a suspected case will be protected from subsequent charges of defamation or breach of confidentiality.[53] Some states such as New York have adult protective legislation but no mandatory reporting requirements. The problem of reporting is even further complicated because cases of self-neglect outnumber neglect of the elderly by their caretakers.[54]

The operation of a geriatric psychiatry clinic, although similar to a general psychiatric clinic, presents special problems such as the following:

Transportation. Many patients require special arrangements for transportation to the clinic. In our program, for example, 47% use taxicabs, 22% use ambulettes, and 31% travel on their own (usually public transportation). This means a great deal of administrative and clerical staff time is spent organizing, telephoning, scheduling, and completing forms just to get the patient to and from the clinic. For those patients who do not have insurance or the means to pay for the special transportation, the clinic must pay for the cost.

Off-site Services. Geriatric programs must provide off-site services to community agencies for case finding, consultation, and education. These off-site services to senior centers, nursing homes, church groups, housing developments, ambulatory medical and inpatient medical programs require a great deal of staff time. These activities are not considered to be patient care services and therefore are not reimburseable.

Social Services. Elderly patients require a wide scope of social services. These include assistance with housing, homemakers, insurance, banking, money-management, entitlements, shopping, and assistance in going from one level of care to another. Whether the services are provided directly by clinic staff or through referral, they also involve a great deal of staff time.

Insurance. As mentioned above, Medicare's coverage of outpatient psychiatric care constitutes economic discrimination with its limitation of outpatient coverage to $250 per year. In addition, Medicare criteria require that the goal of treatment intervention be an improvement in the patient's functions and not merely maintenance of the patient at the same level in order to receive reimbursement. Medicaid provides additional coverage for many other patients. In New York State, hospital-sponsored programs have a negotiated Medicaid reimbursement that is cost based. This means it should reflect the actual cost of providing services (but only on certain approved expenses). The reimbursement in New York, however, was capped a few years ago. Some services, such as home visits and collateral visits, are not reimburseable, although certainly necessary. Programs that are not hospital based receive fixed reimbursements that are less than the hospital rate. Both hospital-sponsored and independently based reimbursement rates are below the actual costs of the services and programs provided and must therefore function under severe financial constraints. For this reason, alternate sources of funding are essential to make up the differences between costs and third-party reimbursements. It is imperative therefore that the geriatric psychiatry service seek outside assistance by securing contracts or grants from governmental and philanthropic agencies.

GEROPSYCHIATRIC INPATIENT CARE

General hospitals increasingly have been assuming a central role in delivering mental health services,[55] including psychiatric treatment of the elderly. As much as 25% of all treatment episodes of the mentally ill elderly occur in the inpatient departments of general hospitals and another 15% of the episodes occur in their outpatient departments.[56] Although acute care psychiatric units have been established in even small general hospitals,[57] the growth of specialized geropsychiatric units have developed at a much slower rate. Those that advocate such acute care geropsychiatric units point out the advantages of having a specialized team of experts who can provide the kind of comprehensive medical, psychiatric, and social intervention that the elderly mental patient requires. They emphasize the importance of specialized assessment techniques[58] and sophisticated neuropsychiatric work-ups in making precise diagnoses. They also stress the expertise need to prescribe safely psychotropic medication to the elderly, many of whom are simultaneously taking other medication for physical ailments. Electroconvulsive therapy is also more frequently used than on a general psychiatric unit because with a number of elderly depressed patients it may be considered safer than antidepressant medication. The geropsychiatrist should therefore be familiar with the June 1985 recommendations of the Census development conference on electroconvulsive therapy.[59] Of course, it is extremely important also to be aware of the particular state statutes regarding ECT.

In New York, electroconvulsive therapy regulations are found in section 27.9 of the Official Compilation of Codes, Rules, and Regulations that are concerned with regulations for surgery, major medical treatment, and the use of experimental drugs or procedures. These last may be administered only with the informed consent of the patient, or of a person authorized to act in the patient's behalf after a full disclosure of potential benefits and harm. If it is not clear whether the adult patient has sufficient capacity to give informed consent, an independent opinion must be obtained from an independent qualified consultant, that is, one who is not an employee of the institution. The director will then decide whether the patient does or does not have the mental capacity to give consent, and will enter his decision and the reasons for this decision in the patient's chart. Each director is mandated to develop standard procedures to evaluate the decisions made on the mental capacity of patients to give consent. An outside qualified consultant must be a member of the review process. If the patient objects to ECT, he or she then has recourse to the procedures in section 27.8 of the same state code, a section that contains procedures through which the patient can object and appeal any form of care and treatment. The practitioner should also be familiar with some of the ethical issues relevant to this treatment modality.[60]

Some comments made by Thomas Arie,[61] an eminent British psychiatrist who helped found the psychogeriatric unit at Goodmayer Hospital in Ilford, Essex, contribute to the discussion about the relative merits of a special geropsychiatric unit. Arie clearly agrees that psychiatric services should be based in general hospitals with collaboration between psychiatrists and geriatricians. However, he does not believe that we should be dogmatic and always have joint involvement of a geriatrician and psychiatrist in predominantly psychiatric patients, nor does he feel that we have a right to be dogmatic about whether or not the psychiatric unit for elderly patients is separate from or part of the ordinary psychiatric ward. According to him, what is important is that "particular psychiatrists should take a special interest in old people."

A question of great consequence is whether or not to admit the elderly patient to the psychiatric ward at all. The abrupt removal of the patient from his or her familiar surroundings may have serious effects. A careful assessment of the patient in the community prior to admission is essential. Individuals who suffer from mild dementia can be worked up in the community and do not necessarily require psychiatric admission. In fact, some states have attempted to discourage dementia as the sole basis for commitment. That the aged are frequently misdiagnosed and can be needlessly involuntarily committed to a psychiatric hospital has been pointed out by Jones, Parlour, and Badger.[62] In a survey of patients admitted to a geriatric psychiatry unit, Turner and Sternberg[63] found that for three quarters of them it was their first admission; more than half were suffering from a functional psychiatric disorder, and a life stress was of etiological importance in 35.8% of the cases. In their study of psychogeriatric hospital admissions, Ross and Kedward[64] found that the elderly living in institutions or living alone in the community were overrepresented, whereas those patients who lived with relatives were at low risk for hospitalization.

Generally, the greatest therapeutic responses in the treatment of geropsychiatric patients occur in the first few weeks after admission. Those patients who need extended hospitalization tend to stabilize at a lower level of functioning than those who could be discharged within 3 weeks, or even to deteriorate.[65] Although good treatment results for the acute problems can be expected, at least one follow-up study of discharged patients indicated that many do poorly later on in the community. A community follow-up program including home visits and a firm institutional liaison is strongly recommended.[66] Bronx-Lebanon Hospital Center operates an acute care general psychiatric ward of 32 beds. Patients are admitted informally, voluntarily, and involuntarily either in an emergency or by two-physician certification. From January 1, 1985 through June 30, 1985 there was a total of 384 admissions to the unit. Ten of these patients were between 65 and 69 years; two were between 70 and 74 years; and only one patient was over 75 years old. When we calculate the elderly mental patient as 60 years and older, this amounts to 15 patients and 3.9% of the total. The breakdown of the grouping is as follows: sex—4 men, 11 women; ethnicity—7 blacks, 7 Hispanics, 1 white; diagnoses—5 organic brain syndromes, 1 primary degenerative dementia, 3 schizophrenias, 2 major depressions, 3 adjustment disorders of late life with depressed mood, 1 epilepsy. Discharge referrals—5 were transfered to the medical ward for further treatment; 8 were discharged to the Services for the Elderly Clinic; 1 was discharged to the Adult Day Hospital Program; and 1 was returned to the nursing home from where she had come. The longest stay on the psychiatric ward was 69 days by a 69-year-old Hispanic male with the diagnosis of primary degenerative dementia who was discharged back to the Services for the Elderly Clinic.

In terms of assessing community needs and program planning, the low percentage of elderly mental patients on a 32-bed ward would not justify establishing a separate geropsychiatric unit at this time, particularly because there is an estimated 25% shortage of acute care general psychiatric beds in the Bronx.

The major legal issues encountered on a geropsychiatric inpatient ward are similar to those on a general psychiatric ward, namely refusal to take psychotropic medication and civil commitment procedures. Additionally, because the elderly mental patient likely suffers from physical ailments, more instances will occur of refusal to take other kinds of medications or to submit to necessary medical-surgical procedures. Geropsychiatrists may also expect to be involved in guardianship procedures and competency proceedings regarding referral of some elderly patients to long-term nursing facilities. For the most part, the assessment of the young mental patient's competency to make treatment decisions is affected by his or her psychotic process. In addition to considerations of psychosis, the elderly mental patient may have to be evaluated for cognitive impairment severe enough to preclude making treatment decisions or research participation. As it relates to the elderly mental patient giving informed consent, cognitive capacity will be dealt with in more detail in a separate section.

GEROPSYCHIATRIC CONSULTATION–LIAISON IN THE GENERAL HOSPITAL

Although geriatric inpatients tend to be referred less often than young patients, the elderly still account for about 25% of the psychiatric consultations in general hospitals.[67,68] Rabins *et al.*[69] found that patients 60 years and older occupied 58.5% of the hospital beds and represented 21% of the psychiatric consultations. The amount of elderly with mental disorders in general hospitals may even be underestimated, as Schuckit *et al.*[70] found that 24% of them were not even recognized as such by their own physicians.

Warsaw *et al.*[71] described two characteristics of the acutely ill elderly that placed them at risk of progressive functional disability and predisposed them to mental health confusion when hospitalized, namely diminished physiological reserve and decreased capacity to adapt to unfamiliar surroundings. The risk of developing these disabilities was found to rise with increasing age. Lipowski[72] believes that about 30% of medical-surgical inpatients seen in psychiatric consultation are aged 65 years and older, and therefore argues that liaison psychiatry and geropsychiatry in general hospitals should integrate their programs. This would necessitate special training in geropsychiatry for liaison psychiatrists.

Most hospitalized elderly patients seen in psychiatric consultation suffer from depression, organic brain syndromes, or both. About 30% to 50% of elderly patients experience delirium during some phase of their hospitalization.[73] Krakowski[74] found the overall percentage of the most frequently diagnosed categories (according to the then used DSM-II) to be the following: organic brain syndrome, irreversible—52.2%; depressions, psychotic and neurotic—24.8%, and organic brain syndromes, reversible—14.2%. That investigator commented on how it was practically impossible to distinguish between psychotic depression and chronic organic brain syndromes without a depressive history. In a retrospective study of patients over 60 years old seen in psychiatric consultation, Rabins *et al.*[75] found cognitive impairment to be present in 54% and depression in 27% of them. In another study of 67 patients 60 years or older who were referred for psychiatric consultation, Ruskin[76] found the most common psychiatric diagnoses to be depression (24%), dementia (19%), delirium (18%), and schizophrenia (16%).

At Bronx-Lebanon Hospital, the present authors have conducted a retrospective chart survey of a total of 549 consecutive cases seen by the consultation–liaison service during the 1984 to 1985 period. Of the total (549) cases seen, 118 (21%) were patients 60 years and older. These cases were divided by age into three groups, and their primary diagnoses recorded (see Table I).

An extremely important differential diagnosis to make, writes Lipowski,[77] is that between transient cognitive disorder (delirium and pseudodelirium) and dementia, a chronic syndrome. Lipowski points out that some 5% to 20% of the diagnosed cases of delirium in the elderly have no identifiable organic factors. In these kinds of cases, where acute functional psychoses may stimulate delirium, the term *pseudodelirium* has been proposed.[78] Elderly patients with a reduction in

Table I. Patients 60 Years and Older Seen by the Consultation–Liaison Service, 1984–1985

Group 60–65 years
  Total number: 34; male: 16; female: 18
  Primary diagnoses by number of cases:
    Adjustment disorder with depressed mood: 18 (53%)
    Organic brain syndrome: 7 (20%)
    Schizophrenia: 4 (12%)
    Dementia: 1 (3%)
    Others: 4 (12%)
Group 66–74 years
  Total number: 46; male: 18; female: 28
  Primary diagnoses by number of cases:
    Adjustment disorder with depressed mood: 16 (35%)
    Organic brain syndrome: 15 (33%)
    Schizophrenia: 4 (9%)
    Dementia: 3 (6%)
    Others: 8 (17%)
Group 75 years and older
  Total number: 38; male: 10; female: 28
  Primary diagnoses by number of cases:
    Adjustment disorder with depressed mood: 15 (40%)
    Organic brain syndrome: 14 (37%)
    Dementia: 7 (18%)
    Others: 2 (5%)

Note. Consistent with the other studies mentioned above, the findings at Bronx-Lebanon Hospital show the high frequency of symptoms of depression and cognitive impairment seen in psychiatric consultations on elderly medical-surgical inpatients.

the reserve capacities of the brain are susceptible to pseudodelirium, which may accompany a functional psychiatric disorder.[79] It is also important to realize that delirium may be superimposed on a dementing illness, further confounding diagnostic acumen. The correct diagnosis of the case is vital to its management. Blank and Perry[80] have found that in the management of the patient during and after delirium, psychological processes also need to be considered. In addition to the usual behavioral and physiological matters, the geropsychiatrist should be concerned with psychological matters.

Another critical distinction that must be made is between pseudodementia and dementia. Pseudodementia has been described as a syndrome in which transient intellectual impairments occur in patients having a variety of psychiatric diagnoses.[81] Wells[82] defines pseudodementia as a "syndrome in which dementia is mimicked or caricatured by functional psychiatric illness." He lists major clinical features differentiating pseudodementia from dementia, and he concludes that with proper attention to these features, differentiation between the two can usually be made. Wells does not believe that pseudodementia is necessarily related to depressive disorders, but he notes that pseudodementia "may occur in a wide variety of psychiatric disorders and lacks diagnostic specificity." McAllister and Price,[83] on the other hand, present four cases of depressive pseudodementia, two that occurred with simultaneous dementing illnesses and two without associated dementing illnesses. These authors point out that profound cognitive impairment can occur in pure depressive illness. This fact may make it extremely difficult to differentiate true dementia from severe depressive pseudodementia even when sophisticated neurodiagnostic testing is employed. McAllister and Price therefore believe that a trial treatment for depression should be considered, and that ECT may be highly efficacious.

INFORMED CONSENT FOR TREATMENT AND RESEARCH

In one university hospital, the most common reason the geriatric patient was referred to the consultation-liaison service was for evaluation of the patient's competence to make decisions about medical treatment or to live at home.[76] The increasing proportion of the elderly in the population, with their varying degrees of cognitive decline, will necessitate correspondingly more involvement by geropsychiatrists in evaluations of competency that, along with adequate presentation of information to the patient and voluntariness, comprise informed consent.

Despite the mass of recent literature concerning competency, Drane[84] notes that the question still remains unsettled because there has been no agreement as to what the standard for competence should be. Drane proposes a solution based not on any one standard but on a sliding scale in which the competency standard changes as the medical decision itself (the task) changes. Roth *et al.*[85] characterize the variety of tests of competency employed by clinicians as comprising five categories: (a) evidencing a choice, (b) reasonable outcome of choice, (c) choice based on rational reasons, (d) patient's ability to understand, and (e) actual understanding. But in their conclusion, they admit that "the search for a single test of competency is a search for a Holy Grail . . . there is no magical definition of competency to make decisions about treatment, the search for an acceptable test will never end." And in a later article, Applebaum and Roth[86] point out that, regardless of the tests of competency used, a number of clinical factors, including the patient's psychodynamics, the kind of information conveyed to and by the patient, changes in the patient's mental status over time, and the setting in which consent is obtained, may profoundly affect the outcome of the competency assessment.

In their 1982 *Making Health Care Decisions*, The President's Commission for the Study of Ethical Problems in Medicine and Biomedical and Behavioral Research argued that in evaluating the patient's capacity to make choices about health care treatment, the determination should be based on three considerations:

> the abilities of the patient, the requirements of the task at hand, and the consequences to the patient that are likely to flow from the decision. The individual must have sufficiently stable and developed personal values and goals, an ability to communicate and understand information adequately, and an ability to reason and deliberate sufficiently well about the choices.[87]

It should be noted at this point that the diagnosis of senile dementia does not necessarily mean that the individual is incompetent to give informed consent for either treatment or research. Stanley[88] specifies five legally derived standards of competency related to giving informed consent to treatment: (a) evidencing a choice; (b) comprehension of consent information; (c) quality of reasoning; (d) appreciation of the situation; and (e) reasonableness of the treatment decision. After describing these standards, Stanley[89] discusses the effect of varying degrees of dementia on the individual's competence according to each of these standards. Stanley recommends these same standards for the determination of the patient's competency to consent to research, although she singles out the final competency test (reasonableness of the treatment decision) as being problematic in the context of research because of the inherent experimental nature of research.

In empirical studies of different groups of elderly patients and groups of geriatric patients matched with young medical patients,

Stanley *et al.*[90] found that the aging process itself may lead to some diminution of competency in elderly medical patients, and that dementia in the elderly can further compromise competence. Although these studies were conducted in the research context, the cognitive capacities required to consent to treatment or research are similar. Significantly, the question of the patient's competency to make treatment decisions is not raised until he or she refuses treatment, whereas in the research context the patient must always be judged competent to participate.

Reisberg *et al.*[91] have also applied the experience they have gained with their Global Deterioration Scale to issues related to giving informed consent to participate in research. They have identified three major clinical phases: (a) an early "forgetfulness" phase; (b) an intermediate "confusional" phase; and (c) a moderately severe to very severe "dementia" phase. Each of these phases is further subdivided into seven stages. The capacity of the patient to give informed consent is affected by the extent of his or her cognitive deficits referable to a phase in the GDS.

The need to facilitate clinical research in the dementing illnesses must always be balanced against the need to protect the individual's right to self-determination. This balance is a highly complex matter involving legal and ethical issues. The National Commission for the Protection of Human Subjects of Biomedical and Behavioral Research publishes a number of reports pertaining to the protection of human research subjects. Their 1978 report contained regulations for the Institutional Review Board (IRB) and for informed consent. Amendments to these regulations were published in the *Federal Register* in January 1981 and again in March 1983.[92] In an effort to address specific ethical and legal issues, the National Institute on Aging (NIA) has published suggested guidelines for conducting research with demented patients. Educational in purpose, these guidelines are intended to provide assistance to IRBs and researchers. They attempt to encourage research with competent patients who are favorably inclined to participate, to assure adequate assessment of competence, and to place special constraints on research involving incompetent patients: court approval would be required for participation of the mentally incapacitated subject to a high risk research procedure that has no associated direct therapeutic benefits.[93]

It is important to recognize that paternalistic overprotection that results in depriving the patient of the right of free choice to participate in a research project constitutes an easily overlooked loss of autonomy for an elderly patient.[94] In the event of a determination of incompetence to make a particular decision, the patient's rights can be continued by a substituted judgment. Ordinarily this judgment would be made by a family member, and would be consistent with what the patient would have wanted. When this information is unavailable (when no one knows what the patient would have wanted), a decision in the patient's best interest would continue the patient's rights.[87] Finally, the rights of the mentally incapacitated could be ensured if the patient had previously made a living will and set up a durable power of attorney.

THE RIGHT TO DIE

The geropsychiatrist's involvement in right-to-die issues centers primarily on evaluating the terminally ill patient's capacity to give legally effective consent. The President's Commission for the Study of Ethical Problems in Medicine and Biomedical and Behavioral Research, in their 1983 report *Deciding to Forego Life-Sustaining Treatment,* stated:

> The Commission has found no reason for decisions about life-sustaining therapy to be considered differently from other treatment decisions. A decision to forego such treatment is awesome because it hastens death, but that does not change the elements of decision-making capacity and need not require greater abilities on the part of a patient.[95]

For the mentally incapacitated, the decision made by a legally authorized surrogate should conform to the principle of substituted judgment, as this is most likely to maintain the patient's self-determination. However, if what the patient would have wanted is unknown, then the surrogate's decision should be based on the standard of the patient's best interests.

In anticipation of future mental incapacity, the Commission recommends the employment of "advanced directives"—either an "instructive directive," in the form of a living will, or a "proxy directive" in which the patient specifies whom he or she wants to make such decisions. Because the power of attorney under common law ordinarily expires when the patient (principal) becomes incapacitated, a durable power of attorney is necessary to allow the attorney (agent) to continue to act even when the patient is incapacitated. The durable power of attorney is primarily a proxy directive but it may also contain some instructive directives. The Commission recommends that advance directives be conferred legal status under state law.

Two handbooks published by the Society for the Right to Die, *Handbook of Living Will Laws 1981–1984*[96] and *The Physician and the Hopelessly Ill Patient,*[97] survey the latest legislative developments on living wills and durable power of attorneys and address some of the legal, medical, and ethical aspects of allowing the terminally ill to die. California was the first state in 1976 to enact a "Natural Death Act" (one of many descriptive phrases for the generic term *living will*). According to the Fall 1985 Newsletter of the Society there has been a marked acceleration of living will legislation; the addition of 13 new states (including Connecticut) has brought the total number of living will states in the United States to 36.

Although the statutes vary somewhat, they do have basic similarities. The statutes acknowledge the right of the patient to control decisions by an advance written directive, provide legal immunity to health care providers, specify what must occur in the patient to begin implementation, specify what is meant by life-sustaining measures. contain a form for the declaration that must be at least substantially followed and establish procedures for both executing and revoking a declaration.[97]

New Jersey, Massachusetts, and New York have not as yet enacted any living will legislation. However, the courts in these states have decided some very important cases related to such legislation. In the landmark Karen Ann Quinlan case, the New Jersey Supreme Court decided to appoint Quinlan's father as her guardian, with the power to choose to remove her respirator as an exercise of the patient's right to privacy if the doctors concluded that there was no reasonable possibility of her coming out of the coma in a "cognitive sapient state."[98] In January 1985 the New Jersey Supreme Court held in *In re Conroy*[99] that termination of life-sustaining treatment, including artifical feeding for the incompetent patient, could be permitted providing certain procedures were followed. Essential to the procedure was a determination of what the patient would have chosen were she competent. In Massachusetts, the Supreme Judicial Court, in *Saikewicz,*[100] issued an opinion on the incompetent's right to die based not on what a "reasonable person" would choose but on "what this particular patient would have done." The court also rejected the role of ethics committees as conceived in Quinlan; it considered only judges to be legally qualified to make life and death decisions. On the other hand, the Massachusetts Appeals Court in *In re Dinnerstein*[101] ruled that the decision to write a do-not-resuscitate order for a 67-year-old woman with advanced Alzheimer's disease and in a coma was a medical and not a judicial decision. The court reasoned that, unlike this case, Saikewicz was a patient who could not expect to live for some time longer in a conscious cognitive state; moreover, he had never been competent and therefore had never been capable of expressing his wishes. In New York, in *In re Eichner,*[102] Brother Fox had previously orally declared a desire not to be maintained by life-sustaining maneuvers if he was in a vegetative state. The New York Court of Appeals ruled that the respirator could be removed because there was clear and convincing evidence of the patient's wishes. *In re Storar*[103] involved a profoundly retarded 52-year-old state facility resident with cancer of the bladder. His mother had requested that blood transfusions be discontinued. The New York Court of Appeals reversed a lower court decision and stated that there was no way to determine what the patient would want done as he had never been competent. This case shares similarities to Massachusetts's decision in the Saikewicz case.

In July 1985, a New York court ruled in favor of the petitioner and for the acceptance of a living will as meeting the standard of "clear and convincing" evidence of what the petitioner had explicitly expressed when the patient was mentally competent. This determination, according to the judge, would permit "death to come in a natural, dignified and peaceful manner, where that is the sacred, rational knowing wish of the terminally ill patient." Mrs. Selma Saunders, the 70-year-old petitioner, was suffering from emphysema and cancer. She had prepared and executed a living will while living in Pennsylvania. When she petitioned the court, she was living with her daughter on Long Island. She wanted legal assurance before entering a hospital that her advance directive would be complied with in the event she became comatose.[104]

In August 1985 the National Conference of Commissioners on Uniform State Laws approved a Rights of the Terminally Ill Act that, if adopted by all the states, would recognize an individual's living will in states other than the state in which it was executed. In 1978, the Yale Law School Legislative Services Project sponsored by the Society for the Right to Die proposed an amendment to the Society's Model Bill to specify appointment of a proxy. This authorized individual would have the power to make treatment decisions on behalf of the living will declarant when he or she is mentally incapacitated. The Society believes that the proxy designation would effectively extend the patient's right of self-determination by being immediately available to make decisions with the physician's recommendations as the patient's medical condition changes. This proxy designation is consistent with the growing trend of appointing surrogates to make treatment decisions in the event of incompetency.[96]

Every state in the country, excluding the District of Columbia, has its own durable power of attorney statute. These statutes do not specifically deal with medical decision making, but many attorneys believe they can be applied for this matter. In New York, although there are no specific provisions on health care matters in the durable power of attorney statute, the attorney general has brought forth an opinion regarding the matter. This opinion is not bound by law, but it indicates the attorney general's view that durable powers of attorney may be used to make known the patient's decision to forgo life-sustaining treatment under certain circumstances. For the durable power of attorney to be used this way, there must be a statement authorizing the agent to communicate these decisions.[97] According to Professor John J. Regan in his presentation, "Withdrawing Life Support from the Elderly," at the Annual Tri-State Chapter Conference of the American Academy of Psychiatry and the Law on January 18, 1986, the New York State durable power of attorney statute has a "gap" regarding health care matters that is being studied by the governor's task force.

In a special article published in the *New England Journal of Medicine* and reprinted in *The Physician and the Hopelessly Ill Patient,* Wanzer *et al.*[105] offered guidelines for physicians treating hopelessly ill adult patients. The authors' considerations were based on two principles:

> The patient's role in decision-making is paramount, and a decrease in aggressive treatment of the hopelessly ill patient is advisable when such treatment would only prolong a difficult and uncomfortable process of dying.

They specify the general levels of care for the hopelessly ill patient that are considered appropriate for each particular stage of the disease process. The general levels of care they specify are: (a) emergency resuscitation; (b) intensive care and advanced life support; (c) general medical care; and (d) general nursing care and effort to make the patient comfortable. The patients who need only the fourth level of care are usually considered to be terminal.

There has been growing concern about the inappropriateness of cardiopulmonary resuscitation on terminally ill patients. Decision making in these cases is basically similar to those considerations of foregoing life-sustaining therapy in general. Because brain death usually occurs in 10 minutes with lack of oxygen there is a presumption in favor of resuscitation when there has been no advance deliberation.[95] Do-not-resuscitate guidelines have been published by the State Medical Societies of Alabama, Minnesota, New York, North Carolina, and by Los Angeles County. These guidelines are advisory only, but serve to encourage hospitals to develop their own DNR policies. Bronx-Lebanon Hospital Center has recently established a DNR protocol to protect the interests of patients and to offer guidance to physicians (Appendix A). Our hospital also intends to establish an ethics committee; among other issues, this committee will undoubtedly advise on matters pertaining to the terminally ill.

RELOCATION AND TRANSFER TRAUMA

Geropsychiatrists need to consider the possible harmful effects on the elderly person of involuntary relocation, whether within the community or to an institution. The physical and emotional harm suffered by the elderly as a result of forced relocation has been termed "transfer trauma." In fact, whether or not transfer trauma actually occurs is uncertain.

A number of conflicting studies have been published regarding morbidity and mortality rates related to relocating the elderly. Borup *et al.*[106] cite six studies that found residential home relocations resulted in increased mortality rates, twelve studies that found such relocations had no significant effect on mortality rates, and one study of relocated elderly that found a significantly reduced death rate. In their own study of 529 patients who had undergone nursing home relocations compared to 453 patients who had not been relocated, Borup *et al.* found that relocation did not increase the mortality rate. One study of the impact of forced residential relocation on elderly hotel dwellers in the community found few adverse health changes.[107] Another study though found deleterious health effects on the elderly who voluntarily relocated in the community and those whose move was involuntary.[108] A relocation from the community to an institution is particularly stressful.[109] The negative effects of temporarily relocating the elderly within the same institution may be avoided by adequately preparing the patient and family for the move.[110]

The issue of transfer trauma was considered by the Supreme Court decision in *O'Bannon v Town Court Nursing Center.*[111] In that 1980 decision, the Supreme Court reversed the Court of Appeals's decision that held the residents of a nursing home had a right to a hearing before the nursing home could be decertified for failing to meet federal standards because such a decertification meant that the residents would be forced to transfer to another facility. In reversing this decision, the majority of the court at least acknowledged that some patients might suffer from the transfer. Justice Blackmun, however, denounced the danger of transfer trauma as speculative. Annas[112] criticizes Blackmun's conclusion, which was apparently based on a single article, the Borup *et al.*[106] study cited earlier. Annas argues that "instead of routinely denying hearing courts should focus on specific individuals who may suffer specific harm." Given the conflicting literature on transfer trauma, it is incumbent on the psychiatrist to conduct as thorough an evaluation of the patient as possible before recommending any kind of a relocation. Depending on the situation. the psychiatrist should employ a variety of assessment scales pertaining to all areas of the patient's functioning. The psychiatrist should enlist the assistance of the social service and nursing department and, most importantly, try to get the patient involved in the decision making. The situation becomes even more problematic when the hospitalized patient is incompetent, and, for example, is in need of long-term placement in a nursing home but insists on returning home. In that case a judicial hearing is held to adjudicate the patient's incompetency. The legally appointed guardian is then empowered to make the decision based on the doctor's recommendation.

THE NURSING HOME

The emergence of the nursing home as a place of treatment for the elderly has challenged these institutions to provide adequate psychiatric services. Although only around 5% of the elderly reside in nursing homes, the lifetime risk of their entering such an institution may exceed 50%.[113] Contrary to previous assumptions, a significant proportion of admissions have short lengths of stay[114]; those patients who stay longer than 6 months are not likely to be discharged.[115] Consequently, it is imperative that correct diagnoses by made on those patients with reversible conditions, and that appropriate treatment be initiated immediately.

The prevalence of psychiatric disorders in nursing home patients has been reported to encompass more than half of the population; cognitive disorder is the most common disorder, followed by depression.[116] Unfortunately, Teeter *et al.*[117] found that in more than three fourths of the patients in skilled nursing homes who had significant psychiatric disorders a strikingly high percentage of diagnoses were missed or misdiagnosed. Sabin *et al.*[115] found when they thoroughly evaluated 111 patients diagnosed as having cognitive impairment, 26 of the 111 actually had treatable conditions, such as normal pressure hydrocephalus, hypothyroidism, low serum vitamin B12 level, or pseudodmentia of the depressive type. These investigators point out the pervasive lack of knowledge and negative attitude among many physicians who are responsible for the care of elderly patients.

A study to find out about conditions in nursing homes for elderly mental patients, many of whom were discharged from state hospitals to nursing and board-and-care homes, was conducted by the Joint Information Service of the American Psychiatric Association and the National Association for Mental Health and published in a volume called *Old Folks At Home*[118] in 1976. Despite the extreme variability of conditions among the facilities visited, certain trends were noted. Although a minority of the facilities visited had a great many discharged state hospitals patients with diagnoses of schizophrenia or, more commonly, organic brain disease, many of the facilities had few patients diagnosed as having a primary psychiatric disorder. Some nursing homes considered "psychiatric patients" to be only those actually discharged from psychiatric hospitals. Patients who displayed only cognitive decline were not considered psychiatric cases. There was practically no difference in the treatment given to schizophrenia patients and those having chronic organic brain syndromes. There was, however, a great deal of tolerance in the nursing homes toward problem behaviors such as wandering about or loud screaming. A great number of nursing homes complained about the difficulty they had getting physicians to come to their facilities and see patients. The researchers estimated that of approximately 1,100,000 nursing home patients, more than three quarters of them had some psychiatric disability and most of them had no contact at all with psychiatrists.

Federal Medicare regulations have mandated that every skilled nursing home must hire a licensed physician to serve as medical director. This physician usually functions as a primary care physician. Intermediate nursing care facilities, on the other hand, are not legally required to have medical directors. Patients in skilled nursing facilities must be seen by a physician at least every 30 days for the first 90 days, and then every 60 days. Patients in intermediate care facilities must be seen by the physician every 60 days.[53] Most physicians treating patients in nursing homes, however, do not keep adequate psychiatric records, optimally prescribe psychotropic medication, or make more visits depending on the patient's psychiatric condition. Nursing staff have also been found to be poorly trained in the identification and management of psychological disturbances in their patients.[117] In this kind of atmosphere, there is usually little appreciation of the full value of psychiatric services. A simple medication review may be the only request made of the psychiatric consultant. However, if this is provided, it may lead eventually to what is really needed, namely, an ongoing educational process in the institution.[119]

At Bronx-Lebanon Hospital Center, two outpatient clinic programs maintain relationships with local nursing homes. The geriatric psychiatry clinic provides case consultation to one nursing home. The alcoholism clinic accepts referrals of nursing home residents for evaluation and group treatment. These relationships serve to foster ongoing educational processes through the provision of both service and psychiatric consultation.

One nursing home near Bronx-Lebanon Hospital Center reports that approximately 75% of the residents are admitted directly from hospitals. These patients, however, are generally older and sicker and require more care than those patients who were admitted to nursing homes 10 years ago. According to that nursing home administrator, the elderly are being maintained longer at home, but once admitted to the nursing home at an advanced age are generally quite sick.

Regarding the issues of termination of life-sustaining treatment and DNR orders for the hopelessly ill, the same New York nursing home has the following policies. It will not participate in or permit the termination of life-sustaining treatment of competent patients nor of incompetent patients without court intervention. If the patient has the capacity and makes the decision to forego life-sustaining treatment, then the patient is requested to discharge himself or herself to the custody of another facility or home. When the resident's mental capacity is in question or is actually judged to be incompetent, medical care is continued. If the family disagrees and requests termination of life-sustaining treatment, the family is advised to petition the court. The living will may serve as a guide to the court, but at this time, is not a binding statement to the nursing home personnel. Concerning DNR orders for residents who have the capacity to decide, the decision is between the resident and the attending physician. When the resident does not have the capacity to decide or where family members, the resident and physician disagree about resuscitation, medically accepted resuscitation procedures are followed.

Advocates of transforming nursing homes into teaching institutions point out that this would simultaneously upgrade patient treatment and educate professionals. Libow[120] is convinced that, just as 40 years ago acute hospitals improved professionally by becoming affiliated with medical education programs, the same process could improve nursing homes today. He foresees at least one nursing home linked to every medical school, and one associated with every teaching hospital, which will make these hospitals major geriatric medical centers. Medical and allied health students will receive required training at these centers. The patients will be admitted directly from their homes for acute care by a geriatric team of specialists whose goal is to return the patient home as soon as possible. Necessarily, the sharing of staff and resources will create a close relationship between the nursing home and the hospital. But the somber realities of the present provide a sharp contrast to this optimistic view of nursing homes of the future.

Nursing home care is expensive, and the government's willingness to pay is in question. The passage in 1966 of Titles XVIII and XIX of the Social Security Law brought Medicare and Medicaid into being, which led to the rapid rise of the "nursing home industry." In 1979, the federal-state Medicaid program spent $8.8 billion, which made up 49% of the total amount spent for nursing homes. Medicare contributed 2%, private insurance paid 1.5%, and the remainder, nearly 50%, was paid by private patients.[121] Usually, the private patient must exhaust his or her own funds before he or she can qualify for Medicaid. Concerning this issue, the *New York Times*[122] reported on March 6, 1986 in a front page article about a 75-year-old woman forced to sue her 77-year-old husband, a resident in a Bronx nursing home, for financial support. When her spouse, who was suffering from severe Parkinson's disease, was hospitalized in a nursing home their financial resources were soon depleted. Medicaid allowed her to keep her own minimum Social Security benefit check and only a small proportion of her husband's income from Social Security and his Post Office pension. The alternative could be divorcing him, but the couple could not bring themselves to that. Interestingly, if the situation had been reversed and the wife was the one in the nursing home, her husband could have refused to contribute any of his Social Security or pension toward the cost of her care. Since the beginning of the Diagnostic-Related Group (DRG) program, patients are being discharged earlier from hospitals, and often require the continuation of intensive care when they enter nursing homes.[123] Medicare limits its coverage for skilled nursing care to a maximum of 100 days. Medicaid payment for skilled nursing care varies from an average of $41 a day in Texas to $90 in New York. The wide range of funding levels of the 1980 Omnibus Reconciliation Act reduced the federal contribution to the Medicaid program and allowed states to set their own rates. Most states did not make up the difference in the lost federal funding.

HOSPICES

The role of physicians in hospices remains unclear because, since the establishment of the first modern hospice at St. Christophers in London in 1967, the prevailing attitude in hospice programs has been one of disillusionment with technology and with the medical model in the care of the dying patient. This attitude, based on the recognition that in the final stages when there is no hope of a cure, what was needed was to provide comfort to both patient and family, resulted in the tendency in hospices to exclude many aspects of the medical model in their treatment of the terminally ill. Since the establishment of the first American hospice in New Haven, Connecticut in 1974, societal recognition of the special needs of the dying patient has led to the opening of an estimated 1,200 additional programs. According to the National Hospice Organization's Directory, 27% of these hospices are affiliated with hospitals, 19% are affiliated with existing home health agencies, 41% are independent, and approximately 12% are connected to multiple agencies.

Concern has been expressed about the impact of Medicare on the hospice movement.[124,125] The passage of Public Law 97-248, the Tax Equity and Fiscal Responsibility Act of 1982 (TEFRA), qualified hospice services under Medicare. There is a cap, however, of only 20% of all hospice patient days that can be spent in a hospital. The maximum payment for hospice patients is limited to 210 days and the cost up to $6,500 during any given year. Only about 200 hospice programs in the country have signed up for the Medicare program.

The remarkable growth of hospice programs in the United States, as Greer[126] points out, is related to "an accurate reading of social trends and the ability to fill a widely perceived void in the medical profession." Greer criticizes the medical profession as narrowly adhering to its "disease orientation and mechanistic approach to illness despite the obvious human and social complexities that confront it." On the other hand, he concludes that geriatricians can choose to expand their range beyond traditional confines and develop more comprehensive social and professional resources.

CONCLUSION

The care of the elderly mental patient, as is the care of geriatric patients in general, is in a transitional stage of development. New programs and approaches are evolving to meet the needs of this increasing population. Dynamic changes in administrative and legal considerations continually affect this process and help shape goals and directions. These considerations are in turn greatly influenced by the social, economic, and political contexts in which they take place. Concurrently, we are facing a period of fiscal constraints and even cutbacks. This development portends, at a minimum, challenging times ahead for medicine in general, and for geriatric services in particular.

The government is determined to reduce Medicare costs. The Gramm-Rudman-Hollings deficit reduction law that was passed to eliminate a \$220 billion federal deficit over the next 5 years[127] went into effect on March 1, 1986. Although the Supreme Court promised to hear arguments on the constitutionality of the law in mid-April and to decide on it before the summer vacation, the first spending cuts have already occurred. The next scheduled cuts are to occur in October 1986 if Congress fails to reduce the federal deficit by a specified amount. According to an *American Medical News* report,[128] President Reagan proposed to legislators a fiscal 1987 budget that would reduce Medicare expenditures by \$5.2 billion in 1987 and by around \$55 billion over the next 5 years. About one fourth of all the cuts in the 1987 budget would come from health programs—Medicare would constitute 15% of the proposed savings. Of these Medicare savings, the bulk (60%) are to be derived from reducing payments to physicians, hospitals, and other providers. About 15% of the cuts would be absorbed by Medicare patients.

Another way the government has attempted to reduce health care expenditures is with the introduction of Medicare's prospective hospital payment system based on diagnosis-related groups (DRGs). Although psychiatric units in general hospitals are so far exempted from DRGs, this situation will likely not last. A study was therefore conducted by a task force of the American Psychiatric Association[129] to ascertain the potential impact of DRGs on psychiatric patients and inpatient psychiatric units in general hospitals. The task force analyzed a large hospital discharge data base and drew the following main conclusion: DRG is a substantially inaccurate method of measuring psychiatric resources used by general hospitals. This could lead to inappropriate discharges and provide financial incentives for hospitals not to admit seriously ill patients but to prefer less ill patients. Interestingly, the task force emphasizes that there has already been a kind of cost incentive Medicare payment system established by the Tax Equity and Fiscal Responsibility Act (TEFRA) and modified by the Social Security Amendments of 1983, which applies to those psychiatric units in general hospitals and psychiatric hospitals still exempted from DRGs. One explanation given why psychiatric care was exempted from DRG is that in fiscal year 1981 approximately only $995 million out of $41 billion Medicare expenditures, which amounted to 2.4%, was paid on psychiatric services. Of this, 83% was spent on inpatient care.

In conclusion, we cannot predict the composition of future geriatric psychiatry programs because of the numerous variables that come into play. But it is clear to us that elderly patients require a great variety of different services at different times. How society is prepared to deal with these needs will ultimately reflect its sense of priorities and values.

REFERENCES

1. Deuteronomy: *The New American Bible.* New York, 1970, 5:16.
2. Kebric RB: Aging in Pliny's Letters: A view from the second century A.D. *Gerontologist* 1983; 23:538–545.
3. Foner N: Old and frail and everywhere unequal. *Hastings Center Report:* 1985; 27–31.
4. Hall GS: Senescence, *The Last Half of Life,* New York, Appleton & Company, 1922.
5. Butler RN: Age-ism: another form of bigotry. *Gerontologist* 1969; 9:243–246.
6. Butler RN: *Why Survive? Being Old in America.* New York, Harper & Row Publishers, 1975.
7. Butler RN: Psychiatry and the elderly: An overview. *Am J Psychiatry* 1975; 132:893–900.
8. Group for the Advancement of Psychiatry: *The Aged and Community Mental Health: A Guide to Program Development,* Vol.8, Series 81, New York, GAP, 1971.
9. Ford CV, Sbordone RJ: Attitudes of psychiatrists toward elderly patients. *Am J Psychiatry* 1980; 137:571–575.
10. Binstock RH: The aged as scapegoat. *Gerontologist* 1983; 23:136–143.
11. Cole TR: The 'enlightened' view of aging: Victorian morality in a new key. *The Hastings Center Report* 1983; 13:34–40.
12. Siegler M: Should age be a criterion in health care? *Hastings Center Report* 1984; 14:24–31.
13. Evans RW: Health care and the inevitability of resource allocation and rationing decisions. *JAMA* 1983; 249:2047–2053.

14. Verwoerdt A: *Clinical Geriopsychiatry.* Williams & Wilkins, Baltimore, Md, 1981.

15. Pitt B: *Psychgeriatrics. An Introduction to the Psychiatry of Old Age.* Churchill Livingstone, Edinburgh, London, Melbourne, and New York, 1982.

16. Weinberg J: Geriatric psychiatry, in Kaplan HI, Freedman AM, Sadock BJ (eds): *Comprehensive Textbook of Psychiatry* ed 3, Baltimore, Md, Williams & Wilkins, 1980. p 3024.

17. Gurland BJ, Cross PS: Epidemiology of psychopathology in old age, in Jarvik LF, Small GW (eds.): *Aging, The Psychiatric Clinics of North America,* Philadelphia, Pa, W B Saunders Company, Vol 5, 1982, p 12.

18. Blazer D: The epidemiology of psychiatric disorders in the elderly population, in Grinspoon L (ed): *Psychiatry Update, The American Psychiatric Association Annual Review,* Washington, D.C., Vol 2, 1983, Chapter 6.

19. Osgood NJ: *Suicide in the Elderly.* Rockville, Maryland, An Aspen Publication, 1985.

20. Atkinson RM: *Alcohol and Drug Abuse in Old Age,* Washington D.C., American Psychiatric Press, 1984.

21. Reifler B, Roskind M, Kethley A: Psychiatric diagnoses among geriatric patients seen in a outreach program. *J Am Geriatric Society* 1982; 30:530–533.

22. Zimberg S, Wallace J, Blume S: *Practical Approaches to Alcoholism Psychotherapy.* New York, Plenum Press, 1978.

23. Shapiro S, Skinner EA, Kessler LG, *et al.:* Utilization of health and mental health services. *Arch Gen Psychiatry* 1984; 41:971–978.

24. Baker FM, Weiner O, Levine M, *et al.:* Utilization of mental health services by the aging. *J National Med Assoc* 1984; 76:455–460.

25. Butler RN, Lewis MI: *Aging and Mental Health.* St. Louis, Mo, C V Mosby Company, 1982.

26. Mumford E, Schlesinger HJ: Economic discrimination against elderly psychiatric patients under Medicare. *Hosp Community Psychiatry* 1985; 36:587–589.

27. Snow DL, Gordon JB: Social network analysis and intervention with the elderly. *Gerontologist* 1980; 20:463–467.

28. Pynoos J, Hade-Kaplan B, Fleisher D: Intergenerational neighborhood networks: A basis for aiding the frail elderly. *Gerontologist* 1984; 24:233–237.

29. Brown R, Lieff JD: A program for treating isolated elderly patients living in a housing project. *Hosp Community Psychiatry* 1982; 33:147–150.

30. Finkel SI: Psychiatric consultation in a community agency serving the elderly. *Hosp Community Psychiatry* 1980; 8:551–554.

31. Wasson W, Ripecky J, Lazarus LW, *et al.:* Home evaluation of psychiatrically impaired elderly: Process and outcome. *Geronotologist* 1984; 24:238–242.

32. Heim P: Psychiatric outreach to the mentally disabled elderly living independently. *Psychiatric Annals* 1985; 15:673–677.

33. Waxman HM, Carner EA, Klein M: Underutilization of mental health professionals by community elderly. *Geronotologist* 1984; 24:23–30.

34. Waxman HM, Carner EA: Physicians' recognition, diagnosis and treatment of mental disorders in elderly medical patients. *Gerontologist* 1984; 24:593–597.

35. Portnoi VA: A Health care system for the elderly. *New Eng J Med* 1979; 300:1387–1390.

36. Kenney RA: Physiology of aging, in Geokas M C (ed): *The Aging Process, Clinics in Geriatric Medicine,* Philadelphia, Pa, W B Saunders Company, Vol 1, 1985.

37. Dagon EM: Planning and development issues in implementing community-based mental health services for the elderly. *Hosp Community Psychiatry* 1982; 33:137–141.

38. Wasylenki DA, Harrison MK, Britnell J, *et al.:* A community-based psychogeriatric service. *J Am Geriatr Soc* 1984; 32:213–218.

39. Reifler BV, Eisdorfer C: A clinic for the impaired elderly and their families. *Am J Psychiatry* 1980; 137:1399–1403.
40. Aronson MK, Levin G, Lipkowitz R: A community-based family/patient group program for Alzheimer's disease. *Gerontologist* 1984; 24:339–342.
41. Sands D. Suzuki T: Adult day care for Alzheimer's patients and their families. *Geronotologist* 1983; 23:21–23.
42. Panella Jr. JJ, Lilliston BA, Brush D, *et al.:* Day care for dementia patients: An analysis of a four-year program. *J Am Geriatr Soc* 1984; 32:883–886.
43. Cicirelli VG: *Helping Elderly Parents. The Role of Adult Children.* Boston, Mass, Auburn House Publishing Company, 1981.
44. Clark NM, Rakowski W: Family caregivers of older adults: Improving helping skills. *Gerontologist* 1983; 23:637–642.
45. Haley WE: A family-behavioral approach to the treatment of the cognitively impaired elderly. Gerontologist 1983; 23:18–20.
46. Mace NL, Robins PV: *The 36-Hour Day.* Baltimore, Md, The Johns Hopkins University Press, 1981.
47. Goldman LS, Luchins DJ: Depression in the spouses of demented patients. *Am J Psychiatry* 1984; 141:1467–1468.
48. Medical News. The elderly newest victims of familial abuse. *JAMA* 1980; 243:1221,1225.
49. U S Congress: *Elder Abuse: The Hidden Problem.* Committee Publication No.96-220, U.S. Government Printing Office, 1980.
50. Steuer J, Austin E: Family abuse of the elderly, *J Am Geriatr Soc* 1980; 28:372–376.
51. O'Malley TA, Everitt DE, O'Malley HC, *et al.:* Identifying and preventing family-mediated abuse and neglect of elderly persons. *Ann Intern Med* 1983; 98:998–1005.
52. Rathbone-McCuan E, Voyles B: Case detection of abused elderly parents. *Am J Psychiatry* 1982; 139:189–192.
53. Kapp MB, Bigot A: *Geriatrics and the Law.* New York, Springer Publishing Company, 1985, p 82.
54. Salend E, Kane RA, Satz M, *et al.:* Elder abuse reporting: Limitations of statutes. Gerontologist 1984; 24:61–69.
55. Keill SL: The general hospital as the core of the mental health services system. *Hosp Community Psychiatry* 1981; 32:776–781.
56. Barton WE, Barton GM: *Mental Health Administration. Principles and Practice.* New York, Human Sciences Press, Inc., 1983, Vol 2, p 452.
57. Glassote RM, Gudeman JE, Beigel A: *The Uses of Psychiatry in Smaller General Hospitals.* The Joint Information Service of the American Psychiatric Association and the National Mental Health Association, 1983.
58. Kane RA, Kane RL: *Assessing the Elderly. A Practical Guide to Measurement.* Lexington, Mass, Lexington Books, 1981.
59. Consensus Conference National Institute of Mental Health and National Istitutes of Health: Electroconvulsive Therapy. *JAMA* 1985; 254:2103–2108.
60. Salzman C: ECT and ethical psychiatry: *Am J Psychiatry* 1977; 134:1006–1009.
61. Arie T: Morale and the planning of psychogeriatric services. *Brit Med J* 1971; 3:166–170.
62. Jones LR, Parlour RR, Badger LW: The inappropriate commitment of the aged. *Bull Am Acad Psychiatry Law* 1982; 10:29–38.
63. Turner RJ, Sternberg MP: Psychosocial factors in elderly patients admitted to a psychiatric hospital. *Age Ageing* 1978; 7:171–177.
64. Ross HE, Kedward HB: Psychogeriatric hospital admissions from the community and institutions. *J Gerontol* 1977; 32:420–427.

65. Neiditch JA, White L: Prediction of short-term outcome in newly admitted psycho-geriatric patients. *J Am Geriatr Soc* 1976; 24:72–78.
66. Sadavoy J, Reiman-Sheldon E: General hospital geriatric psychiatric treatment. *J Am Geriatr Soc* 1983; 31:200–205.
67. Popkin MK, Mackenzie TB, Callies AL: Psychiatric consultation to geriatric medically ill inpatients in a university hospital. *Arch Gen Psychiatry* 1984; 41:703–707.
68. Shevitz SA, Silberfarb PM, Lipowski ZJ: Psychiatric consultation in a general hospital: A report on 1,000 referrals. *Dis Nerv Syst* 1976; 37:300–395.
69. Rabins P, Lucas MJ, Teitelbaum M, *et al.:* Utilization of psychiatric consultation for elderly patients. *J Am Geriatr Soc* 1983; 31:581–585.
70. Schuckit MA, Miller PL, Hahlbohm D: Unrecognized psychiatric illness in elderly medical-surgical patients. *J Gerontol* 1975; 30:655–660.
71. Warshaw GA, Moore JT, Friedman W, *et al.:* Functional disability in the hospitalized elderly. *JAMA* 1982; 248:847–850.
72. Lipowski ZJ: The need to integrate liaison psychiatry and geropsychiatry. *Am J Psychiatry* 1983; 140:1003–1005.
73. Chisholm SE, Deniston OL, Igrisan RM, *et al.:* Prevalence of confusion in elderly hospitalized patients. *J Gerontological Nursing* 1982; 8:87–96.
74. Krakowski AJ: Psychiatric consultation for the geriatric population in the general hospital. *Bibl Psychiatr* 1979; 159:163–185.
75. Rabins W, Lucas M, Tietelbaum M, *et al.:* Utilization of psychiatric consultation for elderly patients. *Am Geriatrics Society* 1983; 31:581–585.
76. Ruskin PE: Geropsychiatric consultation in a university hospital: a report on 67 referrals. *Am J Psychiatry* 1985; 142:333–336.
77. Lipowski ZJ: Transient cognitive disorders (delirium, acute confusional states) in the elderly. *Am J Psychiatry* 1983; 140:1426–1436.
78. Goldney R: Pseudodelirium. *Med J Aust* 1979; 1:630.
79. Roth M: The natural history of mental disorder in old age. *J Ment Sci* 1955; 101:281–301.
80. Blank K, Perry S: Relationship of psychological process during delirium to outcome. *Am J Psychiatry* 1984; 141:843–847.
81. Caine ED: Pseudodementia: current concepts and future directions. *Arch Gen Psychiatry* 1981; 38:1359–1364.
82. Wells CE: Pseudodementia. *Am J Psychiatry* 1979; 136:895–900.
83. McAllister TW, Price TRP: Severe depressive pseudodementia with and without dementia. *Am J Psychiatry* 1982; 139:626–629.
84. Drane JF: The many faces of competency. *Hastings Center Report* 1985; 15:17–21.
85. Roth LH, Meisel A, Lidz CW: Tests of competency to consent to treatment. *Am J Psychiatry* 1977; 134:279–284.
86. Appelbaum PS, Roth LH: Clinical issues in the assessment of competency. *Am J Psychiatry* 1981; 138:1462–1467.
87. President's Commission for the Study of Ethical Problems in Medicine and Bio-medical and Behavioral Research: *Making Health Care Decisions.* U.S. Government Printing Office, 1982.
88. Stanley B: Senile dementia and informed consent. *Behavioral Sciences and the Law* 1983; 1:57–71.
89. Stanley B: Competency to consent to research, in Melnick VL Dubler NN (eds): *Alzheimer's Dementia. Dilemmas in Clinical Research,* Clifton, NJ, Humana Press, 1985.
90. Stanley B, Stanley M, Pomara N: Informed Consent and geriatric patients, in Stanley B (ed); *Geriatric Psychiatry: Ethical and Legal Issues,* Washington D C, American Psychiatric Press, Inc. 1985.

91. Reisberg B, Gordon B, McCarthy M, *et al.:* Clinical symptoms accompanying progressive cognitive decline and Alzheimer's disease, in Melnick VL, Dubler NN (eds): *Alzheimer's Dementia. Dilemmas in Clinical Research*, Clifton, NJ, Humana Press, 1975.

92. Department of Health and Human Services: Final Regulations Amending Basic HHS Policy for the Protection of Human Research Subjects. *Federal Register* 1983; 48(March 4, March 8):9266, 9814.

93. Melnick VL, Dubler, NN, Weisbard, JD, *et al.:* Clinical Research in senile dementia of the alzheimer type. Suggested guidelines addressing the ethical and legal issues. *J Am Geriatr Soc* 1984; 7:531–536.

94. Ratzan RM: Being old makes you different: The ethics of research with elderly subjects. *Hastings Center Report* 1980; 10:32–42.

95. President's Commission for the Study of Ethical Problems in Medicine and Biomedical and Behavioral Research: *Deciding to Forego Life-Sustaining Treatment.* U.S. Government Printing Office, 1983.

96. Society for the Right to Die: *Handbook of Living Will Laws 1981–1984*, New York, Society for the Right to Die, 1984.

97. Society for the Right to Die: *The Physician and the Hopelessly Ill Patient*, New York, Society for the Right to Die, 1985.

98. *In re Quinlan*, 355 A. 2d 647, (N.J. Sup. Ct. 1976), T MDLR 28, 143, Cert denied, 429 U.S. 922 (1976).

99. *In re Conroy*, No.A-108 (N.J. Sup. Ct. Jan 17, 1985).

100. *Superintendent of Belchertown State School v Saikewicz*, 370 N.E. 2d 417 (Mass. Sup. Jud. Ct. 1977).

101. *In re Dinnerstein*, 380 N.E. 2d 134, 135 (Mass App. 1978).

102. *In re Eichner*, 52 N.Y.S. 2d 266 (1981).

103. *In re Storar*, 52 N.Y. 2d 363, 420 N.E. 2d 64, (1981).

104. *In the Matter of Selma L. Saunders v The State of New York*, Index No 10171/85, July 16, 1985.

105. Wanzer SH, Adelstein SJ, Cranford RE, *et al.:* The physician's responsibility toward hopelessly ill patients. *N Engl J Med* 1984; 310:955–959.

106. Borup JH, Gallego DT, Heffernan PG: Relocation and its effect on mortality. *Gerontologist* 1979; 19:135–140.

107. Eckert JK, Haug M: The impact of forced residential relocation on the health of the elderly. *J Gerontol* 1984; 39:753–755.

108. Ferraro KF: The health consequences of relocation among the aged in the community. *J Gerontol* 1982; 38:90–96.

109. Rowland KF: Environmental events predicting death for the elderly. Psychol Bull 1977; 84:349–372.

110. Shamian JN, Clarfied AM, Maclean J: A randomized trial of intra-hospital relocation of geriatric patients in a tertiary-care teaching hospital. *J Am Geriatr Soc* 1984; 32:794–800.

111. *O'Bannon v Town Court Nursing Home*, 100 Sup. Ct. 2467 (1980).

112. Annas GJ: Transfer trauma and the right to a hearing. *Hastings Center Report* 1980; 6:23–24.

113. McConnel CE: A note on the Lifetime risk of nursing home residency. *Gerontologist* 1984; 24:193–198.

114. Liu K, Manton KG: The characteristics and utilization pattern of an admission cohort of nursing home patients. *Gerontologist* 1983; 23:92–98.

115. Sabin TD, Vitug AJ, Mark VH: Are nursing home diagnosis and treatment inadequate? *JAMA* 1982; 248:321–322.

116. Rosner BW, Rabins PV: Mental illness among nursing home patients. *Hosp Community Psychiatry* 1985; 36:119–120, 128.

117. Teeter RB, Gavetz FK, Miller WR, *et al.:* Psychiatric disturbances of aged patients in skilled nursing homes. *Am J Psychiatry* 1976; 133:1430–1434.

118. Glasscote RM, Beigel A, Butterfield Jr A, *et al.: Old Folks At Home.* Washington DC, Joint Information Service of The American Psychiatric Association and the Nation Association for Mental Health, 1976.

119. Herst L, Moulton P: Psychiatry in the nursing home, in Soreff S M (ed): *Alternative to Hospitalization. The Psychiatric Clinics of North America.* Philadelphia, WB Saunders Co, 1985, Vol 8, pp 551–561.

120. Libow LS: The teaching Nursing home: past, present, and future. *J Am Geriatr Soc* 1984; 32:598–603.

121. Rango N: Nursing home care in the United States. Prevailing conditions and policy implications. *N Engl J Med* 1982; 307:883–889.

122. Sullivan R: Nursing costs force elderly to sue spouses. New York Times, Thursday, March 6, 1986, p 1.

123. Wagner L: D.R.G. side effect. Flood of patients puts new strains on nursing homes. Physicians Financial News 3 (10):1,8,9, 1985.

124. Greer DS, Mor V: How Medicare is altering the hospice movement. *Hastings Center Report* 1985; 15:5–13.

125. Bayer R, Feldman E: Hospice under the medicare wing. *Hastings Center Report* 1982; 12:5–6.

126. Greer DS: Hospice: lessons for geriatricians. *J Am Geriatr Soc* 1983; 31:67–70.

127. American Medical News: Budget bill puts Medicare in peril. May hurt care, Dr. Sammons says. Chicago, Ill, AMA, Feb. 7, 1986.

128. American Medical News: Budget cuts Medicare $55 billion in 5 years. Chicago, Ill, AMA, Feb. 21, 1986.

129. English JT, Sharfstein SS, Scherl DJ, *et al.:* Diagnosis-Related Groups and general hospital psychiatry: The APA study. *Am J Psychiatry* 1986; 143:131–139.

APPENDIX
BRONX-LEBANON HOSPITAL CENTER

DO-NOT-RESUSCITATE ORDERS

Principles and Procedures

Introduction

The Medical Board is committed to provide optimum medical care for all patients hospitalized at the Bronx-Lebanon Hospital Center. The Board recognizes that such a commitment should not be confused with a requirement to resuscitate all patients under any circumstances. The protocol described below has been devised to protect the interests of patients and their families and to offer guidance to physicians. As experience is gained it is likely that revisions will be needed to accomplish the goals of the Board and this protocol.

General Principles

1. Resuscitation is therapy. Any competent patient may refuse therapy. Thus the patient's expressed wishes are the most important determinants of whether or not resuscitative efforts are to be instituted. If a patient expresses a wish to be resuscitated or not to be resuscitated at a time when he or she is capable of evaluating his or her situation then the last such request made will be followed. If a living will exists covering the question of resuscitation then that statement may be considered the equivalent of the patient's expressed wishes. If the circumstances of the medical situation change (i.e., the advent of a useful therapy) at a time that the patient cannot participate in the decision then the do-not-resuscitate (DNR) status changes and the DNR order is rescinded.

2. A DNR determination is to be undertaken carefully and deliberately. The mechanisms outlined are designed to encourage full reflection by all parties.

3. A DNR order is to be written only if death is considered proximate and no effective therapy is available.

4. A DNR determination relates to resuscitation for cardiopulmonary arrest only. All other care will be maintained in a manner consistent with the patient's needs.

5. In the absence of a *written* DNR order resuscitative efforts will be vigorous and comprehensive.

6. All data and decisions relating to writing or changing a DNR order are to be clearly and completely documented in the progress notes of the patient's chart.

Procedure

Consent

It is the responsibility of the *attending physician* to insure that a valid informed consent is obtained. Informed consent requires that the physician explain to the patient the state of his or her disease and his or her therapeutic options. The patient must indicate verbally or by sign his or her understanding and agreement to the course of therapy or his or her refusal of such therapy.

If the patient can give consent:

If the patient is both conscious and rational he or she may give consent. Physicians may elect to discuss the choice of DNR with a patient who has not yet broached the subject. If the patient's cognitive functions are doubtful the physician should obtain psychiatric or neurologic consultation.

If the patient cannot give consent:

Patients cannot give consent in the following circumstances:

1. Patients whose cognitive functions are so impaired that they are unable

to understand the nature or state of their illness and lack of available therapeutic options.
2. All minors (below age 18). It should be noted that writing a DNR order for a minor is considered an extraordinary circumstance.

If the patient cannot give consent the attending physician must submit to a Committee, designated by the President of the Medical Board, all facts pertinent to a DNR determination. The Committee should consist of a physician, a lay representative and a nurse; the physician will chair the Committee. The Committee will make a determination that the patient is or is not competent within the meaning outlined above; that he or she is terminal and death is proximate; and that there are no useful therapeutic options. Having confirmed these factors the Committee will decide if a DNR order is appropriate and can be written. If the determination by the Committee of any of the previously described factors is negative, then no DNR order will be written. The Committee will keep a record of its determination and a summary of its reasons for its decisions for submission to the Medical Board.

If the Committee determines that a DNR order is appropriate the attending physician must obtain unanimous agreement from all next of kin in the following order:

1. For minors and unmarried adults:
   a. Surviving parents equally.
   b. Surviving adult siblings equally.
2. For married adults:
   a. Surviving spouse.
   b. Surviving children equally.
   c. Surviving parents equally.
   d. Surviving siblings equally.

If *unanimous* consent is not obtained, a DNR order shall not be written, and a court order can be sought. Familial consent can be withdrawn at any time and requires cancellation of the DNR order.

Documentation and Implementation

After obtaining informed consent (and Committee approval where appropriate) as outlined above, the *attending physician* will write in the progress notes a statement detailing the patient's disease state and therapeutic options, cognitive state, corroborative consultations where appropriate, and the scource of the consent. The attending physician will then enter the following into the patient's order: "*Do not resuscitate.*" or its equivalent. *The order is to be dated and signed.*

Daily notes thereafter shall include a statement about the current DNR status; that is, if the original circumstances and consent are still in force. Any significant change in the medical circumstances or the consent negates the DNR order. Such changes must be documented in the progress notes and the DNR order must be cancelled in the order book. Daily notes can be written by either the *attending physician* or *resident physicians.*

The *nurse in charge* will transfer the written DNR order to the Kardex after verifying that all of the requirements of the DNR protocol have been met. The nurse will include in the nursing notes of that day that the DNR order has been written and that the requirements of the protocol met. Thereafter daily nursing notes will reflect whether or not the DNR order is in effect and this information will be transmitted to each succeeding shift.

In the event a change in the patient's clinical status suggests cardiopulmonary arrest the nurse will call the appropriate house physician, request an assessment of the patient, and inform the responding house physician of the patient's DNR status. The house physician will record in the daily notes his assessment of the patient's medical status and that resuscitation was not instituted in accordance with the DNR order or that resuscitation was instituted and for what reason.

14

Abuse of the Elderly

An Overview of the Problem and Solutions

JORDAN I. KOSBERG

Those in the helping professions, in general, and psychiatrists, in particular, are often in vantage points to detect, prevent, and intervene in cases of intrafamily violence. Treatment of potential or actual perpetrators of violence against others is all too common. Common, also, is treatment of those who harm family members. Although most research, theory, and practice have focused on child abuse and spouse abuse, an increasing amount of attention has been directed to elder abuse. Indeed, whereas the 1960s were called the decade of child abuse and the 1970s the decade of spouse abuse, the 1980s have been referred to as the decade of elder abuse.

Whether in the emergency room, mental health clinic or hospital, private office, or in a patient's home, it is imperative that all mental health professionals be sensitive to the possibility of elder abuse, understand the dynamics, and be knowledgeable as to preventive and interventive resources. The insensitivity to the possibility of abuse of the elderly is not uncommon. As O'Malley *et al.*[1] state:

> Unexplained trauma, neglected medical problems, failure to thrive, malnutrition, and misuse of medications may be manifestations of family-medicated abuse and neglect of the non-institutionalized elderly person. Physicians frequently overlook or misdiagnose this form of family violence.

Intervention into the problem does not occur.

JORDAN I. KOSBERG • Department of Gerontology, College of Social and Behavioral Sciences, University of South Florida, Tampa, FL 33620.

It is the purpose of this chapter to present an overview of the problem of elder abuse and to discuss ethical and legal issues that are related to the problem's detection, prevention, and intervention. Whether the patient is the abused, the abuser, or a collateral, it is important that psychiatrists and other mental health professionals are as knowledgeable about elder abuse as they are about other forms of abusive behavior.

GENERAL VULNERABILITY

There are many reasons why the elderly are especially vulnerable to criminal or abusive behavior. Some of the reasons pertain to social values and attitudes toward the elderly, and others pertain to their physical and economic needs. Still other factors in the vulnerability of the aged are related to their social and psychological losses.

The elderly are likely to live alone, and this is especially true for elderly women. Combining this fact with living in a high crime rate area increases the vulnerability of the elderly and the probability of isolation (anticipating crime and therefore fear of leaving one's dwelling) or actual commission of crime against the elderly.

"Older people have diminished physical strength and stamina; hence, they are less able to defend themselves or to escape from threatening situations."[2] Related to this is the fact that they are likely to suffer from physical disabilities and have impairments affecting hearing, sight, touch, and mobility. The result is a lessened ability to resist abuse, whether the aggressor is a stranger or a relative.

There is a likelihood that the elderly will be vulnerable to crime because they live in high crime rate neighborhoods and are "in closer proximity to the groups most likely to victimize them—the unemployed, teenage dropouts."[2] There are at least two reasons for the fact that elderly live in high crime areas. First, their low incomes necessitate movement to low rental areas of a city. Second, there is a reluctance to leave the neighborhood (and home) where one has lived for many years, even when the neighborhood has greatly changed over the years. The economic, social, and psychological meaning of one's dwelling may be more important than the extent of crime or incongruity of population in one's neighborhood. Moreover, a person may not be able to afford to relocate to a "better" area in the community.

Many elderly persons rely on walking or on public transportation to get around in the community. Accordingly, they are more visible and vulnerable. Walking to and from public transportation, waiting at stops, getting on or off a bus or subway, and being in crowded situations all have ramifications for the possibility of accidents as well as for victimization.

Given a low fixed income, the elderly are especially vulnerable to fraudulent promises and quick wealth schemes. Poor health conditions, coupled with little if any hope for improvement, can make an elderly person vulnerable to health care quackery or schemes to evoke anxiety about health or economic security. Expensive or excessive health insurance, funeral arrangements, cemetery plots, and health devices are but a few of the many gambits to part the elderly from their financial resources. In addition, the loneliness of elderly persons makes them vulnerable to overly friendly and solicitous clerks, salespersons, or strangers. Many elderly have had their savings stolen by a variety of confidence games and unscrupulous salespersons.

"The dates of receipt of mail of monthly pension and benefit checks (and hence the dates when older people are most likely to have cash on their person or in their dwelling) are widely known."[2] Such knowledge results in mail boxes being broken into and checks stolen, robberies, and purse snatchings. Such crimes are especially likely around banks, shopping malls, and grocery stores.

Such conditions pertain to the vulnerability of the elderly, in general, and to criminal activity against the elderly in particular. There are also reasons why the elderly are especially vulnerable to abusive behavior by family, friends, or neighbors.

Elder abuse by family members can occur within the confines of the home and is therefore outside of public scrutiny. Moreover, if a problem is detected by outsiders, it is considered a family affair. Even professionals are reluctant to intervene. Further, the informal care system has been used as a panacea by those in the legal, social service, and health care systems; yet, such individuals turned to for care of often ill and dependent elderly may be unsuited, unprepared, and unmotivated to provide necessary care or may be motivated for all the wrong reasons (i.e., exploitation).

> In the eagerness to find an easy and inexpensive solution to care for an elderly person, those making referrals . . . may turn too quickly to family members without assessing the appropriateness of the family or pressures on the family to be caused by having to care for an elderly relative.[3]

In addition to these factors associated with vulnerability of the elderly to criminal or abusive behavior, there is the social perception of the elderly as a worthless. Such a negative view of the elderly can result, it is believed, in aimless and senseless, nonutilitarian crimes against the elderly. Too often one reads about muggings, beatings, psychological abuse, killings, and the like with no apparent motive. The only conclusion is that the criminal or abusive behavior was based on thrills, taking out aggression on a defenseless scapegoat, or seeking out a (perceived) valueless individual. It is further believed that if an elderly person is also perceived as a deviant (i.e., an alcoholic, drifter, handicapped, mentally ill, etc.), the greater the likelihood of becoming a victim. In American society, being dependent is also an undesirable status, and dependency is related to chronological age.

THE EXTENT OF THE PROBLEM

It is difficult to know the exact extent of abusive behavior against the aged. Little empirical work has been done on elder abuse. Even those studies that have taken place have not reached the medical profession. In 1983, O'Malley *et al.*[1] reported:

> During the past 5 years, no substantive articles and only two editorials on family-mediated abuse have appeared in any major American medical journal. Consequently, few primary care physicians are aware of the studies of this form of family violence.

Given the invisibility and underreporting of the problem of elder abuse coupled with methodological difficulties and differing definitions, conclusions from the limited number of studies on elder abuse preclude definite information about the scope of the problem. This author has concluded that a comprehensive definition of elder abuse should include the following:

1. *Passive neglect,* characterized by a situation in which the elderly person is left alone, isolated, or forgotten; the abuser is often unaware of the neglect or the consequences of the neglect, due to the abuser's lack of intelligence or lack of experience as a caregiver
2. *Active neglect,* characterized as the intentional withholding of items necessary for daily living, such as food, medicine, companionship and bathroom assistance
3. *Verbal, emotional, or psychological abuse,* characterized in situations in which the older person is called names, insulted, infantilized, frightened, intimidated, humiliated, or threatened
4. *Physical abuse,* characterized by the older person being hit, slapped, bruised, sexually molested, cut, burned, or physically restrained
5. *Material or financial misappropriation,* characterized by actions including monetary or material theft or misuse (when not being used for the benefit, or with the approval, of the elderly person)

> 6. *Violations of rights,* characterized by efforts to force an elderly person from one's dwelling and/or being forced into another setting (most often a nursing home) without any forewarning, explanation, opportunity for input, or against the older person's wishes

Self-abuse is a special type of problem, and because it is mainly done without assistance, is not included for discussion. When, however, informal caregivers are cognizant of the self-abuse and either do not intervene or, indeed, knowingly assist in the self-abuse (i.e., buying alcohol or medication, not seeking professional assistance), then the problem can be considered an example of neglectful behavior.

Several studies on elder abuse were undertaken during the late 1970s. Rathbone-McCuan[4] presented information in 1978 on the existence of intergenerational family violence and neglect affecting elderly relatives. Steinmetz[5] was also one of the first to identify and discuss the maltreatment of the elderly by their families. The University of Maryland's Center on Aging[6] undertook a study of battered elderly persons and found that 4.1% of the elderly respondents in their study reported abuse. If projected to a national population of elderly, the researchers concluded that there would be nearly one million cases of elder abuse each year. Other estimates of elder abuse range from 500,000 to 2.5 million cases per year.[7]

In an exploratory study of professional and paraprofessional encounters with abuse in Massachusetts,[8] it was found that 55% of the respondents knew of at least one incident of abuse in an 18-month period. A study in Michigan[9] was based on recollections of professional care providers (i.e., police, social workers, physicians) in five study areas. Although they found little or no direct physical abuse, 50% reported contact with passive neglect. Finally, in a study of abuse of elderly clients at the Chronic Illness Center in Cleveland, Ohio,[10] it was found that 10% of all clients were determined to be abused. Informal caregivers, mainly relatives, were the abusers.

Empirical research findings of abusive behavior by family and friends of elderly persons produced hearings by the United States House of Representatives Select Committee on Aging in June, 1979, and in 1980 a Senate–House joint hearing took place. In 1981, a National Conference on Abuse of Older Persons was held and experts' testimonies were presented to the House of Representatives Select Committee on Aging. In 1981. this Committee published *Elder Abuse: An Examination of a Hidden Problem.*[11] The Committee concluded, after hearing expert testimonies and reading research findings, that "some four percent of the nation's elderly may be victims of some sort of abuse from moderate to severe. In other words, one out of every 25 older Americans, or roughly one million older Americans, may be victims of such abuse each year." The report went on to indicate the abuse was most likely to be a recurring event rather than a single incident, and that elderly victims were likely to be the very old, women, and those dependent on others for care and protection. Generally, the abused older person lives with the abuser, who is most often a relative. The conclusion was reached that no aged person is immune to the possibility of abuse, for it is not associated with social class, educational level, race or nationality, or geographic location.

INVISIBILITY OF THE PROBLEM

Whereas the prevalence of abused elderly should be alarming, it is suspected that the estimates of the problem of elder abuse are gross underestimates. Indeed, the US House of Representatives Select Committee on Aging indicated[11] that whereas one out of every three cases of child abuse are identified and reported, only one out of six cases of elder abuse are detected and reported. The problem of elder abuse is very much invisable to professionals and the general public alike.

Elder abuse is invisible because it occurs mainly within the confines of the home and is, therefore, outside of public scrutiny. As opposed to children, who are required to go to school, etc., the older person may remain in an abusive situation because the individual may not be able to venture out of the home (due to physical and/or mental impairments) and is not required to go anywhere out of the home.

Moreover, even should the problem be detected or suspected by others, it is considered a family affair. Even professionals are reluctant to intervene into a situation that is denied by the abused and the abuser, or is explained as normal family interaction. Both the family and outsiders may believe the problem is "no one's business" but the family's.

As Rathbone-McCuan[4] states:

> Identification of and intervention in cases of physical abuse and purposeful neglect of the elderly within the context of the family are rare; because professionals in the field seem unaware that the phenomenon exists, they fail to recognize cues of willful abuse and neglect.

Indeed, to discuss intervention without first alluding to the need to address a sensitivity to detect the possible existence of abusive behavior against an older person is rather meaningless. "Without some index of suspicion on the part of the physician [or any caregiver], even obvious instances of physical abuse may be misdiagnosed."[1]

Thus, elder abuse generally goes undetected because service providers, and others, are not sensitive to its possibility. When professionals are unaware that elder abuse exists, they fail to recognize the cues of elder abuse. Even when there are obvious signs of a problem (i.e., bruises, contusions, malnourishment, etc.), there is a tendency to attribute the causes of the problems to the characteristics of the older person: poor balance, mental confusion. impaired vision, agitation, etc. Explanations for the condition of an older person, given by a family member (often the abuser), may be accepted without question by a professional. "She bumped into a door. Poor eyesight, you know!" "He forgets to take his medication." "She got confused and fell down the steps."

Those working with and for elderly persons should not expect elderly persons always to verbalize their adversities. Research on elder abuse had found that the elderly do not report their own abuse. In a study of elder abuse, Lau and Kosberg[10] found that denial was the most prevalent reaction of elderly persons, followed by resignation. Reasons for failure to report abusive behavior, and to deny its existence, by an elderly person can include the following[3]: fear of reprisals by the abuser, embarrassment about the behavior of a family member who is the abuser, anticipating that the solution to the problem will be worse than the problem (removal from one's home, institutionalization), fear that legal and criminal action might be taken against the abusing family member, believing the problem to be a family affair, one that should remain within the confines of the family, or feelings of guilt in being dependent and causing tensions and pressures on informal caregivers.

EXPLANATIONS

The causes of abuse and maltreatment of the elderly by informal care providers are complex and varied. Those in the mental health field should be aware of differing explanations for the abuse of elderly persons (mainly by family members), so that appropriate intervention and treatment can be provided. The following is a summary of the major theoretical explanations for elder abuse (or those explanations extrapolated from research findings on child or spouse abuse).

Elderly individuals may be maltreated by individuals who exhibit abnormal or deviant behavior, which includes drug addiction, alcoholism, mental illness, and senility, among others. Steele and Pollock[12] found child abusers to be impulsive, immature, and depressed, and abusive behavior displaced aggressive and sadistic inclinations. Parents may have cared for schizophrenic, retarded, or alcoholic children who became adults. "As aged parents weaken and need care, their adult children become abusing and neglectful caregivers because of an inability to make appropriate judgements and perceptions."[7]

Gelles[13] has discussed sociological interpretations of child abuse that have, it is believed, application to elder abuse. Vulnerability of children to abuse is explained as due to their lack of physical durability to withstand physical force, and the fact that they are not capable of meaningful social interaction, resulting in parent frustration. Further, the infant (or elderly person) may create stress by necessitating economic hardship or interfering with professional, occupational, or educational plans.

Intrafamily violence may be caused by stress-producing conditions. More abusing fathers than would be expected have been found to be unemployed. Another contextual factor is that child abuse is often associated with an unwanted pregnancy; that often the parents had to get married, were ill-prepared to become parents, or the unwanted child caused stress on the family. "The child may be a financial burden, an emotional burden or a psychological burden."[13] So, too, might the elderly parent be an unwanted or unexpected responsibility.

Edwards and Brauburger[14] studied the exchange system between parents and their adolescent children. They found that when exchange between them breaks down, conflict develops. This exchange incorporates a system of rewards, power, the costs of compliance, and reciprocity. Conflict was found related to coercive control techniques used to resolve problems. Whereas for adolescents increased independence from parents can produce tension and conflict, it may well be that increased dependence of elderly parents on adult children can result in conflict. The conflict, in turn, may result in maltreatment or abuse.

Adding to the understanding of exchange phenomena, Richer[15] suggested that the greater the availability of resources and alternative sources of rewards available to children "the less likely parental dictates are to be followed and the more conflict-ridden the relationship with parents will be." It is interesting to speculate whether the relationship problems of earlier times are reproduced in the old age of the parents, or whether a role reversal occurs whereby the elderly parent is expected to comply with the directions of the adult children. In this latter case, the earlier social exchange relationship is altered, if not reversed.

Justice and Duncan[16] discussed stress resulting in child abuse. They focused on life-changing events that require readjustment in the lifestyle of a person or family. When an excessive number or magnitude of such life-change events occur, a "state of life crisis" may be said to exist. The researchers found such states associated with abusing parents. They utilized 43 events requiring some readjustment by the person or family to whom the event occurred. These events include physical illness, occurrence of an accident or injury, and personal, social, economic, or interpersonal changes. Such a multicausal model of abuse focuses on life crises caused by excessive changes as a predisposing factor. The life crises, a series of change events rather than situational day-to-day problems that are often unpredictable, result in abuse. "The end state of the life crisis is a stage of exhaustion, of decreased ability to adjust and increased risk of losing control."[16]

A social-structural theory of family violence focuses, in part, on the socialization of aggression. "The theory states that parents who punish more severely produce children who are more aggressive."[17] This socialization of violence is seen for certain groups with lower educational achievements and lower socioeconomic status as well as among broken families.

The power structure of the family is also considered in relationship to family violence. Each family has a hierarchy of interpersonal relationships with super- and subordinate roles. Each family has different structures, values, and beliefs regarding power relationships. Moreover, the power structure of the larger society affects violence in the family. Societal abuse of children (through world hunger, poverty, poor education)—or any dependent population (including the aged)—can be seen to be more serious than individual abuse. Cultural values support the view that the aged are unimportant and cultural norms toward dependent populations (use of physical force) may sanction abusive behavior.

Less empirically based are efforts that focus specifically on the intergenerational problems between the elderly and their adult children. Tensions between older mothers and adult daughters have been discussed in terms of personality conflicts that are worsened by the passing of years. Failure to redefine family roles with the passing of years has been seen to result in either hostility or overt violence. It has been suggested that conflict between family members and aged relatives is more likely in situations where the family—as individuals or a unit—has difficulty coping with an elderly parent suffering from a chronic disease. Miller[18] has written about the stress to adult children, facing their own aging process and needs of their own children, from having to care for elderly parents. Conflict between generations has also been viewed as reflecting a lack of normative definition with regard to the rights and responsibilities of middle-aged children vis-à-vis their aged parents.

Finally, Garbarino[19] has used the human ecological model to explain child abuse. To oversimplify the use of the model for elder abuse, interpersonal relationships within the family are influenced by the institutional system (economic, political, legal, education, etc.), which is in turn affected by pervasive cultural values. Elder abuse may result from the combination of pressure on families (caused by events and situations that may or may not involve the elderly relative) and the incompetence of the caregiver (resulting from lack of experience, unrealistic expectations, and inability to reorder priorities concerning the gratification of personal needs). The isolation of the family increases the possibility of abusing behavior.

PREVENTION

From the limited research on elder abuse and from information extrapolated from work on child and spouse abuse, we can identify the characteristics of the aged at high risk to abusive behavior, as well as high-risk family members and family systems who are or might be caregivers for frail and vulnerable aged.[20] Prevention includes detection, family assessment, family counseling, and communitywide efforts.

Detection

Elder abuse is best detected in the setting in which it occurs: the home. As was discussed, such detection is difficult because the elderly are often "invisible" within the home and the family is viewed as sacrosanct. Outsiders are therefore reluctant to intervene. Some service providers, however, do have opportunities to visit elderly persons in their homes. Home aides, home health workers, and visiting nurses are three such examples.

Health care workers in clinics, emergency rooms, or in general hospitals may confront elder abuse cases when an older person is brought in for medical care or attention. This can be based on routine medical appointments or for an emergency (which can result directly from the abusive behavior, i.e., broken arm or contusion). Office visits to private physicians provide opportunities to detect cases of elder abuse.

The elderly may also be seen in conjunction with their applying for various forms of benefits or reestablishing eligibility for benefits, or as participants in programs and services for the aged. All agency staff, therefore, need to be sensitive to the signs of abuse.

Increasingly, social and health care professionals utilize systematic screening of elderly clients and patients for signs of elder abuse. Protocols and other screening instruments have been developed[21,22] for professionals to identify physical and behavioral symptoms of elder abuse.

Family Assessment

There is a need to assess families and individuals who are caregivers for impaired and vulnerable elderly or who are being considered to provide care to such elderly. The assessment of caregivers, or potential caregivers, can identify high-risk situations for the older person. There is a great need to assess the characteristics and capabilities of members of an informal support system who are being considered to care for elderly persons.

> Designating a relative as guardian, placing an elderly client with a child, or discharging an elderly patient to the family without adequate assessment may be viewed as a panacea by service providers but may result in great problems for the elderly person.[3]

On one hand are those individuals who are totally ill-suited to be responsible for the care of ill, elderly, and dependent persons: adult children who are alcoholic, mentally ill, senile, or deviant, and so forth. On the other hand are those situations where caring for an elderly relative will result in great pressures or burdens on the caregiver (and family constellation). Alternative placement plans can be made, and—it is hoped— elder abuse may be prevented if proper assessments are made.

Caregiving is very stressful and many caregivers "burnout" due to excessive demands placed on them. The costs of caregiving are not only financial but also social, physical, and emotional. Instruments measuring actual or anticipated family burden, such as the *Cost of Care Index* (CCI),[20] are extremely useful in working with families. The CCI systematically explores the caregiver's perceptions of the actual or potential consequences of providing care to an elderly relative by focusing on the five following areas: (a) personal and social restrictions, (b) physical and emotional health consequences, (c) the value of providing care, (d) the older person as provocateur, and (e) economic costs.

In assessing a family situation involving elder abuse, or to prevent placing an older person in an abusive situation, it is important that the assessment focus on the relationship of the aged person to the caregiver. When abuse has been detected, professionals should record every abusive incident reported or witnessed in detail. There are checklists available, such as the *Comprehensive Index of Elder Abuse* developed by Sengstock and Hwalek,[21] that provide an organized and thorough means of recording case notes so as to ensure that the interviewer will not fail to focus on all possible dimensions of abuse and neglect.

Family Counseling

In order to intervene in an abusive situation, the therapist or counselor must try to understand the roots of the problem and must focus the assessment on the caregiver and the factors that precipitated the abusive situation. It is important to know what type of relationship has existed with the patient over the years, how the caregiver was identified, to what extent their expectations have matched reality, and how well they have been able to accept their role. One also needs to understand the exact nature of their care-providing duties and whether the amount of required effort matches their skill level and physical and emotional capabilities.

Counseling should be made available during the time a family is making a decision regarding care for an elderly relative. Families can face great pressures and strains from caring for impaired individuals and need to be made aware of possible problems and the consequences on the family system, including the vulnerable older person. Some families, motivated to provide care out of guilt or pressure from others, need to recognize that such motivations do not translate necessarily into proper care needed by the older person. The family also needs to be made aware that the demands for care may increase as the condition of the older person gets worse.

Professionals need to recognize unrealistically optimistic expectations of families regarding the needs of an older person. On the other hand, some families may have overly pessimistic views that result in a hesitancy to care for an older person. Such families need to be familiarized with community resources that can assist them in the care of their elderly relative.

It is unfortunate that some families who will be caring for an impaired older person do not receive counseling on what to expect in the present or in the near and distant future. Assistance in making the decision to be or not to be a caregiver will be beneficial to the family and older person, and may in fact prevent the occurrence of elder abuse.

Communitywide Efforts

Other community efforts to be undertaken can include the requirements of family assessments, follow-ups after placement of an older person, periodic visits to older persons, and screening for abuse of elderly clients and patients.

Increased public education on the problem of elder abuse is vital, so as to sensitize professionals and the general public to the problem, to increase the probability that family members will consider the potential implications of their providing care to an elderly relative, to encourage families undergoing pressures resulting from providing care to seek assistance, and to encourage abused elderly to seek assistance.

Finally, communities need to consider a comprehensive and coordinated approach for dealing with elder abuse.[9] Such an approach should combine attention to referral systems, verification of problems, and in-home, community-based, and institutional services for the aged victims and for informal caregivers.

PROTECTIVE SERVICES AND MANDATORY REPORTING

Whereas efforts for detection, assessment, counseling. and communitywide organization may or may not exist, there are two prevalent mechanisms that have erroneously been considered as effective in dealing with elder abuse: protective services and mandatory reporting of elder abuse legislation.

Protective services for adults exist in most states in America and generally "includes the provision of both voluntary and involuntary services to a physically or mentally infirm older person in order to assist that person in carrying out the activities of normal living."[23] Protective services seeks to aid the vulnerable adult from abuse, maltreatment, self-destructive behavior, and other forms of adversity. Legal authority permits—in extreme cases—intervening in family situations, temporary removal of an older person into a shelter, and initiation of guardianship proceedings

Mandatory reporting of elder abuse is another mechanism utilized in many states. Basically, mandatory reporting legislation requires that all persons, not only health and social service workers and law enforcement officers, report suspected or known cases of abuse, neglect, or exploitation, to a central state agency. Often there is a statewide toll-free number. Mandates for an agency to investigate and arrange voluntary and involuntary services are usually included in these provisions. Procedures for provision of involuntary service to clients through court intervention vary widely. The authority vested in the investigating agency, the criteria for emerging intervention, and the protecting of due process rights are not standardized among states. The legislation often provides legal action against anyone who, knowing of elder abuse, fails to report it.

Nineteen states had abuse reporting in 1984 as a component in their laws dealing with adult protective services.[24] A number of states identify certain social services to be made available, such as respite care, alternative housing, and counseling to elderly victims of abuse. Certain elder abuse laws provide for specific funding for implementation of the law. Other statutes provide for protective placement or guardianship. As part of the complaint-investigation process, some statutes provide for petitioning the court to allow entry to the home of an older person believed to be abused. Due process safeguards, such as the right to a hearing following complaint investigation, have been included in some statutes.

In Florida, for example, a protective services worker must investigate a complaint within 24 hours of receipt of the complaint. If forcible entry is required in a life-threatening situation, a law enforcement officer accompanies the worker. If it is considered necessary, the abused person may, when authorized by a court order, be taken into custody and taken to a medical or protective services facility. Temporary placement can be ordered for 4 days by the court.

CRITICISMS OF THESE MECHANISMS

Both protective services and mandatory reporting of elder abuse presuppose the detection of a problem and available resources once abuse has been identified. As will be discussed, both the identification of the problem and actions following the identification are hardly guaranteed. Mandatory reporting and protective services have been attacked for two basic reasons: as being ineffective and as an infringement of personal liberties.[23,24,25]

As to the ineffectiveness of mandatory reporting legislation, the invisibility of the problem has been discussed (due to the sanctity of the family, privacy of the home, lack of detection by professionals, and denial by the aged). Further, such legislation has been criticized as an inexpensive and therefore politically attractive way for the state to deal with the problem.

Other critics have noted that mandatory reporting of elder abuse policies should have, as a conceptual basis, adequate community resources that can be mobilized and utilized for the benefit of the abused (and the abuser). Community resources for the aged who are abused can be seen more as an ideal than a realistic solution. As protective services and adult abuse workers can testify, often needed resources are not in existence, are inaccessible, or have lengthy waiting periods. (There will be dire consequences of the Gramm-Rudman act; the cutback of community resources will impede efforts to deal with elder abuse.) Given the official identification and labeling of abuse, and the absence of substantial community service resources, one can speculate whether the older abused person's plight might be exacerbated by the lack of treatment resources coupled with a retaliating abuser. Faulkner[23] has pointed out that mandatory reporting legislation, as a mechanism to identify new cases of abuse, has been ineffective, and cites studies that have found identified cases of abuse had already been reported to local police, welfare agencies, or the courts.

It has been found that although a state may have mandatory reporting of elder abuse legislation, many professionals in the state claim ignorance about the policy. Others appear indifferent, at best, or indicate that reporting abusive behavior would be an infringement upon the privileged communication with a patient. The fear of litigation has kept many professionals from reporting suspected cases of abused individuals. Although many states have a fine, or other penalties, for negligence in reporting suspected cases of abuse, this writer is not aware of any cases of prosecution for failure to report an elderly person who is abused or suspected of being abused.

Mandatory reporting legislation and protective services have been criticized for the involuntary intrusion into the life of an older person.

> Specifically, the legislation being proposed or adapted does not suggest the mandatory reporting of cases of suspected abuse of persons adjudicated incompetent or impaired, but proposes instead requiring such reporting for all persons over a specified age.[23]

As such, the legislation infringes on individual rights of older persons to privacy and self-determination. Further, it can be viewed as an example of ageism. Should the older person refuse assistance, competency proceedings may be initiated and result in institutionalization (and complete loss of independence).

When the abused older person is competent and understands the consequences of the adversity and resources for elimination of the problem, but refuses to leave the abusing situation, nothing can or should be done. Unlike child abuse, when the victim has no right to remain in an abusing situation, a competent older person can decide to remain in an abusing situation by refusing assistance.

CONCLUSION

This chapter has presented an overview of the problem of elder abuse. The problem is widespread, invisible, and affects all segments of society. Those who abuse the elderly are either ill-suited or incompetent to be caregivers, or they become overwhelmed by the demands of care provision. Prevention by detection, family assessment, counseling, and community coordination has not yet been effective. Protective services and mandatory reporting legislation have been criticized for being not only ineffective in preventing the existence and reoccurrence of elder abuse, but also as fostering ageism and the infringement on self-determination. What can be done about the problem of elder abuse, a problem projected to increase, as the number and proportion of the oldest of the elderly increases?

There are those who might believe that stricter penalties against those who abuse the elderly will act as a deterrent. Unfortunately, very few cases get to either domestic or criminal court for adjudication. Although statistics do not exist to document the proportion of all cases of elder abuse by those who act out of ignorance, who are mentally or emotionally impaired, who act out of impulsive or spontaneous "bursts" of frustration, who are deviants (i.e., alcoholic), it is doubtful whether stricter penalties will preclude the victimization of the aged by such abusers. There is anecdotal information on the skeptical responses of lawyers and probate court judges, among others, to the reliability of the testimonies of older persons in legal proceedings.

It seems clear that elder abuse will continue to exist, as long as ageism and violence remain on the American scene. In the meanwhile, it is imperative that those in the helping professions, especially those in the mental health field, are aware of the possibility of elder abuse, familiar with the dynamics that produce the problem and the resulting consequences, and consider—thoughtfully—the placement of frail and vulnerable elderly.

REFERENCES

1. O'Malley T, Everett ED, O'Malley HC, *et al.:* Identifying and preventing family-mediated abuse and neglect of elderly persons. *Ann Inter Med* 1983; 98(6):998–1005.
2. Goldsmith J, Thomas N: Crimes against the elderly: A continuing national crisis. *Aging* 1974; June-July: 10–13.
3. Kosberg JI: The special vulnerability of elderly parents, in Kosberg JI (ed): *Abuse and Maltreatment of the Elderly: Causes and Interventions.* Boston, Mass, John Wright-PSG, 1983, 263–276.
4. Rathbone-McCuan E: Elderly victims of family violence and neglect. *Social Casework* 1980; May: 296–304.
5. Steinmetz SK: Battered parents. Society 1978; 15:54–55.
6. Block M, Sinnot J (eds): *The Battered Elderly Syndrome: An Exploratory Study.* College Park, University of Maryland Center on Aging, 1979.
7. Rathbone-McCuan E, Hashimi J: Isolated Elders: Health and Social Interventions. Rockville, Md, Aspen Systems, 1982.
8. O'Malley HC, Segars H, Perez R, *et al.: Elder abuse in Massachusetts: A survey of professionals and paraprofessionals.* Boston, Mass, Legal Research and Services for the Elderly, 1979.
9. Douglass RL, Hickey T, Noel C: *A Study of Maltreatment of the Elderly and other Vulnerable Adults.* Ann Arbor, Mich, University of Michigan Institute on Aging, 1980.
10. Lau EE, Kosberg JI: The abuse of the elderly by informal care providers. *Aging* 1979; September-October: 10–15.
11. US House of Representatives, Select Committee on Aging: *Elder Abuse: The Hidden Problem.* US Government Printing Office, 1981.
12. Steele B, Pollock C: A psychiatric study of parents who abuse infants and small children, in Helfer, RE (ed.): *The Battered Child.* Chicago, Ill, University of Chicago Press, 1968.
13. Gelles RJ: Child abuse as psychopathology: A sociological critique and reformulations. *Amer J Orthopsychiatry* 1973; 43:611–421.
14. Edwards JN, Branburger MB: Exchange and parent–youth conflict. *J Marriage and Fam* 1973; 30:462–466.
15. Richer S: The economics of child rearing. *J Marriage and Fam* 1968; 25:462–466.

16. Justice B, Duncan DF: Life crisis as a precursor to child abuse. *Public Health Reports* 1976; 91:110–115.
17. Lystad MH: Violence at home: A review of the literature. *Amer J Orthopsychiatry* 1975; 45:328–345.
18. Miller DA: The 'sandwich' generation: Adult children of the aging. *Social Work* 1981; 26:419–423.
19. Garbarino J: The human ecology of child maltreatment: A conceptual model for research. *J Marriage and Fam* 1977; 39:721–735.
20. Kosberg JI, Cairl R: The cost of care index: A case management tool for screening informal care providers. *Gerontologist* 1986; 26(3):273–278.
21. Sengstock MC, Hwaleck M: *Comprehensive index of Elder Abuse.* Detroit, Mich, Wayne State University, 1985.
22. Quinn M, Tomita S: *Elder Abuse and Neglect: Causation, Diagnosis, and Intervention Strategies.* New York, Springer, 1986.
23. Faulkner LR: Mandating the reporting of suspected cases of elder abuse: An inappropriate, ineffective and ageist response to the abuse of older adults. *Family Law Quarterly* 1982; 16(1):69–91.
24. Salend E, Kane R, Satz M, *et al.:* Elder abuse reporting: Limitations of statutes. *Gerontologist* 1984; 24(1):61–67.
25. Reagan JJ. Protective services for the elderly: Benefit or threat, in Kosberg, JI (ed.): *Abuse and Maltreatment of the elderly: Causes and Interventions.* Boston, Mass, John Wright—PSG, 1985, 279–291.

15

Divorce and the Elderly

LARRY H. STRASBURGER AND SUE S. WELPTON

INTRODUCTION

The aging of the American population may be the most dramatic demographic change of the 20th century. Our neighbors are growing noticeably older, even if we choose to ignore our own aging. We can expect to live longer than our grandparents. Only recently, however, have we begun to appreciate the significance as well as the magnitude of this change. A longer life expectancy is appreciated by most, yet it brings with it both new problems and old ones in a new context. One problem that our older population can be expected to encounter is divorce. A painful problem of younger couples, divorce has some uniquely difficult characteristics for the elderly.

It has been projected that divorce will end nearly half the marriages of today's young adults.[1] Although divorce rates for the elderly are not expected to rise to this level, older persons will be more at risk for divorce in coming years than they used to be. Because divorce is a stressful and stigmatizing event, it and marital separation rank as the second and third most demanding life events listed on the Holmes and Rahe Schedule of recent events.[2] Both the stigma and the stress of divorce appear to be greatest among the elderly. One woman summed it up by saying, "My generation doesn't like it."[3] When this generation of elderly were married, there were fewer divorces. For most older people divorce is not an expectable outcome of marriage. Both the phe-

LARRY H. STRASBURGER • McLean Hospital, Belmont, MA 02178; Department of Psychiatry, Beth Isreal Hospital, Harvard Medical School, Boston, MA 02115.   **SUE S. WELPTON** • Brookline Council on Aging, 61 Park Street, Brookline, MA 02146.

nomenon of elderly divorce and its complications have not received widespread recognition.[4]

Part of the reason that the problem of divorce among the elderly has not been generally recognized is attributable to the psychological denial with which we defend against loss. Divorce adds to other clearly expectable losses that occur with increasing age. These include loss of work identity, social identity, family members, friends, physical capacity, economic power, and mental acuity. Divorce intensifies these losses and adds to them. Ageism, as well, contributes to the nonrecognition of the phenomenon in a way that is similar to the ignoring of sexuality among the elderly. Surprisingly little has been written about elderly divorce,[5] but there is sufficient information available to indicate that the area merits psychiatric attention.

INCIDENCE/PREVALENCE

The incidence of divorce, as well as the absolute number of divorced persons, is lower among the elderly; however, every indication is that this situation is rapidly changing. The incidence of divorce among the elderly was relatively stable during the 1960s but almost doubled between 1970 and 1975, going from 1.2 to 2.2 per thousand.[6] By 1979 approximately 3.3% of both males and females over the age of 65 were divorced. Both the incidence and the prevalence are expected to rise so that by the year 2000 there will be over one million divorced elders.[5]

There are five reasons to expect this rise. First, there is the statistical projection of current trends into the future. Second, remarriages are increasing, and remarriages are more vulnerable than first marriages to divorce. A third reason the rate will rise is an increasing acceptance of divorce among the elderly as a solution to unhappy marriages. The liberalization of divorce laws has made them easier to obtain. There is more exposure to the idea. More of their peers have divorced. A child may have divorced. A fourth reason to expect a rise in the incidence and prevalence of divorce among the elderly is the increased economic independence of women, which alters a long-standing barrier. Increasing numbers of women are in the labor force, and they are there for a longer time, allowing more access to pensions and social security. In addition, recent tax laws allow divorcing spouses to share social security and pension benefits without financial penalty. A fifth reason for the increase is the general decrease in mortality rates, which means that fewer marriages will be terminated by death prior to old age.[7]

CAUSES

The psychological developmental changes of later life provide a serious challenge to relationships between older spouses. Adapting to them requires recognition and acceptance of changes in role patterns and changes in the nature of sexuality for elderly persons. Older males shift from an investment in production, competition, and a stance of active mastery to an interest in more collaborative supportive relationships, increased tolerance for dependency, and an increasingly diffuse sexuality. Women become more aggressive in later life, less affiliative, and more managerial or political.[8] The reversal of these culturally determined role patterns in both men and women can provide a substantial challenge for adaptation. The husband who used to support his wife both physically and emotionally may now become more dependent on her as a kind of maternal authority.

An additional challenge comes from change in the role of the sexual drive as masculine sensuality becomes more diffuse and its significance shifts away from procreation and the maintenance of masculine identity.[9] Increasingly with age sex for both men and women becomes

> the opportunity for the expression of passion, affection, and loyalty; affirmation of the value of one's body and its functions, a means of self-assertion and affirmation of life, the pleasure of being touched and caressed, the defiance of the stereotype of ageing, and a continuing search for sensual growth and experience.[10]

These meanings must be assimilated by both partners after the children leave home, otherwise serious tensions will affect the relationship.

Dysfunction seems to be both cause and consequence of divorce among the elderly, though the subject has not been adequately studied. Demographic information is sparse, but one sample[4] indicates that elderly divorced persons are generally urban residents of low occupational status who have few assets and weak religious ties. Among this group it appears that divorce results from a long-standing lack of emotional gratification aggravated by some type of a precipitating event.

A typology of divorces among the elderly may include some of the following: (a) One of the marital partners becomes involved with a third party for whom he or she wishes to leave the marriage. (b) The increased stress on the relationship from the obligatory togetherness of retirement. "For better, for worse, but not for lunch" is not simply a joke. (c) Differential aging among marital partners may give rise to life-style differences that become intolerable. (d) "Mid-life crisis revisited" may occur as men, having sold businesses, find themselves with substantial assets. (e) A spousal caregiver may become exhausted after an extended period of caretaking for a mentally or physically incapacitated partner. (f) The divorce may simply formalize a long-standing de facto situation of tacitly agreed separation. This separation may have included actual partition of living quarters, residence in different houses, or even different cities. (g) A "Medicaid divorce" may be sought through which personal assets are sequestered from depletion when one partner must enter a nursing home. (h) "Terminal divorce" is a phenomenon we have recently encountered in which a dying partner resolves that despite rapidly approaching death he or she will terminate the union. This gross statement of rejection has great psychological importance for both members of the couple. "If it's the last thing I do," may be the final statement in a long-standing tragic family dialogue.

CONSEQUENCES

The consequences of divorce can be separated into social, economic, health, and psychological categories.

Social Consequences

Divorce has been both unmentioned and unmentionable for older Americans. A divorce results in a loss of social identity for the elderly couple. Couples are often seen as a single entity, and the loss of that entity may mean loss of all of its connections. The loss of status within the community is particularly difficult for women, who appear to be victims of a form of sexism. Marriage as work constitutes another loss for women particularly. Marriage is twice invested by women in a society that measures personal worth by social utility. A woman who has given her life to child rearing and homemaking becomes disenfranchised from her work identity by a divorce. A large segment of the elderly female population has had no work identity other than marriage. The low paying, low status jobs some held were not invested with work identity. They were only a way "to make ends meet."

Loss of a kinship network occurs among the elderly divorced. There is a high likelihood of reduction of both the amount and the quality of contact with family members.[3] Accessibility both to children and grandchildren, as well as in-laws, may be reduced in a divorce. The tradition of younger daughter or daughter-in-law as caretaker has been a norm in the support for an elderly individual. Divorce may result in that caretaker being alienated or overwhelmed by having to split time between two parents whose needs have increased.

Friendships may be less available to elderly divorcees. The loss of the work place makes new friends hard to find. The loss of particular friends of the spouse may be painful. All these losses of connections have social consequences as friendships reduce dependence of elderly people on formal support services. Active friendship tends to obviate reliance on social agencies.

Economic Consequences

There is a severe decline of economic well-being for elderly divorcees, with a resultant increase in demands for social welfare support. Costs increase for food, shelter, health insurance, and caretakers. Legal fees for the divorce can be unimaginably large. The sale of a house or household goods can result in a substantial investment loss. Furnishing separate living quarters adds to the bill. Health insurance may be difficult to obtain or more expensive although recent changes in law allow divorced spouses to continue to be covered by a single medical insurance policy.[11]

The economic hardships are particularly severe for women. Many elderly women have never handled a checkbook or credit before and are particularly vulnerable to the effects of inexperience. Only rarely can women effectively enter the job market over the age of 60 except as baby-sitters or minor clerical workers.

As a final blow, access to quality long-term care may be diminished by low assets. Trusts are frequently written to preserve assets by excluding payment for chronic care. Nursing homes are often unwilling to accept patients without a guarantee of one to two years' resources for private payment.

Health—Physical and Mental

The enormity of the stress of divorce is reflected in the increase of both morbidity and mortality afterwards.[10] Individuals who have been separated and divorced have a higher admission rate to psychiatric facilities, more suicide, more homicide, more alcoholism, and more automobile accidents.[12] In 1960 the death rate for male divorcees was 33% greater than for married men, and the death rate for female divorcees who were over 65 was up 7%.

There is a markedly increased mental illness rate among the elderly divorced,[13,14] as is true across the age spectrum. One observer notes that three quarters of divorced women and two thirds of divorced men had or currently have a psychiatric disorder.[15] Some psychological symptoms, of course, reflect long-standing characteristics of individuals rather than phenomena that have been induced by the divorce itself. Nevertheless, it is clear that the psychological toll of divorce in the latter years is enormous. The greatest stress appears to be for those individuals who have resisted the divorce.[4] Perhaps they had the most to lose. Widowhood followed by remarriage and divorce is the most stressful sequence.[16] The stress of divorce is greater than that of retirement,[17] and dissatisfaction with life is worse for the elderly divorced than for elderly entering into widowhood.[18]

It is clear that coping skills of all kinds may be overtaxed as a result of elderly divorce. Managing finances, elementary housekeeping, personal care, nutrition, and transportation may be overwhelming. The stigma of divorce is a part of the psychological stress for the elderly. Women suffer social censure in this regard more than men do.[5] This may be partially true because men are more likely to remarry. Another stressor may be a profound religious alienation for an older person whose religious institution does not approve of divorce. This may mean the loss of an important element of social support.

Terrible feelings of aloneness await elderly persons who divorce. Women appear to suffer even more than men in this respect. Divorced women do not go alone to restaurants or theaters.[19] Traditionally, elderly women do not go alone on vacation. This aloneness can lead to a marked atrophy of social skills.

An inability to detach from the former way of life is accompanied by the loss of a sense of future for divorced elders. These people cannot imagine what life will be like in the future. There is an uncertainty as to what to do next. It has been observed that 77% of divorced men over 50 cannot project one year into the future in order to imagine what life will be like.[19] Realistic planning becomes difficult.

There is a lower morale among the elderly divorced.[19] This lower morale has a number of sources. Reality issues are combined with loss of feelings of self-worth where such individuals look to their own personal shortcomings that caused the marital failure. Self-blame and self-pity can become important sources of decreased self-esteem. A lower satisfaction with relations in general may be the result of depression and an inadequate social support network. Our society's lack of role models and behavioral standards for normal aging is more problematic for the elderly divorced, for whom there are few behavioral models or guidelines. These uncertainties add to the confusions of aging. Ordinary situations become ambiguous in the new context of divorce, and there is a loss of perspective. Such a "betwixt and between" position is discouraging to older people, who require structure and tolerate ambiguity less well than they did earlier in life.

ROLE OF CHILDREN

There is a commonly circulating joke among divorce attorneys in which a couple in their nineties is awaiting the award of their divorce decree. Before handing it to them, the judge asks them why they waited so long to get divorced, and the couple responds by saying, "We had to wait until the children died." This story illustrates both the stigma experienced by elderly divorcees and the intergenerational conflict sparked by late-life divorces. Frequent side taking and perpetuation of dysfunctional family dynamics occurs. There may be pronounced intergenerational alienation, with attorneys reporting that children often withdraw from both partners. The divorce may interfere with middle-aged children's persisting need to idealize the parents' marriage. In other cases, children squabble over heritable assets and exhibit great concern that a parent may enter into another late-life marriage that might divert assets outside the family.

The question of who needs support is an important one. The issue of custody is reversed in elderly divorces. The issue becomes frequently the child's custody of the parent rather than vice versa. Who will take care of whom? There is a change in the "dominant generation." This probably occurs when adults of age 50 or thereabouts have parents in their seventies and children in their twenties. If a marriage dissolves after this shift in generational dominance occurs, it is a different matter and probably has substantially different effects than if it occurs when children have not become the dominant generation. The gamut of caretaking responses can range from children and stepchildren refusing to take any responsibility to care for their parents to premature assumption, shouldering a caretaking role leading to compromised functioning of a child's family, which is then blamed on the elderly parent.

ROLE OF THE CLINICIAN

The forensic clinician can become involved with elderly divorcing persons in a number of ways. The clinician has an opportunity to provide valuable services in anticipating the painful consequences and in making a preventive intervention. The counseling process should be invoked for elderly divorcing people in much the same way that premarital counseling is routine. Divorce counseling for the elderly should take into consideration the special issues unique to the elderly. The clinician may well find himself or herself treating the mental disorders that so frequently follow on the heels of an elderly divorce if they have not proceeded it. Although mediation is an area primarily undertaken by attorneys, clinicians may find a systems approach useful to the mediation of disagreements occurring within the divorce process. Familiarity with the family systems theory may enable constructive interventions using a mediation/family systems approach instead of an adversarial one.[20]

COMPETENCY AND ITS DETERMINATION

The most common entry point of the clinician into the divorce process for elderly individuals occurs when there is a suspicion that one member of the couple is not competent to proceed. In such circumstances the court will usually appoint a guardian who will act for the incompetent party. In some states the role of the guardian will be to offer a substituted judgment, an attempt to provide input to the process that the incompetent individual himself would have provided had he been able.

The legal standard for competency to divorce is quite vague in many states. In Massachusetts it is defined as "insanity,"[21] with little in the way of case law to further elucidate what the standard should be. "The judge who hears the divorce case is required to make a determination of the facts as to the insanity."[21] An action may be brought by a plaintiff committed to a mental institution.[22] States are divided on whether a suit may be undertaken on behalf of an incompetent plaintiff by a guardian, but an incompetent defendant will not be a bar to a proceeding, providing a guardian has been appointed.[23]

Competency is a socially defined threshold for a minimal acceptable level of adult performance.[24] We propose that an ideal standard should involve the evaluation of both cognitive and affective abilities. There should be cognitive knowledge as well as affective appreciation of the nature and quality of the act of divorce. The cognitive and affective abilities should function unimpaired in three specific areas:

1. Knowledge that an individual is being divorced
2. The duties, obligations, and privileges that are being relinquished with the divorce
3. The expectable consequences of the divorce including, social, psychological, economic, and health considerations

The clinical evaluation of competence to divorce will be not unlike the evaluation of other competencies. A careful clinical history should be taken, including any history of physical and/or mental illness. Physical and laboratory findings pertinent to a diagnostic understanding should be included. A careful mental status examination should be performed that will incorporate a cognitive and affective inquiry regarding the specific issues of the divorce. Where a mental or physical disorder prevents competent participation in a divorce process, the clinician must be careful to elucidate the causal linkage between the psychiatric findings and the legal standard that we have just elaborated. This process will obviate the all too frequent determination of competency made on the basis of ill-defined allegations by family members, the family physician, or the attorney.

CONCLUSION

We have seen that divorce among the elderly is and will continue to be on the increase, exacting a terrible toll from those caught up in it. The subject is much in need of further research and study. To date, clinicians have made a very limited contribution to both the needs of these individuals and the needs of the legal system. It is hoped that by careful elaboration of a new standard for competence, as well as a heightened awareness of the emotional problems that elderly divorced persons face, social workers and forensic psychiatrists may increase their future contributions to this difficult and challenging area.

REFERENCES

1. Norton AJ: Family life cycle: 1980. *J. Marriage and Fam* 1983; 45:267–75.
2. Holmes TH, Rahe RH: The social readjustment rating scale. *J. Psychosom Res* 1967; 11:213–218.
3. Johnson ES: Other mother's perceptions of their child's divorce. *Gerontologist* 1981; 21:395–401.
4. Wilson KB: Causes and consequences of divorce in late life. *Dissertation Abstracts Intl* 1983; 44(A):298.
5. DeShane MR, Wilson KB: Divorce in late life: A call for research. *J Divorce* 1981; 4(4):81–91.
6. Carlson E: Divorce rate fluctuation as a cohort phenomenon. *Popul Stud* 1979; 33:523–536.
7. Uhlenberg P, Myers M: Divorce and the elderly. *Gerontologist* 1981; 21:276–282.
8. Gutman D: The cross-cultural perspective: Notes towards a comparative psychology of ageing. In Birren JE & Schaie KW (eds): *Handbook of the Psychology of Ageing*. New York, Van Nostrand, 1977.
9. Hildebrand HP: Psychotherapy with older patients. *Brit J Med Psychol* 1982; 55:19–28.
10. Butler RM, Lewis ML: *Sex After Sixty*. New York, Harper & Row, 1978.
11. Massachusetts General Laws, Chapter 175, Section 110 I.
12. Chiriboga DA, Brierton P, Krystal S, *et al.*: Antecedents of symptom expression during marital separation. *J Clin Psychol* 1982; 38:732–741.
13. Gove W: Sex, marital status and suicide. *J Health Soc Behav* 1972; 13:204–213.

14. Gove W: Sex, marital status and mortality. *Am J Sociol* 1973; 79:45–67.
15. Briscoe CW, Smith JB, Robins E, *et al.:* Divorce and psychiatric disease. *Arch Gen Psych* 1973; 29:119–125.
16. Lopata H: *Women as Widows: Support Systems.* New York, Elsevier, 1979.
17. Chiriboga DA: Adaptation to marital separation in later and earlier life. *J Gerontol* 1982; 37:109–114.
18. Gubrium JF: Marital desolation and the evaluation of everyday life in old age. *J Marriage and Fam* 1974; 36:107–113.
19. Kitagawa EM, Hauser PM: *Differential Mortality in the United States: A Study in Socioeconomic Epidemiology.* Cambridge, Mass, Harvard University Press, 1973.
20. Jacobs JW: Divorce and child custody resolution: Conflicting legal and psychological paradigms. *Am J Psychiatry* 1986; 143:192–197.
21. Massachusetts General Laws: Chapter 208, Section 15.
22. *Turner v Bell,* 198 Tenn 232, 279 SW2d 71, cert den 350 US 842, 100 L Ed 751, 76 S Ct 83.
23. 24 Am Jur 2d. Divorce and separation. Sec 265–6.
24. Kuypers JA, Bengtson VL: Social breakdown and competence: A model of normal aging. *Hum Dev* 1973; 16:181–201.

16

Non Compos Mentis [1]

The Psychiatrist's Role in Guardianship and Conservatorship Proceedings Involving the Elderly

ROBERT LLOYD GOLDSTEIN

Sophocles, in advanced old age, was brought before a court by his son, who claimed that the elderly tragedian was incompetent to manage his own affairs. Sophocles' sole defense was to recite to the jurors a number of passages from a play he had recently completed. As a result, the case was dismissed, the son fined and the old man allowed to continue to manage his assets (and presumably to continue to write his plays).[2] Such a dramatic example of "poetic justice" strikes a deep chord within us and appears to vindicate our feelings about the dignity and autonomy of the elderly, as well as to reaffirm our commitment to the sacred right of every individual, regardless of age, to the possession and control of his person and property, free from all restraint and interference by others (unless it is established by clear and unquestionable authority of law that he is no longer able to do so).

I propose in this chapter to discuss guardianship and conservatorship proceedings for the elderly, the role psychiatrists are assigned in such legal interventions, and the criticisms and proposals for reform that have been generated in view of a widespread perception that such proceedings often oppress the elderly under the guise of affording them protection. As one commentator observed: involuntary guardianship

ROBERT LLOYD GOLDSTEIN • Department of Psychiatry, The College of Physicians and Surgeons, Columbia University, New York, NY 10032; Member of the New York State Bar.

269

and conservatorship proceedings, as they are administered in many states today, are a glaring anachronism in a society that values independence and individual self-determination.[3] We will examine our present system in the light of our Sophoclean heritage, weigh its shortcomings, and attempt to draw attention to the deeper philosophical problem presented by the emergence of a large elderly population.

THE NATURE OF GUARDIANSHIP AND CONSERVATORSHIP PROCEEDINGS

In what way do guardianship and conservatorship proceedings differ from each other? Although it is true that governing law is fragmentary and that a coherent national pattern has not yet emerged, a few general statements can be offered that apply to most jurisdictions.[4] In general, guardianships (also known in some jurisdictions as committeeships) are usually established after a judicial finding of incompetency. A formal adjudication of incompetency results in the appointment of a guardian, who thereby gains custody and control of both the property and the person of one who is found unable to care for himself. Conservatorships are said to avoid the stigma that accompanies a formal adjudication of incompetence. The principal object of conservatorship statutes is to provide protection for the property of persons who by virtue of some disability are unable to manage it by themselves.[5] The conservator's protective powers are thus typically limited to the property of the conservatee and do not extend to his person.[6] Such proceedings are viewed as an application of the *parens patriae* doctrine,[7] under which the judge becomes, in effect, a substitute parent. Because of the ostensibly beneficent and protective nature of such proceedings, usually initiated by concerned friends or relatives (or often by the state itself[8]) who seek no advantage for themselves other than the security of knowing that their loved one will not fritter away his assets and will be properly cared for, informal procedures and lax standards of adjudication have usually been the case in order to expedite the "rescue" attempt and to avoid discouraging the efforts of the petitioners. The literature suggests, however, that such a sanguine perception of conservatorship procedures is illusory and that assumptions that the proceeding is always in the best interest of the proposed ward or conservatee must be seriously questioned.[9] These concerns will be addressed in detail in a later section of this chapter. A discussion of the specific legal mechanisms and steps that are followed in both a typical guardianship proceeding and a typical conservatorship proceeding are somewhat beyond the scope of this chapter and can be found elsewhere.[10]

Although the statutes differ greatly from state to state, the gist of their provisions is similar: if advanced age, mental illness, physical incapacity, or some other statutorily enumerated disability or combination of disabilities (a) results in the individual's inability to manage his own property, or (b) renders it likely he will not be able to care for himself or his property, or (c) renders it likely that as a result his property will be dissipated, or (d) renders it likely that he might be deceived or imposed upon by artful or designing persons, then a court may be petitioned to appoint a guardian or a conservator for that individual.

THE ROLE OF THE PSYCHIATRIST

The imposition of an involuntary guardianship or even the supposedly less stigmatizing conservatorship is always a demoralizing event, one that suspends the elderly person's power to create legal relationships with others and reduces him to the status of a child in the eyes of the law.[11] He can no longer make any legally binding decisions or transactions. Aside from the total deprivation of control over property, it may also result in deprivation of a number of recognized liberty interests as well.[12]

Although protective intervention may relieve concerned relatives or mental health professionals of guilt or the feeling that nothing is being done and the situation is out of control, in fact such intervention has been shown to be unlikely to protect the health or promote the welfare of the proposed ward or conservatee.[13] Protective services for the elderly

> often means segregation from their communities, consolidation of large numbers into confined settings, and imposed supervision. Meanwhile, proponents of protective services pay requisite lip service to rights of self-determination and individual autonomy.[14]

Against such a backdrop, the psychiatrist is called upon to render an opinion as to whether the elderly individual in question is competent to take care of himself and manage his property. Much of the decision making as to incompetency has shifted from the court to the psychiatric expert. The question of guardianship and conservatorship has become more and more of a psychiatric than a legal question.

According to a typical statutory classification, an incompetent is one who, by reason of old age or mental illness, is incapable of caring for himself and/or managing his business affairs or providing for his family, or is liable to dissipate his property or become the victim of designing persons.[15] The court is supposed to make two findings in such a case: (a) that the person has the requisite mental condition (e.g., primary degenerative dementia) and (b) as a result of said mental condition, he is unable to care for himself and/or manage his property. Very often, the reliance on a psychiatric opinion overemphasizes the mental condition by itself, without giving proper attention to the individual's functional capabilities. The court often fails to make a meaningful distinction between a psychiatric diagnosis *per se* and the individual's residual capacity to be self-reliant. For example, an elderly person who manifests one or more of the characteristics of an organic brain syndrom (e.g., forgetfulness) may find these translated into the wider conclusion that he is mentally ill and therefore in need of a guardian or conservator.

In *In re Coburn,*[16] the appointment of a guardian was approved even though the elderly respondent had the mental capacity ordinarily found in a man of his years as well as a clear grasp of his current situation. The court believed that the presence of a mild form of dementia would likely progress to the point that future difficulties were likely to ensue. Thus, even in the absence of any past or present indication of dissipation of property or other difficulties, a court may not hesitate to act on the rationale that guardianships/conservatorships are intended to be preventive in nature. In *In re Sigel's Estate,*[17] the court conceded that the proposed conservatee had displayed an adequate degree of knowledge and understanding in regard to her property and investments, enabling her to manage them in a sound fashion; nonetheless, the court focused on her chronic paranoid symptomatology in determining that she was incompetent. The court explained its decision as follows:

> Appellant's chief contention is that, because her hallucinations have existed without dissipation of her estate, her ability to manage her own affairs has been demonstrated. But the very purpose of the Act is preventive and prospective in nature. It was intended to operate prospectively in order to shield mentally defectives [sic] against their own improvidence. Proof of specific instances that appellant in the past dissipated her property is not necessary.[18]

As a result, the mere existence of a psychiatric condition may often be determinative of the issue, without making further inquiry as to the effect of said condition on the elderly individual's ability to care for himself or manage his property. Psychiatrists must be alerted to the important consideration that their input in such proceedings should address the full set of statutory criteria; it is not sufficient to determine that the individual in question has the requisite mental condition. Specific scrutiny of the effect of the elderly individual's mental condition on his ability to function satisfactorily (in regard to caring for himself or managing his property) must always be undertaken as well, lest the court automatically assume that the mere presence of the mental condition suffices to make a finding of incompetence. The psychiatrist should share his decision-making process with the court, explain the basis of his conclusions in regard to the functional capacity of the elderly individual, and avoid the pitfalls of conclusory testimony based solely on diagnostic labels or on speculative predictions as to future conduct.[19]

I will refrain from a detailed discussion of each of the various mental disorders in the context of a psychiatric examination for a guardianship or conservatorship proceeding. Such issues are too basic to consider in our discussion, which is aimed at an experienced and sophisticated audience of forensic psychiatrists.[20] Suffice it to say that whatever the mental disorder in question in the elderly patient, the yardstick for incompetence would be the same: Does whatever mental disorder that is present interfere with the patient's ability to care for himself and/or manage his property? If it does, then in what specific ways does it impair functioning in this area? *The presence of mental illness by itself is insufficient to warrant a finding of incompetency requiring the appointment of a guardian or a conservator.* This is analogous to the proposition that all schizophrenics are not thereby incompetent to stand trial (or the proposition that all involuntarily committed patients are not thereby incompetent to refuse treatment).[21] The psychiatrist who engages in such determinations primarily as a proponent of a philosophical or political point of view, for example, to protect the elderly against state "oppression" at all costs because in his view preservation of civil liberties is paramount to all other considerations, is in danger of losing sight of his proper professional role and responsibilities.[22] There is no substitute for a thorough clinical evaluation and a detailed assessment of functional incapacities on an individualized case-by-case basis.

WHO BENEFITS FROM GUARDIANSHIP AND CONSERVATORSHIP PROCEEDINGS?

Many commentators would conclude that we as a society have been untrue to the classical example of Sophocles that was offered at the beginning of this chapter. Critics of the system allege that despite strong statutory language that guardianships and conservatorships are to serve the best interests of the elderly individual (which is one justification for the use of informal, nonadversary proceedings), in reality the protection of the elderly individual's *assets* appears to be the most important factor.[23] The appointment of a guardian or conservator usually assures preservation of the estate for the potential beneficiaries of the ward's (or conservatee's) affluence. Not infrequently, such incompetency proceedings are vehicles through which potential heirs preserve the assets that they hope to inherit.[24] As a result of this situation, there is concern that reforms in current procedures are called for in order to protect the alleged incompetent from his alleged protectors.[25] Suggestions for modifications of existing law that would provide for due process protections that are often lacking or insufficient would include (a) expanded notice provision, that is, giving notice to the proposed conservatee of the nature and purpose of the proceeding and a warning of the possible serious legal consequences, [26] (b) the presence of the proposed ward or conservatee in court when the issue is resolved,[27] (c) the right to effective counsel,[28] and (d) the right to considered, reviewable decisions and to actual periodic review.[29] Equally important suggestions for reform encompassing substantive changes in current laws would include deletion of all *status* considerations in incompetency proceedings (e.g., old age)[30] and the substitution of nonmedical functional tests turning on the individual's ability to provide himself with food, shelter, clothing, and medical attention and his ability to manage his property.[31]

The goal of suggested reforms in this area is simply stated: to enable the elderly individual to continue to enjoy the greatest amount of freedom consistent with his physical and mental faculties[32]; to deprive him by law of his independence to a limited extent only, that is, only to the extent that is required to prevent a substantial danger of serious harm to himself or others[33]; and to provide services of a supportive nature designed to prolong or restore his functional independence.[34]

PROMOTING THE BEST INTERESTS OF THE ELDERLY

The number of aged in the United States is growing rapidly. In 1900, the elderly (defined as those 65 years of age and older) numbered 3 million (or 4% of the population); in 1970 they numbered 20 million (or 10% of the population); by 1985, the number of persons 85 years and older alone was estimated to be almost 2 million.[35] As people are living longer and living more often into the oldest ages, the incidence of chronic disease, long-term illness and disability, which are the characteristic health problems of the elderly, also increases, along with their vulnerability and dependency on others.[36] What is the reaction of our youth-oriented culture likely to be, presented with the emergence of an ever-expanding elderly population? One writer, commenting on the oppression that is latent in many of the present systems of guardianship and conservatorship, wrote:

> The number of aged persons will certainly rise, but it is not at all certain that the value a youth-oriented culture places on its elderly citizens will similarly appreciate. The aged no longer have strong ties to the nuclear family. They no longer produce in a society that prizes and rewards productivity. Their

role is vague and superfluous. These low-value factors make it easy to relegate the interests of the elderly to the bottom of society's list of priorities, to bury them in institutions, or to hand them over to managers.[37]

In contrast, psychiatrists have shown a deep commitment to the care and treatment of the elderly. They can play an important role in guardianship and conservatorship proceedings as well and promote the best interests of the elderly by protecting such individuals from neglect, exploitation, and abuse when indicated. Likewise, they can encourage the development of maximum self-reliance and independence in the elderly when that is indicated. The court looks to psychiatrists in these proceedings for assistance in reaching the most appropriate disposition. The psychiatrist should be able to identify those individuals who are seriously impaired to the extent that they should be found incompetent. They will require a full range of protective services in view of their patent inability to care for themselves or manage their assets.[38] Equally importantly, the psychiatrist should be able to distinguish those individuals who are only partially disabled and still competent, provided that they are afforded some form of limited assistance. They should be found competent and provided the assistance they need for independent functioning.[39] Such an approach is premised on the idea of the least restrictive alternative, that is, no individual should be deprived of his independence to a greater degree than is necessary.[40] Such a considered approach would do much to promote the healthy functioning of many elderly individuals who otherwise would be unnecessarily and unjustly stigmatized by a determination of incompetency, with its attendant loss of civil rights and social rights.[41]

REFERENCES

1. The term *non compos mentis* is an ancient one, dating back at least to the time of Lord Coke. Interestingly, a leading psychiatric dictionary defines the phrase in legal terms ("mentally incapable of managing one's affairs"; Hinsie LE, Campbell RJ: *Psychiatric Dictionary*. New York, Oxford University Press, ed 3, 1960, p 501), whereas the most widely used legal dictionary defines it in psychiatric language ("embracing all varieties of mental derangement. See Insanity" [under "Insanity" appear a number of diagnostic labels at one time part of the psychiatric lexicon]; *Black's Law Dictionary*. St. Paul, MN, West Publishing Company, ed 4, 1968, p. 1200). A bewildering compendium of terms also used to denote incompetency to manage one's property or affairs has been utilized in the various state statutes. These terms include the following: incompetent, insane, lunatic, unsound mind, mental weakness, improvidence, advanced or old age, idiocy, mental illness, disease or disorder, mentally retarded, deficient or defective.
2. Hamilton E: *The Greek Way*. New York, W.W. Norton and Company, 1943.
3. Alexander GJ: Who benefits from conservatorship? *Trial Magazine* 1977; 13:30–32.

4. Rohan PJ: Caring for persons under a disability: A critique of the role of the conservator and the "substitution of judgment" doctrine. *St. John's Law Rev* 1977; 52:1–41.
5. "Formal proceedings to protect and preserve the property of mentally disabled persons substantially predated provision for their institutional confinement and care. Although the ward's property has always been well cared for, his body, depending upon the temper of the times, has at worst been subjected to tortures designed to exorcise the spirits possessing it, and at best cared for privately by friends and relatives." Horstman P: Protective services for the elderly: The limits of parens patriae. *Missouri Law Rev* 1976; 40:215.
6. Although the statutes copiously enumerate the fiduciary obligations of a guardian concerning the ward's property, rarely does such legislation impose any specific duties on a guardian towards the ward's person beyond that of "care and custody." (One possible reason for this failure to impose more specific duties is that the ward is often placed in an institution, e.g., a hospital or nursing home.)
7. "The King, as the political father and guardian of his kingdom, has the protection of all his subjects, and of their lands and goods; and he is bound, in a more peculiar manner to take care of those who, by reason of their imbecility and want of understanding, are incapable of taking care of themselves." Shelford L: *A Practical Treatise on the Law Concerning Lunatics, Idiots, and Persons of Unsound Mind, Vol 2*. London, S. Sweet, 1933, p 9.
8. The state of New York, for example, is the single largest petitioner for guardianship.
9. The protection of the assets of the ward or conservatee appears to be the most important factor. The elderly person's management of his assets comes under scrutiny at a time when he usually undergoes a major shift in spending patterns. With diminishing life expectancy and the absence of minor dependents, the elderly person may shift the emphasis from saving to spending assets and income. Thus the critical question is: *for whom* is the court attempting to safeguard these assets? *See,* Comment: The disguised oppression of involuntary guardianship: Have the elderly freedom to spend? *Yale Law J* 1964; 73:676. Unusual spending patterns may be utilized as proof of deteriorated mental capacity.
10. Regan JJ: Protective services for the elderly: Commitment, guardianship, and alternatives. *William and Mary Law Rev* 1972; 13:569–622; Hall HG: Appointment of a conservator: A useful tool. *N.Y. State Bar J* 1986; 58:29–33.
11. Every year, courts place thousands of elderly adults under involuntary guardianships and conservatorships because they are incompetent to manage their own personal or financial affairs. The legal disabilities resulting from an incompetency adjudication in a guardianship proceeding may include removal of the elderly individual's right to contract, sue or be sued, make gifts, marry, vote, hold public office, and drive. Conservatorships are often preferred as a means of avoiding the stigma that attends such a formal adjudication of incompetency. Although the appointment of a conservator does not constitute a formal determination that the conservatee is mentally incompetent, nonetheless a conservatee is limited in his contractual powers. (In some states all of his contracts are void *ab initio;* in others, his contracts are voidable at the option of the conservator.)
12. Guardianship empowers the guardian to make decisions regarding day-to-day matters, e.g., where the ward will live, with whom, what medical treatment he will have, how his money will be spent, and so on. The relationship between guardian and ward has been aptly described as analogous to the parent–child relationship. A determination of incompetency is often used in conjunction with social service programs intended to provide care and protection to the elderly; it is often the first step in the forced displacement of the elderly into undesirable custodial facilities.

13. The assumption that there is a positive effect from services to the elderly appears doubtful. Those who are the "beneficiaries" of social services, the "protectively placed" elderly, may be at even higher risk of injury or death. Confinement in an institution usually means curtailment of both the enjoyment and length of life, loss of self-esteem, of freedom, and of meaningful activity.

14. "programs and policies produce the opposite of their original intentions. . . . Social legislation is introduced with the avowed aim of improving the health and well-being of citizens; it brings instead a system of dependence and powerlessness." Wolin R: Carter and the new Constitution. *New York Rev of Books,* June 1, 1978, at 16, col. 1.

15. "Consider also the phrase, 'likely to be deceived or imposed upon by artful or designing persons.' The cueing that takes place in that phrase instructs the judge to consider not only the proposed ward and those who petition for conservatorship but also the probability of deception by other people who can be seen as artful or designing. What does that trigger? I'm not sure, but I have always supposed that it was designed to focus on Daddy's young girlfriend or Mommy's hairdresser or the new religious faith or charity that either of them have found." Alexander, *supra* note 3, at 30–31. In regard to the problem of those who are preyed upon by cults, *see* Cults, deprogramming and guardianships. *Columbia J Law & Social Problems* 1982; 17:163–286.

16. *In re Coburn,* 165 Cal. 202, 131 P. 352 (1913).

17. *In re Sigel's Estate,* 69 Pa. Super. Ct. 425, 82 A.2d 309 (1951).

18. *Id.* at 311.

19. Forensic psychiatrists, in order to maintain their credibility as scientists with something of value to contribute to the legal system, "must be able to substantiate their conclusions on the basis of the available data in such a way that their own decision making process can be clearly appreciated by the factfinder." Goldstein RL: Castles in the air: The credibility of forensic psychiatrists. *Newsletter of the American Academy of Psychiatry and the Law,* 1986; 11(2):35–37.

20. It goes without saying that close attention must be paid to such areas as reality testing, cognitive impairments, memory deficits, lapses in concentration, psychotic features, judgment, and so on. For a more detailed discussion of the impact of various specific psychiatric disorders in a related context, i.e., in the evaluation of testamentary capacity, *see* Goldstein RL: Last will and testament: Forensic psychiatry's last frontier? in Rosner R, (ed): *Critical Issues in American Psychiatry and the Law, Vol 2.* New York, Plenum Press, 1985, pp. 107–117.

21. Goldstein RL: "The fitness factory," part I: the psychiatrist's role in determining competency. *Am J Psychiatry* 1973; 130:1144–1147; Brooks A: The constitutional right to refuse antipsychotic medications. *Bull Am Acad Psychiatry Law* 1980; 8:179–221.

22. Taking an ideological and absolutist position may result in disservice to the elderly patient. One patient may need the full range of protective services in order to survive, whereas another could function quite adequately with minimum assistance and should not be found incompetent and deprived of dignity and independence thereby.

23. "[B]eneficiaries find themselves in the cynical position of being forced to plead in court for surrogate management premised on benefit to the object of the proceeding rather than, candidly, benefit to themselves." Alexander, *supra* note 3, at 30.

24. *See* note 7 and accompanying text.

25. *See* Mitchell AM: The objects of our wisdom and our coercion: involuntary guardianship for incompetents. *Southern Calif Law Rev* 1979; 52:1405–1449; Regan JJ: Protective services for the elderly: commitment, guardianship, and alternatives. *William & Mary Law Rev* 1972; 13:569–621; Hall HG: Appointment of a conservator: A useful tool. *N.Y.State Bar J* 1986; 58:29–33.

26. Notice only of the fact of a hearing is inadequate: the proposed ward or conservatee

should be advised of the serious personal and legal implications of the determination, the need to prepare a response to the allegations, the evidence to be introduced, and so on. Notice should be sent to the object of the proceedings sufficiently well in advance to permit adequate preparation.

27. In the absence of the object of the proceedings, they are usually brief and one-sided. Only the petitioner and his attorney are present and there is no cross examination, no hearsay objections, and no opposing witnesses. Proposed wards or conservatees are too often "excused" because of alleged health problems, thus permitting a hearing to be conducted and an adjudication made in their absence. The presence of the proposed ward or conservatee should be mandatory unless impossible for health reasons.

28. Without representation by counsel, the hearing becomes a meaningless formality. The prospective ward or conservatee is destined to lose without counsel to advise him of available rights and options and to prepare the best possible case to resist the imposition of a guardian or conservator.

29. One study of over 600 guardianship cases in New York State failed to discover a single appeal of the trial court finding. Quoted in Horstman, *supra,* note 4 at 260 n. 209. If the condition that led to the guardianship or conservatorship was old age, mental deterioration, or mental illness, it is unlikely that an appellate court would ever be persuaded that the reason for the guardianship or conservatorship had disappeared. In practice, the opportunity to terminate a guardianship or conservatorship, although provided for in most states' statutory scheme, is illusory. Such procedures are rarely used to restore legal competency.

30. "Now, instead of justifying intervention in a person's life by assuming he is irresponsible due to a diagnosed mental disorder, the statutes [in California] authorize the imposition of a guardianship over *anybody* who is incapable of proper conduct and responsible choices." Mitchell, *supra* note 23 at 1431–1432.

31. *See* Alexander, *supra* note 3 at 32.

32. For a discussion of the concept of "limited guardianship" and the philosophy of encouraging utilization of the guardianship process to discern and utilize guardianship alternatives that least restrict the rights of the elderly disabled person, *see* Jost DT: The illinois guardianship for disabled adults legislation of 1978 and 1979: Protecting the disabled from their zealous protectors. *Chicago-Kent Law Rev* 1980; 56:1087–1105.

33. *Ibid.* The "limited guardianship" is premised upon the idea of the least restrictive alternative.

34. *See* Horstman, *supra* note 4.

35. In a Los Angeles study conducted by the National Senior Citizens Law Center, it was discovered that 80% of the people named in conservatorship proceedings were over 65. Ninety-three percent of them were not present at trial. Ninety-seven percent of them were not represented by counsel! Quoted in Alexander, *supra,* note 3 at 32.

36. It is well known, for example, that disabling psychiatric disorders such as primary degenerative dementia affect between 2% and 4% of the population over age 65 and that the prevalence increases with increasing age, particularly after age 75.

37. Regan, *supra,* note 23 at 622.

38. *Ibid.*

39. Jost, *supra,* note 30 at 1097–1099.

40. *Ibid.*

41. Brakel R, Rock R: *The Mentally Disabled and the Law.* Chicago, Ill, University of Chicago Press, 1971, p 264.

17

Rights of Nursing Home Residents

STANLEY R. KERN

> *Don't fear the law and judges. When you know your rights the laws work for you, not against you.*
>
> Anonymous

Our country is gradually and inexorably being populated by increasingly older individuals. In 1980, 2.2 million people were 85 years and older, more than double the number in 1960. It is estimated that by the year 2000 there will be 4.9 million people over 85 and by 2030 the figure will go to 8 million. These carry important meanings for nursing homes and their residents. Nearly 25% of the over-85 population are in nursing homes compared with 1.5% of the 65 to 84 age group. About 2.1 million people will be in nursing homes in the year 2000, compared with 1.2 million 5 years ago.[1] Consequently, resident legal rights in nursing homes will affect more and more people as time goes on.

The term *nursing home* as used here refers to two types of nursing facilities: intermediate care facilities and skilled nursing facilities. The skilled nursing facility is eligible to participate in Medicare and Medicaid programs and offers a high level of nursing care. Intermediate care facilities offer a relatively lower intensity of nursing care to a less sick population and may only participate in its state's Medicaid program.[2] In

---

STANLEY R. KERN • Department of Psychiatry and Mental Health Science, New Jersey Medical School, University of Medicine and Dentistry of New Jersey, Newark, NJ 07103; Rutgers Law School, Newark, NJ 07102.

279

1974 federal regulations[3,4] were enacted requiring skilled nursing facilities to establish patient rights policies and in 1976 regulations were adopted extending similar rights to residents of intermediate care facilities. As of 1985, 30 states have enacted nursing home patient rights legislation to expand the rights guaranteed by federal law.

As an example of the type of regulations enacted by states over the past 10 years, I have chosen two states, California and New Jersey, to illustrate the rights that should be enforced.

CALIFORNIA (TITLE 22, DIVISION 5, SECTION 72527 OF CALIFORNIA NURSING HOME REGULATIONS)[5]

Each person admitted to a nursing home has the following rights, among others:

A. To be fully informed, as evidenced by the patient's written acknowledgment prior to or at the time of admission and during the stay, of these rights and of all rules and regulations governing patient conduct.

B. To be fully informed prior to or at the time of admission and during the stay of services available in the facility, and of related charges including any charges for services not covered by the facility's basic per diem rate or not covered under Titles XVIII or XIX of the Social Security Act.

C. To be fully informed by a physician of his or her medical condition, unless medically contraindicated, and to be afforded the opportunity to participate in the planning of his or her medical treatment and to refuse to participate in experimental research.

D. To refuse treatment to the extent permitted by law and to be informed of the medical consequences of such refusal.

E. To be transferred or discharged only for medical reasons or for his or her welfare or that of other patients or for nonpayment for his or her stay; to be given reasonable advance notice to ensure orderly transfer or discharge. Such actions must be documented in his or her health record.

F. To be encouraged and assisted throughout his or her stay to exercise his or her rights as a patient and as a citizen, and to this end to voice grievances and to recommend changes in policies and services to facility staff and/or outside representatives of his or her choice free from restraint, interference, coercion, discrimination, or reprisal.

G. To manage his or her personal financial affairs or to be given at least a quarterly accounting of financial transactions made on his or her behalf should the facility accept his or her written delegation of this responsibility subject to the provisions of Section 72557.

H. To be free from mental and physical abuse, and to be free from chemical and (except in emergencies) physical restraints, except as authorized in writing by a physician for a specified and limited period of time or when necessary to protect the patient from injury to himself or herself or to others.

I. To be assured confidential treatment of his or her personal and medical records, and to approve or refuse their release to any individual outside the facility except in the case of his or her transfer to another health facility or as required by law or third-party payment contract.

J. To be treated with consideration, respect, and full recognition of his or her dignity and individuality including privacy in treatment and care for his or her personal needs.

K. Not to be required to perform services for the facility that are not included for therapeutic purposes in his or her plan of care.

L. To associate and communicate privately with persons of his or her choice and to send and receive his or her personal mail unopened unless medically contraindicated.

M. To meet with and participate in the activities of social, religious, and community groups at his or her discretion unless medically contraindicated.

N. To retain and use his or her personal clothing and possessions as space permits unless to do so would infringe upon the rights of other patients and unless medically contraindicated.

O. If married, to be assured privacy for visits by his or her spouse, and, if both are patients in the facility, to be permitted to share a room unless medically contraindicated.

P. To have daily visiting hours established.

Q. To have members of the clergy admitted at the request of the patient or person responsible at any time.

R. To allow relatives or persons responsible to visit critically ill patients at any time unless medically contraindicated.

S. To be allowed privacy for visits with family, friends, clergy, social workers, or for professional or business purposes.

T. To have reasonable access to telephones—both to make and receive confidential calls.

NEW JERSEY (P.L. 1976 — CHAPTER 120)[6]

Every resident of a nursing home shall:

A. Have the right to manage his own financial affairs unless he or his guardian authorizes the administrator of the nursing home to manage such resident's financial affairs. Such authorization shall be in writing and shall be attested by a witness that is unconnected with the nursing home, its operations, its staff personnel and the administrator thereof, in any manner whatsoever.

B. Have the right to wear his own clothing. If clothing is provided to the resident by the nursing home, it shall be of a proper fit.

C. Have the right to retain and use his personal property in his immediate living quarters, unless the nursing home can demonstrate that it is unsafe or impractical to do so.

D. Have the right to receive and send unopened correspondence and upon request to obtain assistance in the reading and writing of such correspondence.

E. Have the right to unaccompanied access to a telephone at a reasonable hour, including the right to a private phone at the resident's expense.

F. Have the right to privacy.

G. Have the right to retain the services of his own personal physician at his own expense or under a health care plan. Every resident shall have the right to obtain from his own physician or the physician attached to the nursing home complete and current information concerning his medical diagnosis, treatment, and prognosis in terms and language the resident can reasonably be expected to understand, except when the physician deems it medically inadvisable to give such information to the resident and records the reason for such decision in the resident's medical record. In such a case, the physician shall inform the resident's next-of-kin or guardian. The resident shall be afforded the opportunity to participate in the planning of his total care and medical treatment to the extent that his condition permits. A resident shall have the right to refuse treatment. A resident shall have the right to refuse to participate in experimental research, but if he chooses to participate, his informed written consent must be obtained. Every resident shall have the right to confidentiality and privacy concerning his medical condition and treatment, except that records concerning said medical condition and treatment may be disclosed to another nursing home or health care facility on transfer, or as required by law or third-party payment contracts.

H. Have the right to unrestricted communication, including personal visitation with any persons of his choice, at any reasonable hour.

I. Have the right to present grievances on behalf of himself or others to the nursing home administrator, State government agencies or other persons without threat of discharge or reprisal in any form or manner whatsoever. The administrator shall provide all residents or their guardians with the name, address, and telephone number of the appropriate State governmental office where complaints may be lodged. Such telephone number shall be posted in a conspicuous place near every public telephone in the nursing home.

J. Have the right to a safe and decent living environment and considerate and respecful care that recognizes the dignity and individuality of the resident.

K. Have the right to refuse to perform services for the nursing home that are not included for therapeutic purposes in his plan of care as recorded in his medical record by his physician.

L. Have the right to reasonable opportunity for interaction with members of the opposite sex. If married, the resident shall enjoy reasonable privacy in visits by his spouse and, if both are residents of the nursing home, they shall be afforded the opportunity, where feasible, to share a room, unless medically inadvisable.

M. Not to be deprived of any constitutional, civil or legal right solely by reason of admission to a nursing home.

Patient rights regulations and legislation include basic human rights, such as the right to be treated with respect and live in dignity.

Other rights include constitutional rights due all citizens. Finally, nursing home residents have special rights because of their particular situation and circumstances.

Various authors have described and classified these rights in numerous ways. I have found the following categories and some examples of each category most useful, though there may be some overlapping:

A. The right to know:
   1. To be fully informed about one's rights
   2. To be fully informed about the rules and regulations governing the facility
   3. To be informed about the services available and the charges for services
   4. To be informed about his or her medical condition
B. Privacy rights:
   1. Privacy of visitation
   2. To receive and send unopened correspondence
   3. Confidentiality of personal and medical records
C. Property rights:
   1. To retain and use personal property, including money and clothing
   2. To be free to manage one's affairs
D. Rights of transfer and discharge: transfer within a facility or between facilities, or discharge, should be reasonable, fair and justifiable
E. Rights of association and communication:
   1. Reasonable visitation by friends, family, and clergy
   2. Reasonable access to telephones to make and receive calls
   3. Correspondence rights (as previously noted)
F. Rights of safety:
   1. To be free from mental and physical abuse
   2. To be free from any unreasonable restraints, chemical or physical
G. Autonomy rights:
   1. To participate in decisions about one's care and treatment
   2. Not to be required to perform services for the facility
H. Procedural rights:
   1. To be able to voice concerns without fear of reprisal and to have complaints resolved
   2. To be encouraged and assisted to exercise his or her rights as a patient and citizen
   3. To participate in social, religious and community groups, including resident councils

   I. Rights of human dignity:
      1. To have a safe and decent living environment
      2. To be free of any type of discrimination
      3. To be treated with consideration, respect and full recognition of his or her dignity and individuality

CATEGORIES

The Right to Know

It is interesting to note that the first three rights listed among the California regulations, as well as the federal regulations, concern themselves with the right to know, and New Jersey does list it in its regulations. It is the responsibility of the facility at the time of admission to make sure that the resident and his family are fully informed about his rights, as well as the rules and regulations of the institution. Preferably the resident should be given a printed list in large type and the admitting official should review it with the resident and the facility. The American Health Care Association[7] recommends that the admission agreement be signed by the resident and an individual responsible for the resident verifying that they have been informed of their rights and responsibilities.

Information regarding available services and their cost should be provided in writing. It should be reviewed when inquiries are made concerning the possible admission of a resident and at the time of admission. Also, it should be provided when requested by a resident or family member before any changes in cost are made and in response to any inquiry by an official or an interested party. If a resident is personally responsible for any charges, he should receive an itemized monthly bill.

If an individual is to be free to make decisions regarding his medical treatment, it is imperative that he be informed about his medical condition. Therefore, the resident should have access to his medical records and be provided with information about his condition upon request, or when deemed necessary by the administration of the institution. Policies regarding the giving of information should be developed by the facility and attending physicians should be made aware of these policies. If a question of the competency of a resident arises, the resident should be evaluated by a physician, preferably a psychiatrist, and legal competency procedures should be initiated, if indicated by the physician's findings. In situations where the resident has been adjudicated to be incompetent, the individual responsible for the resident should be informed about the medical conditions so that reasonable decisions can be made.

Privacy Rights

As a part of basic human rights every resident has a right to privacy and both California and New Jersey list various privacy rights. Privacy rights include personal privacy, such as not being exposed when dressing or undressing, and the right to privacy with a spouse. If possible, spouses should be allowed to share a room and residents should be allowed to invite willing members of the opposite sex into their rooms.

Conversations with visitors should be private and visitors should have ready access to residents unless it is not in the best interests of the resident. Residents also have the right to refuse visitors or to terminate visits. Residents should be free to communicate privately by mail or by telephone and it is the responsibility of the institution to provide easy access to mail and telephone facilities. Both states make mention of this.

All personal and medical records should be held in confidence and only revealed with the resident's written permission or when required by law. Nursing home facilities should develop procedures to reflect this policy and staff members should be fully informed regarding the confidentiality of information regarding residents.

Property Rights

Every resident has the right to wear personal clothing and retain personal property when reasonable. Residents should have reasonabl access to personal items held in safekeeping and the facility should maintain a list of such items available for review when requested. Any restrictions regarding the possession of personal items should be made clear at the time of admission and any restrictions imposed following admission should be reasonable and documented.

Competent residents should be free to manage their affairs, including financial matters, and both California and New Jersey include this right in their regulations. In fact, New Jersey lists this first among their regulations and New Jersey requires that if the resident or guardian authorizes the administrator of the nursing home to manage his financial affairs, such authorization shall be in writing and witnessed by a person not connected with the nursing home. However, California makes provision for an accounting if the resident requests that the facili ty manage his affairs. Arrangements for management by others of a resident's finances should be in writing, including the scope of such management and any charges for the service.

Rights of Transfer and Discharge

On occasion a resident is involuntarily transferred or discharged and this potentially could be traumatic to the individual. Therefore, transfers and discharges should be made only for a good cause, such as for medical or psychiatric necessity or for nonpayment of legitimate fees and charges. Federal regulations require that appropriate advance notice be given to the resident except in an emergency.

An exception to the general prohibition against involuntary transfers should be noted. In a 1982 case[8] the United States Supreme Court ruled that a nursing home may discharge or transfer a Medicaid resident to a lower level of care, without prior notice, based upon a utilization review committee's decision, without violating the resident's constitutional right to due process. The Court found that such decisions are "medical judgments made by private parties according to professional standards" and with due consideration of medical ethics. Because the decision of the utilization review committee depends on documentation supplied by the treating physician, it is incumbent upon the physician to appraise the committee of any adverse change in the resident's condition that might be expected to occur in response to an involuntary transfer.

Rights of Safety

As part of the right to be treated with respect and human dignity, all residents should be free from mental and physical abuse. The concept of abuse is quite extensive and ranges from inferior care and treatment, such as an inadequate diet or a resident spoken to in a harsh or threatening manner, to actual physical mistreatment.[9,10] Though not as frequent today as in the past, abuse is still a fact of life and every now and then the administration or staff member of a nursing home is charged with abuse of residents. Treating physicians should be cognizant of potential abuse of residents and report it, when found, to the proper authorities. This may mean reporting a staff member to the administration or, if abuse appears to be a general policy of the nursing home, to legal authorities. New Jersey has a separate statute that requires "any caretaker, social worker, physician, registered or licensed practical nurse or other professional" to report any suspicion or belief of abuse to the office of the Ombudsman for the Institutionalized Elderly.[11]

Because the rights of one resident cannot be sacrificed for the rights of another resident, it may be necessary to temporarily restrain an individual who may be threatening or making unwanted sexual advances. The use of chemical or physical restraints should only be instituted for the good of the resident or the protection of others and never for the convenience of the facility. Chemical restraints generally refer to the use of psychotropic medication to control a resident rather than being used as a therapeutic agent. Therefore, if it is necessary to prescribe psychotropic medication, one should choose the medication that is safest, has the fewest side effects and provides the resident with the least restrictions. If it seems that restraints may be required, be they chemical or physical, one should consider other possible alternatives first. Justification for the administration of medication or the use of physical restraints should be documented, reviewed at a reasonable frequency and, of course, lifted as soon as possible.

Autonomy Rights

Every competent resident has the right to make decisions regarding his care and treatment and he should be encouraged to participate in these decisions after being fully informed as to his condition. He has the right to refuse medication, diagnostic and surgical procedures, and to participate in any experimental program. If the resident's condition precludes him from being fully informed or able to participate in the decision-making process this should be documented and justified. In the absence of an emergency one should consider the institution of competency proceedings and the appointment of a guardian or conservator when indicated.

No resident should be required to perform services for the facility. Small chores, such as bed making, may be performed voluntarily, but no one should be forced to work or be punished for refusing to work. If the resident is given various tasks as part of a therapeutic program, this should be documented and explained.

Procedural Rights

Every facility should permit residents to voice their complaints and concerns, and procedures should be developed to resolve any complaints. This may be done informally or through a resident council. New Jersey requires that the telephone number of the State governmental office where complaints may be lodged be posted near every public telephone in the nursing home. Participation in various organizations should be encouraged as part of a therapeutic program as well as a right. However, residents may refuse to participate in an activity, if they so desire, without fear of coercion.

Rights of Human Dignity

The right to a safe and decent living environment, as well as to be treated with consideration and respect, underlies all the above mentioned rights. No facility should discriminate against anyone because of age, sex, race, or handicap. Although a nursing home may not have the facilities to care for all types of residents, it should take steps to ensure that it does not discriminate. For instance, it cannot refuse admission because an individual has a prosthesis or is deaf.

Not all rights have the same significance and one instance of physical or mental abuse is much more serious than one instance where a resident is not permitted to wear a piece of clothing or receive a single phone call. However, if minor infractions of these rights occur frequently, they may be deemed to become a serious violation of the law and might be dealt with harshly. It is the responsibility of family and friends of nursing home residents, as well as state officials and interested parties (such as the American Association of Retired Persons volunteers)[12] to protect the rights of those who may not be capable of looking out for their own best interests. Attending physicians carry a greater burden: first, as citizens, and second, as professionals who care for the welfare of others. I would hope that with greater knowledge and understanding of the rights of nursing home residents we will be better able to carry out these responsibilities.

REFERENCES

1. Aging in New Jersey Sixty-five and Over. New Jersey Department of Community Affairs, August 1985.
2. Kapp B, Bigot A: *Geriatrics and the Law*. New York, Springer Publishing Co, 1985.
3. *Federal Register* 1974; 39 (May 1):15231.
4. *Federal Register* 1975; 40 (March 4,):8957.
5. California Department of Human Services: *A Consumer Guide to Nursing Homes*, 1985.
6. NJSA 30:13, 1 to 11. P.L. 1976-Chapter 120. Nursing Home Patients Bill of Rights.
7. American Health Care Association Policy Paper. Patients' Rights, September 1981.
8. *Blum v Yaretsky*, 102 S.Ct. 2777 (1982).
9. Gapen P: Nursing home faces loss of license over patient deaths. *American Medical News*, Jan 17, 1986, p 10.
10. Gapen P: Physician refutes prosecution testimony. *American Medical News*, Jan 24/31, 1986, p 10.
11. NJSA 52:27, G–Z. P.L. 1983-Chapter 43. Mandatory Reporting of Adult Abuse.
12. American Association of Retired Persons. Highlights, January–February 1985.

18

An Overview of the Elderly in the Criminal Justice System

Mental Health Perspectives

NAOMI GOLDSTEIN

The criminal victimization of the elderly is of great concern to society, is widely publicized in the media, and addressed sympathetically in a fairly extensive literature. It is the subject of criminological research and of community programs for prevention. But the subject of the elderly in the criminal justice system, that is, the elderly as criminals, has received relatively little attention except in those places where the numbers are sufficient to demand attention, such as in retirement communities or in the prisons that must care for those few who are ultimately committed to their care.

The justice system is not geared to the elderly: from initial ambivalence about making the arrest, to the use of handcuffs on frail old people, or to the meaning of prosecuting an elderly tax evader who will almost invariably be placed on probation. What is the significance of bail for a person of limited mobility or no resources? What is the meaning of deterrence and rehabilitation for the first-time elderly offender who has frequently led a law-abiding, productive life?

This chapter will present an overview of the elderly in the criminal justice system with particular emphasis on mental health or psychiatric issues, from initial contact through ultimate disposition. Of special in-

NAOMI GOLDSTEIN • Department of Psychiatry, New York University School of Medicine, New York, NY 10016.

terest is the nature of the elderly criminal and the crimes committed, the processing through the system, the role of psychiatry and other expert services in evaluating and diverting out of the justice system, and finally, the management of the elderly prisoner. Many thousands of elderly commit crimes, less are arrested and convicted, very few go to prison to join their elderly brethren already serving time, and eventually, an even smaller number will be released to the community.*

In 1984 there were almost nine-million arrests, of which 1.6% were of persons age 55 to 59, 1% were of persons 60 to 64, and 0.9% were over 65, altogether about 315,000 elderly persons.† As of 1981, there were approximately 8,853 persons 55 years and older in state and federal prisons in the United States.[1] The total prison population was then in the range of 250,000 persons, not including jails or detention facilities, which clearly process many more elderly criminals but on a transient basis. The definition of elderly for the purpose of the criminal justice system varies, but is often defined as 55 or 60 years of age and up, sometimes 50 years and older, reflecting changes in social status, retirement, patterns of criminality, and the beginnings of chronic health problems that require more institutional resources. The absolute numbers appear to be increasing whereas the percentages of elderly may be going down or may have stabilized. The increases may reflect the "graying" of America. but the numbers are still small. There is a real increase in the number of all incarcerated prisoners, reflecting a trend toward more punitive sentencing. More inmates are serving longer sentences before being eligible for parole. This includes the present elderly and will result in more inmates growing old within the system. It is difficult to determine whether the increase in the number of incarcerated prisoners generally is a reflection of an increased crime rate or increased reporting. What is patently clear is that the criminal behavior of the elderly is a growing social problem.

The elderly criminal is an embarrassment to the elderly themselves and the elderly prisoner is more often than not lost to society and a burden. There is no real constituency for this group except from within the justice and correctional systems, perhaps from departments of aging and from religious leaders. Occasionally, a graduate student or a sociologist may become involved in programming and research.

The general literature on the subject of the elderly criminal is growing. An excellent new book by Newman & Newman, *Elderly Criminals,*

---

*The author is indebted to the staffs of the New York State Department of Correctional Services and of the Federal Bureau of Prisons for reviewing programs, problems, and data on the elderly in New York State and Federal institutions.

†Data on arrests for 1984 was provided by the National Criminal Justice Reference Service.

reviews all aspects of the subject. A number of nonmedical texts deal with the stress and coping mechanisms of aging and there are some useful articles that deal with various aspects of the subject from research to diversion.[2–5] The psychiatric literature on aging, although it discusses mechanisms of defense, psychopathology, and treatment at length, is noteworthy for its almost total lack of attention to the emergence of antisocial behavior in the elderly. One can look in such a major text as *The Comprehensive Textbook of Psychiatry*[6] in its most recent edition and in other excellent volumes such as the *Psychiatric Clinics of North America on Aging,*[7] and will not find the word sociopathic or antisocial in the indexes or chapters on aging, whether as a reflection of organic deterioration, as a new form of coping, as a defense, or breakdown of defense systems in previously law-abiding citizens. Only Busse,[8] as early as 1969, devotes a small section in his text to criminal behavior in the elderly. It was a relief to find it there. Of interest is a preliminary report[9] from the Forensic Psychiatry Clinic in New York (and an unpublished follow-up) attempting to describe some of the demography and diagnoses of the elderly defendants who are examined at the clinic.

THE CRIMES AND PSYCHIATRIC DISORDER

What kind of crimes do the elderly commit? My first evaluation at a lower court clinic in 1961 was of a 55-year-old depressed woman shoplifter with no prior history of mental illness or arrest, very typical, I thought, of a small group of middle-aged depressed women who shoplift. Today she would be included in this group of elderly. Her somewhat unusual arrest at that time, which very much disturbed the court, is now more typical. Until recently elderly violators were underarrested and often diverted to non-criminal justice alternatives for such property offenses as shoplifting and check fraud.

Petit larcency, that is, shoplifting, is the major crime of the elderly and has always been probably the most underreported. Shoplifters are not always indigent persons. Although shoplifting is often a reflection of the subsistence existence to which many of our elderly have been reduced, shoplifters may be middle-class individuals or others with adequate resources. The elderly steal food products, followed by clothing and medical related articles often worth very little. It is a problem of particular concern to states, such as Florida, with high concentrations of elderly. There has also been an increase in fraud, check forgery, counterfeiting, writing bad checks, and embezzling. Our elderly are enterprising and desperate.

Elderly violators today are likely to be arrested and prosecuted for serious crimes against persons. The FBI reports an increase in so-called index crimes (more serious and violent crimes) of the elderly. According to Newman,[7] depending on definitions, 23% to 47% of elderly arrests are for forceable rape, robbery, aggravated assault, and murder, crimes generally associated with a more youthful population. A significant number of homicides are committed by the elderly, often associated with chronic alcoholism and intoxication at the time of the incident, or with paranoid thinking. Frequently the victims are relatives or friends, often suggesting long-standing conflicts and escalating tensions and a breakdown in interpersonal relations—preventable homicides, for the most part.

Sexual crimes, which are well-known among the elderly, are generally immature, rather passive sexual activities, such as indecent exposure or fondling a child well known to the perpetrator, without threats or force. However, rape does exist amongst the elderly. There is a suggestion that many of these late-life sex offenders have histories of earlier sex offenses and of sexual difficulty. These crimes are also thought to be underreported.

Among the elderly there are some professional thieves who have matured and survived, more efficient than ever, and con men, who continue better than ever into old age with sophisticated skills, perhaps well-groomed gray hair, and considerable audacity. There are even elderly imposters with incredible records of impostering going back to adolescence. Tax evasion is not uncommon, either through deliberate action or inaction or frequently through inability to cope with the process. More sophisticated white collar crimes, such as fraud and embezzlement, may be seen in later life when the individual is in a position to command the resources or organizations that are defrauded. There is also a small contingent of leaders of organized crime, major leaders, the godfather types, who have evaded prosecution until the later years of life. Increasingly, although in small numbers, these persons will go to prison for the first time.

Crime, of course, is not to be equated with psychiatric illness, perhaps a mistake made in the earlier years of this century when badness and madness became confused with each other. However, there is obvious psychopathology in some instances. The roots of criminal behavior in the elderly may be no different than in younger criminals and theories are endless, but we need to be particularly concerned with persons who exhibit criminal behavior for the first time in later life, or for the first time in many years. These are generally people who have internalized societal values and in many instances who have had good prior psychological resources. In the face of diminished social and emotional resources, coping with the later years of life may place extraordinary demands on the individual while limiting outlets, so that previously effective coping strategies, both internal and external, may no longer be adequate.

Not really that much is known about how adaptive styles change relative to the earlier years of life, although there is considerable discussion of the stresses and defenses. We do know, of course, that there is often a loss of status, financial insecurity, difficulties adjusting to retirement, elimination of socially acceptable roles, loneliness, loss of friends, families, and support systems. There are theories of disengagement, and of withdrawal from mainstream society. Do we attribute shoplifting to economic factors, not always justified by the facts, or to lowered frustration tolerance, impulsivity, feelings of inferiority, need for attention, excitement, anger and irritability, or boredom and a search for excitement, and poor problem solving? Or do we attribute it to some deeper pathology? For instance, how many of these elderly shoplifters have a clinical depression? How many first-time elderly offenders have significant, or more likely, subtle organic deterioration, contributing to the loss of superego controls and problem solving skills, to confusion, to impulsivity, and to paranoia?

Interestingly, antisocial personality disorder is less prominent among the elderly violent than among the young. But alcoholism apparently is more prevalent.[7] It appears that a significant number of those arrested for more serious violent crimes do suffer from functional mental disorders such as schizophrenia, manic depressive illness, and paranoid psychosis. In one study it was estimated that 75% of the elderly violent have either organic psychosis or functional disturbances, with a high incidence of paranoid symptomatology regardless of diagnosis, and unfortunately, often associated with acute alcoholic intoxication.[7]

Statistics are difficult to obtain as there is considerable selectivity before arrest, and before referral for evaluation. In many instances there is a paucity of historical information available when the individual is referred or screened.

FROM ARREST THROUGH SENTENCING

The police have been reluctant to arrest the elderly, often seeking the family or taking the person to a civil hospital, but this is changing, a reflection of a more demanding society, a law-and-order approach that leaves the police with less discretion. That is, there is increasing pressure to arrest, particularly for drunken driving, and shoplifting, which are very costly to all of us, and for incidents of family violence. A noninterference policy in family violence is changing, as a result of pressure from womens' groups. There is a traditional judicial reluctance to break up families. For example, a judge was reluctant to charge and detain an assaultive wife whose assaulted husband was completely dependent on her. He referred the matter to family court. There is much sympathy with the elderly, it is offensive to incarcerate old persons, and this is perhaps a denial, a disbelief in their criminality. There is also a sense of futility and frustration. What is the meaning or purpose of arrest? Questions of deterrence and rehabilitation seem almost meaningless, which is not the case with a very comparable group, the adolescents, who have their whole lives to lead. What seems more important is the availability of community resources, social service, and supervision.

Some courts do understand the mental and physical needs of the elderly and accommodations are made following a concept of the least restrictive alternative.[7] A summons is used to avoid the trauma of holding an individual in jail. Pretrial release may be arranged without bail unless escape is a real possibility. Pretrial diversion is arranged with supervision and counseling. Alternative sentencing through diversionary programs may be recommended, such as in Broward County, Florida, which established the Broward County Diversionary Program, a senior intervention and education program for first-time offenders.[7] Prosecutions may be deferred. If the individual must be prosecuted and is guilty, probation will be encouraged unless the individual poses a serious threat to the community. In the latter instance incarceration may be mandatory.

In spite of attempts to divert and considerable prosecutorial discretion, many crimes must be prosecuted to conclusion. Where resources are available and lawyers and families care, attempts will be made to intercede, to evaluate and dispose of, to prevent heavy sentences and prison time, attempts initiated by the court and prosecutors as well as by defense lawyers. This is part of a winnowing out. The role of psychiatry here is quite significant, at least in more sophisticated communities. Experts will be asked to evaluate prior to arraignment or pleading, for competency to give confessions or to stand trial, and for the defendant's mental state at the time of the crime, either as to responsibility or diminished capacity. Finally, evaluations may be requested as to competency to be sentenced or special needs on sentencing and the potential impact of incarceration. Medical and neurological assessments are often obtained either to determine whether the defendant is well enough to come to court, to testify, or well enough to go to prison. Judges have grave concern about incarcerating elderly and infirm persons. But only to a point. A disappointed judge who had just sentenced an elderly man convicted of murder and a gun charge, was reported on a recent news broadcast to have said, "he's been getting away with all kinds of things because of his age . . . he should have had a stiffer charge and a heavier sentence."

Students of geriatric criminology have suggested that the elderly should have a special court of their own, parallel to the special courts for children, a geriatric court, with a protected, understanding environment. In this ideal setting there would be an adequate number of social workers, psychological and other counseling services, and the individual would be addressed in a more intimate fashion with much closer attention to his or her real needs. Another major legal, philosophical, and social issue is whether the elderly should be considered incompetent by virtue of age, as are little children, or irresponsible by virtue of age; the issue is a veritable Pandora's box. Such ideas do not seem to be popular or practical, and are patronizing of the elderly. In practice, they may actually deprive elderly citizens of their due-process rights. The family court model has failed notoriously.

CASE VIGNETTES

As a psychiatrist, I have been asked to evaluate a number of elderly persons arrested for serious criminal behavior. The cases below illustrate some of the special issues already mentioned. They will be presented in such a way as to highlight these particular issues, with a little editorializing on the part of the author.

Sally R. was a 58-year-old divorced, psychiatric social worker who was charged with a rather bizarre stock fraud. She was referred for general psychiatric evaluation in anticipation of a renewed effort to negotiate a deferred prosecution that had already been denied in the absence of known psychiatric history. She was arrested on the job, her first involvement with the law. She was competent to stand trial. Psychiatric evaluation was difficult because she was so defensive, and she appeared to be extremely paranoid, possibly alcoholic, and severely emotionally disturbed in spite of the fact that she was working. There was a history of poor interpersonal relationships, particularly with members of her family. She denied all difficulty, including the charges. Following psychiatric examination a deferred prosecution was granted, Mrs. R.'s job was protected, and efforts were made to refer her for alcohol counseling.

Guiseppi G. was referred for competency evaluation. He had never been arrested in his 82 years, nor was there any prior psychiatric history. It appeared that there was some early senile deterioration as reported by the family, with irritability, personality change, and some forgetfulness. He had regressed to the use of Italian although he had a good knowledge of

English. Mr. G. habitually visited the commercial bakery he had founded, now owned and managed by his son. He irritated the employees with his interference, with his own contrariness, and with inappropriate attempts to manage. The employees had been taught that they must not get angry with him and that they should tolerate him.

One day Mr. G. arrived at the bakery, only to find that a full-blown investigation of the bakery premises by officers of the Drug Enforcement Administration was in progress. The agents were actively searching for guns, cash, and drugs, apparently part of a major drug investigation. At that point Mr. G. very clearly, in the presence of other witnesses, offered a substantial sum of money to an agent if they would all just go away, and within a short period of time, he produced $10,000. He was charged with bribing a federal agent, a very serious charge.

Repeated attempts were made by Mr. G.'s lawyers to discuss these charges with him. He was confused, tearful, and uncooperative. He could not remember the charges from visit to visit with the lawyers, nor did he recognize the lawyers. His family was frantic, believing that his cooperation was necessary to proceed with his case, while noting progressive senile deterioration at home. He required increasing supervision and his daughter, with whom he lived, was afraid to allow him to leave the house. She was in tears over his deterioration. Attempts to work with him and to talk to him about the importance of cooperation were futile. He was very depressed and confused.

During the competency evaluation, Mr. G. was extremely cooperative, at least for the first interview, answering questions in Italian and looking to his daughter for reassurance. He was tearful and agitated at times but appeared to be trying to please until he became irritable and fatigued. On the second visit to my office, he failed to recognize me in spite of his extremely full attention and cooperation on the first visit. During the second interview he was confused and preoccupied, unable to focus, and there was no meaningful discussion of his legal situation as there had been during the first interview. Psychological testing was obtained and confirmed the depression and the presence of an organic mental syndrome.

It was my opinion that Mr. G. was not competent to stand trial, that is, I did not think he could grasp the legal situation and cooperate rationally as a result of organic deterioration and depression. There was a very difficult distinction to be made between dementia, pseudodementia secondary to depression, and the possibility of malingering. The diagnostic conclusions were of great importance prognostically for judicial management. The prosecutor was incredulous that a man could deteriorate this way in the 3 or 4 months from the time of arrest. He had so clearly perpetrated the bribery attempt for which he was charged and that did require some very clear, logical thinking. In my opinion the prognosis was poor for recovery within a reasonable period of time and the case to my surprise was actually dismissed. Fortunately, I was not asked the very difficult issue of responsibility. I suspected that Mr. G.'s behavior in the bakery was a reflection of great commitment to his son and of early deterioration, of impaired judgment, a loss of subtlety, and perhaps some impulsivity that led this law-abiding citizen into this embarrassing predicament. This deterioration probably would have amounted to a diminished capacity (which might not have been applicable by law).

Joe S., a 74-year-old black widower, was referred for competency evaluation by the court. There was no prior psychiatric history nor arrest record. He was charged with killing his much younger girlfriend with a knife, a woman who evidently was taking care of him. He had no family by then, lived alone, and had become extremely paranoid about this woman, who he thought was abusing his funds and whom he had suspected of having a young boyfriend. Mr. S. was alcoholic, and alcohol appeared to have played a significant role in the homicide. In spite of an early organic mental syndrome, Mr. S. was found to be competent after full psychiatric, neurological, and psychological evaluation in a forensic hospital. He was convicted, and was eventually sent to prison.

Anthony A., a retired pizza maker of 75, was charged with conspiracy in a major drug scheme in which his son, daughter, and son-in-law, as well as others, were co-conspirators. Mr. A.'s son was considered to be the major perpetrator and Mr. A.'s participation was not clear, although it was suspected that he had some degree of real culpability and $800,000 in cash had been found in the ceiling of his home, in addition to drugs and drug paraphernalia. Mr. A.'s son was a fugitive, and the father, son-in-law, and daughter were in federal prison awaiting trial.

Mr. A. was referred for evaluation of his competency by the court with consent of both defense lawyers and the prosecution. When Mr. A. met with the lawyers he stared at the ceiling blankly and refused to participate in any of the legal discussions. He claimed not to know English, which he was known to speak, and he appeared to be extremely preoccupied and depressed as well as somewhat irritable, even nasty, in the prison. He did cooperate with evaluation with the services of an Italian translator that I felt he really did not need. On interview he gave a number of adequate answers, but some responses were confused, forgetful, and hesitant. After considering his functioning prior to incarceration, remarks made on arrest, and the inconsistency of his responses, I felt that these were reflections of malingering, but they might also have been evidence of depression, which had to be considered. It was clear that he was terribly upset, bursting into tears when I asked him about it. He said he was afraid of dying in prison. The government prosecutors were concerned that if released he might flee to Italy, which he had the resources to do. It seemed to me that he was in a terrible position, in a close-knit Italian family, as it appeared that if the fugitive son had been apprehended, the father might have been released. He was very angry but would not talk.

Mr. A. was aware of the charges against him, but he expressed his conviction that the government would not be able to connect him with the money in the ceiling of his house and that he was not guilty. This was more or less the only allusion he made to the charges. Part of the evaluation included psychological assessment by a psychologist who spoke the same Italian dialect. We both thought that he was competent, although depressed. The court also found him to be competent, and he was acquitted at trial, possibly by a sympathetic jury in the absence of the fugitive son.

Seventy-one-year-old Dolores R. killed her physically abusive, alcoholic husband. His abuse was well documented and had led to hospitalization for injury on several occasions over the years and to the family court 10 years before. The court had, nevertheless, denied an order of protection. Mrs. R. also suggested that the same woman judge had denied her a divorce. Mrs. R. had multiple medical problems, high blood pressure, diabetes, serious cardiac disease, and arthritis. There was no substance abuse history. The year before the incident she had been psychiatrically hospitalized for the treatment of an atypical psychosis, a first psychiatric hospitalization around the age of 69.

Immediately after the incident, Mrs. R. was rehospitalized in a civil psychiatric unit, as she was actively psychotic. After a number of months of psychiatric hospitalization, she was discharged on her own recognizance to an intermediate care facility, still in the community, principally because she was thought to be unable to take care of her medical needs, and because there was still some residual psychosis. She was never incarcerated following the homicide. The psychosis cleared and she remained in the health related facility, where I was asked to visit her to evaluate her competency to take a plea to a much reduced charge of manslaughter. Her lawyers believed that she was competent but wanted this assessed by a psychiatrist.

The competency evaluation was done in the nursing facility after reviewing the original psychiatric records and discussing Mrs. R.'s progress with the nursing-home staff. She was amnestic for the time of the incident although it was not determined whether this was due to psychosis, conscious supression, or repression. She was not psychotic when evaluated, but there was evidence of a mild organic mental syndrome. Nevertheless, Mrs. R. was competent to take the plea. Nobody connected with the case seemed to want to remove her from the nursing facility, where she was doing well. After taking the plea, she was allowed to remain in this health facility, sentenced to probation.

Frank T. was a notorious, powerful, "Mafia" godfather of 79, who was charged with racketeering. This involved a major fraud, highly publicized, but it was also suspected that Mr. T. was responsible for a number of murders over the years. He had evaded prosecution through his life. Following conviction on the racketeering charge he was referred to me by his own lawyers for an evaluation of competency to be sentenced and to sustain a prison sentence.

Mr. T. had multiple serious medical problems, even a history of surgically treated cancer, and was basically a sick old man when psychiatrists had found him to be incompetent to stand trial. There was no psychiatric history. There was a full medical evaluation. The judge was incredulous that he should be incompetent for trial and found him to be competent after hearing, as among other evidence, the judge had witnessed tapes of this man looking fit, prancing down the sidewalk with his mistress on his way to a wedding, shortly prior to the arrest.

By the time I evaluated Mr. T. in my office after many months of trial, he appeared to me to be clearly senile, required supervision at home for personal decency and toileting, for general functioning. It was clear that every need of his was immediately taken care of by family and bodyguard. He was on multiple medications for his medical problems. He was cooperative, sustaining a long interview, but he answered my questions with some confusion. He repeatedly denied his guilt. He told me that he wanted to tell the judge that he was innocent "so help me God." His lawyers were very upset with my conclusion that he was competent to be sentenced, that I felt that he understood what was facing him, although he was not very articulate. He was a very proud old man who wanted to make a last stand. On the other hand, I felt that if he were to be removed from his home and placed in prison where he would be in a strange environment, where he would be disoriented and confused, where he would require excessive nursing help that was not available, he would deteriorate rapidly and would die quickly.

At the sentencing, the judge prohibited any psychiatric testimony, telling the court angrily that he was very disturbed by the earlier psychiatric testimony at the competency hearing, and that he could see for himself the extent to which Mr. T. had deteriorated in the months of the trial process. He had come to court initially on his own power, but over the months, he was reduced to a crumpled heap in a wheelchair. Nevertheless, Mr. T. was sentenced to 10 years in prison. This, of course, was immediately appealed by the lawyers, and he was allowed to stay at home during the appeal. Mr. T. died before the appeal was completed, almost as if to say that he would never serve time in jail with God's help.

Mordechai A., a 69-year-old ultra-orthodox Jew, was referred by the court itself at the recommendation of the probation department for a sentencing evaluation following a conviction for fraud. He had a record of prosecuted and unprosecuted fraud dating back to 1945. In 1976 and 1977, around the age of 60, he had been given a deferred prosecution on an equally sophisticated fraud charge, following a psychiatric evaluation. Facing charges in 1976, with no psychiatric history, he had gone to a psychiatrist's office, laid on a couch with his eyes closed, and acted in general in a very bizarre fashion. The psychiatrist, who was a general psychiatrist and not forensically oriented, described him as "utterly confused, disoriented" with severe memory impairment, and gave him a diagnosis of organic brain syndrome, as well as schizophrenia. Mr. A. did not cooperate with treatment once he received a deferred prosecution. It was clear to a subsequent examiner that he was malingering. There was no further evidence of psychiatric disturbance. There was some medical history of significance, but Mr. A. remained active in business.

When Mr. A. appeared in my office for evaluation on the new fraud charges, he whispered, he was hesitant, he could not or would not answer questions, and he was very withdrawn, tired, and depressed. He was accompanied by a grandson who translated from English to Yiddish for him, but it was known that he could speak English. Unfortunately, while the legal proceedings were ongoing, and throughout the presentence evaluation, Mr. A.'s wife, with whom there was an ambivalent but long relationship, was dying of cancer, a brain tumor for which surgery had been done and every treatment made available. She was at home in a hospital bed, with full nursing care, and the home was very disrupted. This state of affairs was confirmed by the investigating probation officer; I contacted the treating physicians. It was necessary in the psychiatric evaluation to distinguish between malingering, possible organic mental syndrome, and depression. In my opinion Mr. A. was depressed, and he was also attempting to malinger, although there was no bizarre behavior as noted in 1976. A great deal was known about Mr. A.'s sophisticated and knowledgeable fiscal dealings and about his role in the charges. It was he who was considered the major perpetrator without a question, rather than his son, who was a co-conspirator. Mr. A. was sentenced to 10 years in prison following his wife's death.

The importance of careful clinical evaluation, and of good medical, neurological, and psychiatric diagnosis, in these elderly defendants cannot be stressed sufficiently. Many can be treated and restored to some level of better functioning, and from the point of view of the court, the prognosis has great import. Unfortunately, resources are not always available. If the evaluation accompanies the individual to probation services or to prison, it will have further value. The experts have a heavy responsibility, as recommendations in this age group have particularly great impact.

The group of cases just summarized, all with serious charges, are rather typical of cases referred for psychiatric evaluation, although one would expect at least one elderly sex offender, a somewhat higher incidence of prior psychiatric history, a few more prior arrests, and perhaps an insanity defense. Typically, very few are actually incompetent to stand trial and insanity defenses are rare, and very few are sent to prison. A dying or seriously ill spouse at home is not unusual and may have an impact on the sentence, but not always. A history of chronic emotional problems, such as depression and anxiety, and a degree of medical impairment, is not sufficient to protect the individual against incarceration in more serious crimes. Although this sample is not statistically meaningful, as I considered the other elderly defendants I have evaluated, very few actually do go to prison, and it is usually for a very

 301

serious crime. According to one experienced prosecutor, the elderly tend to cooperate with investigation, to provide information so as to get more favorable sentencing. They do not want to die in prison, having a very real sense of their options, and no time to waste. It is noteworthy that the one recidivist in the group, Mr. A., had a history of evading prosecution through psychiatric intervention.

THE ELDERLY IN PRISON

Joe S., who murdered his girlfriend and had no family support system, and Mordechai A., the conniving embezzler with a vast and attentive family, both went to prison, joining a small group of aging individuals. Some of these elderly are serving long sentences, perhaps they are lifers who grew old in prison, some are recidivists, and some of them are first-time offenders. In fact, there are relatively few elderly individuals in the state and federal prisons.

In New York State, for example, typical of statistics reported for other states, in February, 1986, out of approximately 34,997 inmates, 1,100 were between the ages of 50 and 64, or about 3.2% of the prison population.* Those over 65 were about 0.3% of the total or 99 individuals. These are typical statistics. In 1981 in New York there were 57 over the age of 65,[5] in 1985 there were 72, or about 0.2% of the prison population.† It is likely that the real numbers will increase although not necessarily the percentages, but it is the real number that is meaningful to the prison. It is this number that requires special services and puts an added burden on institutional resources.

Of those individuals committed in New York State, a relatively high percentage are serving time for homicide, more than twice the percentage in the general population. This is also true in many other states.[4] Newly convicted elderly recidivists, however, actually constitute the highest number of elderly in prison. For example, in New York again, most of the elderly were already over 60 on commitment and had prior arrest records. I would speculate that relatively few of these recidivists had been referred for psychiatric evaluation. Thus it is a very different group than those who are seen in court clinics, in forensic hospitals, or in private evaluation.

Among these elderly prisoners there is a high incidence of alcoholism, a low incidence of drug abuse, and a relatively low incidence of

---

*Data provided by the New York State Department of Correctional Services, Department of Records and Statistical Analyses.
†Data provided by the U.S. Bureau of Prisons, Public Information Office.

mental illness. In one study,[3] the prior psychiatric hospitalizations were thought to be mostly alcohol related. Histories are also very difficult to obtain, even when there is a screening on admission, as the newly committed elderly inmate may be ill, confused, vague and forgetful, or too anxious and frightened to give history. There is more often than not little family to supplement the history.

There is almost no statistical information available on elderly women in prison, as there are so few. It comes as a surprise to realize that Jean Harris, who killed her physician-lover when she was in her mid-fifties, is clearly in the group of elderly. Although she had been teaching actively in her women's prison, a recent heart attack has placed special demands on the institution. One elderly New York State inmate was committed in 1931 following a murder conviction when she was 24. She was subsequently paroled and was returned to prison, later refusing further parole.[5]

Most states have some intake screening and elderly inmates will ultimately be placed in special units if available and if needed. Most of the special units have few extra services except that they may be more protected, and generally have greater access to nursing or medical services. Unfortunately, the better units for chronic illness or disability are often in minimum security facilities, thus precluding many prisoners, including the violent elderly. These individuals must then remain in the general population. Some states may put the elderly in special units with the handicapped, retarded, or the mentally ill. A few states have developed special units just for the elderly. Little is done of a unique nature for the elderly, even by the Federal Bureau of Prisons, which does have some other special housing with additional services for which a few elderly inmates may qualify.

The prison environment is unsuitable for the elderly. Diet is generally poor, and special diets are rarely available although nutritionists are sometimes assigned to instruct the prisoners. There are cold walls and steep steps to dormitories as well as long walks to the dining rooms. Rehabilitation and recreation programs are geared to the young and elderly inmates are poorly motivated and rarely participate, thus reducing eligibility for parole. They may become quite dependent and isolated, and may be overlooked by staff. The elderly are often victimized, raped, and robbed, and they become quite fearful and withdrawn. Some are placed in protective custody units, which although restricting freedom, provide security. In these units food is brought to the cell so that the prisoner does not have to walk the long distance to the dining rooms. On the other hand, many of the elderly fit in well, gain status in the prison, find a niche, may find peers, and often serve their time quietly and passively, comfortable in their relatively uncomplicated and stable environments. Little is known about incidence of psychopathology, except again, for a relatively high incidence of alcoholism.

The pattern appears to be one of placing the individual in the unit where he is able to function, consistent with custody level and medical need. My impression is that this is done case by case, and that elderly and infirm individuals may be looked after by individual staff members and other prisoners. There is an interesting debate as to whether it is better to isolate or segregate elderly inmates in special units for their protection, or alternatively to integrate them, to keep them scattered throughout general population as a stabilizing influence on the rest of the inmates. They are less violent and there is relatively little sexual acting out.

Many of the elderly reject parole, feeling that the world has passed them by. "Project 60" in Pennsylvania is a rare program for elderly prisoners that attempts to involve community and families, but generally there is little public interest and few resources. Families are often gone or rejecting, or lost to contact, if they existed at all. It is quite common for older prisoners to come from broken homes or to have no family contact in the first place. They are also poor candidates for parole. The elderly generally have not involved themselves in prison programs, which were not intended for them in the first place, and are not employable in the community. Most are indigent. Whereas a 60-year-old man facing prison for the first time may be terrified of dying in prison, those who are in for many years often want to remain there, to die there; it is truly home, the only home they have known in a long, long time. A Federal Bureau of Prisons administrator reported that the major problem in management of the elderly has been in getting them to leave.

When they do leave, there are virtually no resources for these elderly. It can only be hoped that the prison administration will have had the time and resources to anticipate support, housing, and medical or mental health needs for each of the few elderly who do leave, and to have made adequate arrangements in advance.

CONCLUSION

It is evident that the problems of the elderly in the criminal justice and correctional systems are manifold and that there is a need for a comprehensive examination of the entire issue. Research is needed. It appears to me that the handling is chaotic. It is not at all clear that the system routinely abuses the elderly; it seems to try in a clumsy way, but there is no consistent policy and few resources.

The role of the psychiatrist and other mental health professionals is not different from that role with younger individuals, and can be ambiguous, but the need for proper evaluation, diagnosis, treatment, and appropriate diversions at all levels of the system is imperative. The problem of alcoholism in the elderly, both acute and chronic, has received relatively little attention. It must be recognized as a serious, if embarrassing problem for society, as is the elderly criminal him or herself. There is fairly clear evidence that alcohol plays a major role in violent crime by the elderly, and the literature suggests that it plays a major role in much of the lesser crime as well. The lack of resources to do these evaluations for prevention and program planning, as well as for treatment and disposition, is readily apparent. It is an issue which must be addressed by the policy makers as the numbers increase, if there is to be any meaningful input into the courts, any protection for society and for the elderly. Much elderly crime appears to be preventable, and the elderly in prison need more adequate and appropriate services. The author hopes that psychiatrists and other mental health workers will take the lead and be available and sensitive to the complex problems generated by this population.

REFERENCES

1. Newman E, Newman D, Gewirtz, ML: *Elderly Criminals*. Cambridge, Mass, Oelgeschlager, Gunn and Hain, 1984.
2. Malinchak A: *Crime and Gerontology*. Englewood Cliffs, NJ, Prentice-Hall, 1980.
3. Walter BL: *Aged as criminals—Pennsylvania*. Pittsburgh, Pa, University of Pittsburgh Press, 1980.
4. Krajick K: Growing old in prison. *Corrections Magazine* 1979; 34:32–43.
5. Grossman J, MacDonald D: *Survey of Inmates 65 years old and over, February 1981*. Albany, New York, New York State Department of Correctional Services Division of Program Planning, Research and Evaluation, 1981.
6. Kaplan HI, Sadock BJ: *Comprehensive Textbook of Psychiatry/IV*. Baltimore, Md, Williams and Williams, 1985.
7. Jarvik L, Small G: *Aging, The Psychiatric Clinics of North America*. Philadelphia, Pa, W. B. Saunders Company, 1982, vol 5.
8. Busse E (ed): *Behavior and Adaptation in Late Life*. Boston, Mass, Little, Brown, 1969.
9. Rosner R, Wiederlight M, Schneider M: Geriatric felons examined at a forensic psychiatry clinic. *J Forensic Sci* 1985; 30:730–740.

19

Ethical Considerations in Geriatric Forensic Psychiatry

JAY E. KANTOR

Certain issues concerning professional obligations come up in the practice of forensic psychiatry that do and should leave the psychiatrist in a quandary. These are problems that cannot be resolved simply by consulting medical textbooks, or lawbooks, or professional codes of ethics. Moreover, they are problems that cannot be avoided by any psychiatrist who conceives of himself as an ethical professional. Such problems are common occurrences in medical practice in general, and constant occurrences in the practice of forensic psychiatry. There is a growing consensus that the best way to deal with these problems is to have some familiarity with formal ethical theory and ethical analysis. Although such a familiarity may not always be sufficient in itself to help resolve the issues, it will always provide a reasoned and structured method of approaching solutions.

The forensic psychiatrist working with geriatric cases must begin with training in general psychiatry. I think it is useful to begin on a parallel course with ethics. Thus, I will start by discussing some ethical issues that occur in psychiatry in general, move on to some ethical issues that occur in the general practice of forensic psychiatry and, finally, reach the special geriatric issues.

JAY E. KANTOR • Department of Philosophy, College of Liberal Arts, New York University, New York, NY 10016; Department of Psychiatry, School of Medicine, New York University, New York, NY 10016; Forensic Psychiatry Service, Bellevue Hospital, New York, NY 10016; Franklin Delano Roosevelt Hospital, Veterans Administration, Montrose, NY 10548.

305

The first step in treating an ethical problem in medicine is to see the presence of the problem. That is difficult at times because an ethical issue may be hidden. Sometimes an ethical issue may appear in the guise of a medical issue. Sometimes a practitioner may disguise an ethical issue as a medical issue in order to avoid the frustration that ethical dilemmas present.

A step after diagnosis might be an inquiry into the etiology of the problem. Let us begin with diagnosing and examining the causes of some ethical issues in an imaginary case. We can then use the specific case as the basis for a more general discussion.

A 79-year-old woman is admitted to the hospital with gangrenous toes caused by circulatory problems. Her prognosis would be good if amputation were to be done, poor if it were not done. She refuses the surgery, and the attending physician calls in a liaison psychiatrist to evaluate her mental capacity to make the decision. The psychiatric interview reveals that she has been leading an active life. She has two adult children. She has a number of friends, and a seemingly good relationship with her children. She lost her husband to cancer the year before. Although clearly in pain and somewhat rambling in her conversation, she seems to understand her medical condition and the consequences of refusing the surgery. The psychiatrist suggests that she speak to her children, with whom she claims a close relationship. She refuses and asks that her family not be told anything about her condition. She says that if God is ready to take her, then she is ready to go "to join her husband." A suggestion is made that she speak to clergy. She refuses, saying "It is all in God's hands now." In the course of the interview, she says that the psychiatrist reminds her of her son and, after eliciting a promise from the psychiatrist not to tell anyone, she strongly hints that she had "made the way easier" for her husband by agreeing to his request for an overdose of pain medication.

What should the psychiatrist do? He seems faced with a number of issues. First, there are the issues relating to her refusal of treatment. On the one hand, he feels that she should have the surgery and that perhaps he has an obligation to do anything he can to make sure the surgery is done. He thinks that perhaps he should break confidentiality and tell her family so that they can exert pressure on her to change her mind. If she were to ask later why they were told, he can always lie and claim that he did not tell them. He thinks too that perhaps he should acquiesce to her treating physician's request, and evaluate her as mentally incapacitated so that a surrogate can be chosen who will sign for the surgery.

On the other hand, although her decision is not one that he would make if he were in her position, he thinks that in fact she understands what is going on and, given her understanding, he feels that he has no right to impose his own opinion on her.

While mulling this over, he begins to wonder what he ought to do about her near confession that she had helped her husband commit suicide. Should he keep confidentiality, or should he report the admission to someone?

Moral dilemmas usually occur because we find ourselves pulled by a very few ethical theories that often seem to have contrary implications about the "right thing to do." Let us break down some of the moral conflicts faced by the psychiatrist in this case, and try to find their theoretical underpinnings. In thinking that he ought to help force the patient into treatment, he is operating from one of two major ethical theories that form our collective superego—utilitarian theory. Utilitarian theory holds that a morally right action or policy or law is one that produces more happiness than any other action or policy or law would have produced. The utilitarian moral actor considers the probable consequences of his possible actions. He is then obligated to choose the action that he reasonably believes will have the consequence of producing more happiness than any available alternative. If the alternative courses of actions all have dismal consequences, then he is obliged to perform the action that has the least dismal consequences. Telling the truth to patients, obtaining informed consent, keeping patients' confidences, each and all are utilitarian obligations if and only if they will have the consequence of producing more total happiness than would lying to patients, or treating without consent, or "spreading abroad what is heard and seen in the course of treating the ill." Utilitarian theory has been used as the justification for what has been called medical paternalism. That is the belief that the physician–patient relationship should be based on a parent–child model. The physician acts in the best interests of the patient, but is no more obligated to let the patient make a treatment decision than is a parent obligated to let his 3-year-old child decide whether candy or broccoli should be served for dinner. Ethical medicine used to be based on the utilitarian-paternalistic model. In spite of recent major reevaluations of the nature of the physician–patient relationship, the model is still widely used.

When the psychiatrist questions whether he has a right to impose his beliefs on the patient, and questions his right to break confidentiality, he is operating from the second of the important theories of moral obligation—Kantian, or autonomy theory.

Whereas utilitarian theory focuses on the production of happiness as the source of all obligations, autonomy theory focuses on respect for persons. Unlike lower animals, which Kant, at least, believed behave only according to instincts and the pleasure/pain principle, normal adult humans have the ability to overcome their instinctual drives and rationally choose their courses of action. They can consider different courses of action and choose those that they believe are morally right. That is, the normal adult's behavior is not bound by the laws of his biology. He is able to create his own laws. The normal adult is "autonomous"—self-governing and morally responsible for his actions. For autonomy theory, this ability to be a responsible moral agent gives the agent dignity and makes him worthy of respect. It gives him the status of a "person" with infinite value rather than that of a "thing" to be valued simply for its usefulness. A person's ability to contribute to society, his age, his social status as rich man, poor man, beggar or thief, are all secondary to his primary status as a person equal to all other persons. His being a patient or a physician are roles overlaid on and limited by his personhood. As an autonomous person, he has the right to make his own decisions. To deny him that right is wrong. At the worst, he may be denied his autonomy by being used without his consent in order to benefit others (e.g., as a victim of crime or as an unconsenting research subject). Seemingly more benign, but just as wrong for autonomy theory, attempts may be made to make decisions for him for his own benefit. That is, he may be treated paternalistically. Clearly, paternalistic treatment of rational beings is anathema for autonomy theorists. It shows a lack of respect for the person's own ability to make decisions and denies him the dignity of full responsible personhood by infantilizing him. The autonomy theorist does believe that we have obligations to produce happiness, but never by using a person without his consent. (Utilitarian theory, in principle at least, does not entail that limitation.)

The recent thrust in medical law and ethics has been towards an acceptance of autonomy theory as the ground for moral decisions in medicine. Many of the ways that the physician-patient relationship is presently defined flow from a growing acceptance of the patient's right to autonomy.*

For example, the right to informed consent may be looked upon as corollary to the recognition of autonomy. A person cannot exercise his autonomy unless he has enough information to make a decision, and also the right to make a voluntary decision. The right to refuse treatment is a recognition of the autonomy theorist's claim that we cannot paternalistically impose our own view of what is beneficial on another person.

After a few hours of soul-searching, our imaginary psychiatrist decides to evaluate the patient as mentally capable of making the decision to refuse surgery. He thinks to himself: "She is old, has lived a full life, and hasn't that much longer to go anyway." As for the second issue—the question of whether or not to report her admission about her husband,

---

*Though, as we shall see, utilitarian theory also may provide justifications for some of these concepts.

he decides to forget about it. He thinks "Her husband was dying anyway, and why put her through the stress of legal proceedings. Reporting her statements could benefit no one and could only harm her." He has used utilitarian grounds to make his decisions. In regard to the patient's refusing treatment, he judged* that the patient had had her life's quota of happiness. His decision not to report her statements about her husband's death was made on the basis of the probable consequences; no one would have benefited if he had reported, and his patient probably would have been harmed if he did report.

However, things are really not that simple for the utilitarian. The theory would require the psychiatrist to perform a more rigorous analysis. The psychiatrist would have to consider not only the immediate effects of his decision on the patient, but also the possible effects on other staff, the patient's family, and on society in general. Applying utilitarian theory is never an easy matter, for it is really an application of inductive principles to make predictions about the total effects an action or policy will have upon society in the future. Because of the complexity and size of the data, it is very easy to pick just enough data to rationalize a predetermined decision. For example, physicians who want to avoid the emotional difficulty of confronting a patient with a poor prognosis will often use a utilitarian rationalization for lying or withholding the truth—"Telling the patient will make him depressed." That is likely to be true; however, a complete utilitarian analysis would require evidence for that conclusion, and would also require consideration of the consequences of lying on the future plans of the patient, the effect of trying to sustain the lie on the rest of the health care team, and the effects that a policy of lying would have on the rest of society.

Every action has a ripple effect, and utilitarian theory requires an analysis of the extent of ripple effects. The psychiatrist's decision may have been the same after such an analysis, but nevertheless he is obligated to perform the full analysis.

An autonomy theorist would have made the same decisions in regard to this patient, but for very different reasons. In regard to the patient's refusal of treatment, he would have taken that as one of her rights as an autonomous being. In regard to revealing the information about her husband, the autonomy theorist would have said that once the psychiatrist made the promise to keep quiet, he had an absolute obligation to keep that promise, regardless of the consequences.

There are many instances in which the two theories do converge in their conclusions. In fact, sometimes an ethical dilemma seemingly caused by conflict between these theories can be resolved after a full and

---

*I have not used the term *judged* by accident here.

careful analysis of the implications of each theory about the issue. Unfortunately, that does not always seem to happen, and there are also many instances in which the theories seem to diverge. Suppose, for example, our imaginary case were a bit different. Suppose that the patient, after eliciting a promise of silence from the psychiatrist, said that she was going to kill her otherwise healthy husband. The utilitarian psychiatrist might have an obligation to break the promise and report. The autonomy theorist psychiatrist would still have to keep silent.*

The case raises issues that include paternalism, informed consent, the right to refuse treatment, and confidentiality.

Each of these have become standard categories of issues in biomedical ethics. Each poses unique problems for the general psychiatrist and very difficult problems for the forensic psychiatrist. I will first briefly sketch the difficulties, and then discuss them more fully.

Autonomy Issues

To begin with, a psychiatric patient is, by definition, less than autonomous. Patients come to treatment or are brought to treatment for the very reason that their abilities to control their own lives and decisions are at least somewhat impaired. Thus, those clinicians who stress the importance of recognizing patient autonomy confront a dilemma at the very onset of treatment—how to respect the autonomy of persons who are less than autonomous. To put the problem another way, How much lack of autonomy in a person justifies treating him paternalistically?

Double-Agency Issues

With some aspects of mental health care—involuntary commitment, the management of patients who act out, and much of the work of forensic psychiatrists, to name a few, we are not even sure that the psychiatrist is involved in providing health care. The ethical issue of double-agency is especially cogent here.† More than any other type of health care professional, the psychiatrist is asked to not to treat his "patient," but to protect the state, or the public, or individuals, from his "patient."

---

*To those who do not like this implication of autonomy theory, the theorist would say "Be careful about the promises you make!"

†For excellent discussion of this issue, see the articles by Samuel Gorovitz[1] and Ruth Macklin[2] in Volumes 1 and 2 of this series.

Conceptual Issues

The problem of providing objective, value-free definitions of mental health and mental illness is always a real presence in any discussion of the ethical issues relating to mental health care. It is another issue that is central to the work of the forensic psychiatrist. They face other, related conceptual issues too. The forensic psychiatrist who steps into an insanity defense puts himself knee deep into the ancient philosophical issue of free will versus determinism. Underlying his clinical judgment and testimony about a defendant are conceptual judgments about the criteria for calling actions voluntary and for calling persons morally responsible.

AUTONOMY ISSUES AND THE GERIATRIC CASE

The elderly person who comes in contact with the mental health care system often presents with distinctive ethical issues. First, there are the general psychiatric ethical issues.

Let me begin anecdotally: my mother was an immigrant to this country, arriving here with her parents from czarist Russia at the age of nine. She grew up in the low end of the lower middle class. She became an adult well before autonomy theory had begun its capture of medicine and law. Her notion of a good physician was probably little different from other persons of her era, culture, and class. A physician was judged not so much for his medical competency (the assumption was that most physicians were medically competent), as for his "good bedside manner." A good bedside manner meant that he was kind, caring, and available when needed. Of course, that included a willingness to do housecalls, even though he was obviously "a very busy man." I remember my mother pausing from ministering to our childhood illnesses in order to clean house in preparation for his housecall.

Her loyalty to him was absolute, and no neighbor's physician could stand up to comparison.

In short, the good physician was the paternalistic physician. He was what Aristotle called a "Great Souled Man"[3]—knowledgeable, wise, and, although not overly immodest, certainly aware of his practical virtues and his standing in society. His judgment and advice were to be taken as law.

When my mother was in her eighties it became necessary to find a cardiologist for her. Being a medical ethicist of Kantian leanings, I set out to find someone who was not only up-to-date in cardiology but was also up-to-date in medical ethics. I found such a physician. The results were, or should have been, predictable. On the way back from the office visit she volunteered to me that she had not liked him very much. When I asked why, she replied "He told me too much. I didn't want to know all that."

My mother was not an exceptional instance. She was, as I said, probably very typical of persons of her age and class. She was assimilated into American society. In her later years she became active in a senior citizen center. There she dutifully, willingly, and regularly, attended the lectures that informed her of her rights as patient and consumer. She had even marched for the rights of senior citizens. Perhaps all that is what fooled me. I imagine that just as old dogs let loose from the leash have a hard time fending for themselves, those who have spent years looking upon physicians as gods have a hard time with the new existential freedom of autonomy theory.

Too, she was raised in the still-standing tradition that the old are to be taken care of. She felt that she had paid her dues to family and society, and now deserved the right to be a passive recipient of care, rather than having to work actively at making informed choices.

Still another obstacle to her autonomy was her "membership" in the public health care delivery system. The physical strain demanded by modern public health care delivery—making appointments at various clinics, traveling to those clinics, waiting in lines, and filling out forms, is not conducive to active participation in decision making for those who fatigue easily. For many of the old dependent on the public health care bureaucracy, taking care of their health has become full-time and difficult labor.

Certainly my mother's attitude toward physicians extended to her attitude to authority in general. Just or unjust, there was no question in her mind about the dire consequences of disobeying the policy, or her landlord, or her social worker, or the Medicare and Medicaid systems. That early attitude learned in a czarist ghetto, must have reemerged and been bolstered in her old age as she became more feeble and more dependant on the state for her income and her health care.

Her diagnosis, although not dire by today's standards in cardiac care, was dire by the standards she had grown up with, and knowledge of it cut away denial of her advanced age and closeness to death. Moreover, she was an active person, and the increasing realization that her powers were weakening depressed her. Side effects of some of the multitude of her medications made her confused, and that increased her fears about her loss of powers and her increasing closeness to death. One of the physicians who saw her recommended a referral to a psychiatrist. I thought that was a good idea, if only because I believed that she needed a reevaluation of her polypharmacy of medicines. (My attempts to suggest that to her inevitably failed because, after all, if the prescriptions were wrong or had bad side effects, the physicians would not have prescribed them.)

The suggestion of referral to a psychiatrist had its own predictable effect. In spite of her intellectual awareness that psychological therapy had become run-of-the-mill for many obviously sane people, she was still a product of her generation and class. Her gut reaction was that referral to a psychiatrist was an accusation that she must be crazy. No talk of changing medication or of helping her with sleep problems worked. She adamantly refused to see a psychiatrist. The implications of seeing a psychiatrist were strong enough to make her assert herself and go contrary to the physician's recommendation.

The moral of the story is complex. It certainly did not end with my miraculous conversion away from autonomy theory. But it did cause me to reappraise the application of the theory. Although the law, and codes of ethics, and many ethicists, have been calling out, "Haven't you heard, Paternalism is dead!," many patients have not come to the forest to hear them.*

The educated, middle-class patient may be comfortable enough in the presence of physicians to assert his rights. The young, street-wise person may be able to assert his rights because he is capable of going against, or around, authority. But there is a large residue of patients that do not have those assertive capabilities. Numbered among these are certainly many of the elderly.

The psychiatrist attempting to deal with the aged in an ethical manner has to sail a narrow and rocky strait. On the one side, in spite of the law's warning that persons are competent until shown otherwise, he will see that the elderly are often treated as incompetent until shown otherwise. In the worst, but not uncommon instances, they are treated as incompetent even when they are clearly competent. There are still many physicians that are surprised when they find they can hold a rational conversation with a 75-year-old. In spite of that, many of the same physicians will walk away from the bedside of such patients and, without a second thought, allow the family to make treatment decisions for the patient.

That is not to make the claim that all elderly patients are competent. Nor is it to say that there is an absolute obligation to recognize the autonomous self-determination of every patient. The requirement of autonomy ethics is that all those who are capable of autonomous decisions should be treated as autonomous. No one who believes in an auton-

---

*Of course many physicians have heard the call but seem to have ignored it.

omy-based ethic would deny that we have the right, if not the duty, to be paternalistic to those who are clearly incapable of self-determination.[1] But that incapability must be proven, not just assumed.

There is a wide range of patients that are neither delusional or demented enough to be clearly incompetent, nor "together" enough to be seen as clearly autonomous. It is with these patients that ethical dilemmas frequently emerge. Their lack of autonomy may have various causes.

First, there are those patients who, like my mother, are capable of self-determination but seem to prefer to be passive and choose paternalistic treatment.

What should be done with the patient who rejects his own autonomy? For example, the patient who is unwilling to hear the results of the diagnostic tests or his prognosis? Or the patient who, as an attempt to get informed consent for a procedure begins, says "I leave it up to you, doc. Just give me a pen and I'll sign."

Then there are patients that are passive for subtler reasons, having to do with transference reactions. Issues relating to transference raise difficult pragmatic and conceptual problems for psychiatric ethics.

Pragmatic Issues with Transference

On the pragmatic side, the psychiatrist must be wary of the possibility that the elderly patient (or, elderly evaluee) may unwittingly become his parent. He must also be wary of negative tranferences. It is easy to become impatient with the slower thought processes of many of the elderly, and easy to become impatient with their tendency to beat around bushes, and easy to become impatient with their impaired hearing.* Such impatience, immoral in itself, can also have immoral consequences. For example, it may easily lead to a hasty judgment that the treatment decisions for the patient should be made by someone other than the patient. Any ethicist who has dealt with medical cases involving the elderly has seen that scenario acted out a number of times in hospitals. On a semiconscious level, the psychiatrist simply may not want to take the extra time and care that might be needed to interview the patient, or to explain procedures or treatment plans to the patient, or to adjust medications until the patient regains a capacity that was dulled by medication. He may rationalize that impatience by saying he has no time for such "niceties that have nothing to do with treatment." On the unconscious level, the psychiatrist may be emotionally reacting to agendas

---

*Remembering William James' claim that we are sad because we cry, I sometimes wonder if we can get angry (at the deaf) because we yell.

that have more to do with his own unresolved feelings towards his parents than with anything that is immediately and presently happening with his elderly patient. The psychiatrist who reacts in these ways and ignores the patient in the treatment decisions has disguised the ethical issues concerning consent and truth telling as a medical issue concerning mental incapacity. The pragmatic issue is soluble. The solution can be expressed in different ways; in medical terms, we can say that it requires that the psychiatrist conform to good psychiatric practice, which includes a constant awareness that transference problems always lurk. In ethical terms, we can say that the psychiatrist owes his patients a duty of respect and owes himself the Socratic duty of self-knowledge. In Kant's terminology, he has an obligation to treat his patients with "practical love" rather than with "pathological love." That is, he has an obligation to be dispassionately empathetic to all patients whether or not he feels emotional empathy towards them.[5] To lose that "close distance" would be to unfairly favor some patients, unfairly disfavor others and, as well, probably interfere with good psychiatric treatment. That is not to say that the psychiatrist should be callous. That is another extreme to be avoided, although it seems to be a common danger faced by those constantly working with seriously ill patients. Presumably not a new danger either, as Kant said in writing of the immorality of callousness: "In England butchers and doctors do not sit on a jury because they are accustomed to the sight of death and hardened."[4]

Conceptual Issues with Transference

Even if the psychiatrist is alert to and aware of the transferences, the ethical issues relating to transference still will not be totally resolved.

Many psychiatrists believe that effective psychiatric treatment excludes the possibility of full patient autonomy. I am not speaking of physicians who think that the current stress on patients' rights is an infringement on their own right to benign despotism. More subtle problems are involved even for the more democratic psychiatrist. First, the nondirective analyst who has formed a conclusion about the source of a patient's problems may believe that effective therapy requires that he withhold his conclusions from his patient. That is, his function is not to tell the patient what is wrong with him and why, but instead to lead the patient to draw the conclusions himself.

Related to that are theoretical problems relating to transference which may entail an ethical paradox. Many believe that successful therapy depends on a successful transference being achieved during some stage of the therapy. If that is so, then there is a stage in successful therapy in which the relationship between psychiatrist and patient is not the adult-to-adult relationship between physician and patient required by autonomy theory. On the contrary, in this model at least, successful therapy seems to require a paternalistic stage. That is, a stage in which the relationship between psychiatrist and patient is more likely to be a parent–child relationship, rather than an adult–adult relationship. During this stage, it may be said that the patient has unconsciously ceded his autonomy.

Hence, the psychiatrist may be faced not only with problems of what to do with the patient who prefers to be passive and thus consciously cedes his autonomy, but also faced with the problem of what to do with the patient who unconsciously cedes autonomy.

We should also add to those two problems the derivative conceptual problem of differentiating between the two. That is, the problem of determining whether the patient who throws his fate into the hands of the psychiatrist is making a conscious and truly voluntary choice, or whether his waiver of autonomy is coming from the transference.

Conceptual Issues about Mental Health

We can see the paradox revealed in another way if we take a broader look at the goals of mental health care. The concepts of mental health and mental illness have been undergoing a major change over the past hundred years. Mental illnesses used to be ascribed to persons because they exhibited behavior that deviated from the statistical or accepted norm or else because their behavior deviated from what was believed to be normal and natural behavior for our species. Insanity and incompetency were determined in the same way. The former belief still persists in the lay mind. The latter belief, which has its foundation and justification in the Natural Law theories of Aristotle[3] and St. Thomas Aquinas,[6] still persists in many ways, places, and laws among laity, clergy, and health care professionals.* The newer criteria are leaner, and may simply boil down to one: An individual is mentally ill to the degree that he lacks the internal ability to govern his own life. That is, mental health is internal autonomy, mental illness is a lack of internal autonomy. We can see this reflected in the definitions given in DSM-III.[7]†

We can now see a connection among the problems faced by the psychiatrist who accepts autonomy theory as both a theory of obligation

---

*Thus, Pope John Paul II, in accord with the Natural Law influence on Catholicism, called the practices of homosexuality, masturbation, and sex outside of marriage "moral disorders."

†This, of course, is oversimplified. For comprehensive treatments of this issue, see Boorse,[8] Culver and Gert,[9] and Moore.[10]

and a theory about the goals of psychiatric treatment; how can he recognize and respect his patient's autonomy rights if (a), his patient is, by definition, less than autonomous and if, (b), his goal as psychiatrist is to restore an autonomy that the patient lacks, and if, (c), that restoration depends on encouraging and taking advantage of a stage in which the patient is less than autonomous?

Put still another way, how can the psychiatric patient be treated as an adult if the theoretical framework of treatment presumes that he is a child?

Double-Agency

Problems are compounded for the forensic psychiatrist who works with the elderly.

There are the pragmatic problems already mentioned having to do with objectivity. On the one hand, it may be easy to feel sympathy toward some elderly defendants. One may feel that the impoverished elderly defendant arrested for shoplifting deserves an extra break. Or that it would be abetting cruel and inhuman punishment to help send an elderly defendant away for a long prison term. On the other hand, it may be easy to feel antipathy toward other elderly defendants. The psychiatrist may lose his "practical love" and allow himself to indulge his feelings of disgust and impatience with an elderly alcoholic street person who is filthy and reeking and verbally abusive. With that, he may find such evaluees getting short shrift in evaluations or examinations.

I have already mentioned the "easy" solutions to those pragmatic problems. That is, that professionalism in health care requires self-awareness and dispassion. However, I should be fair and admit that there are ethical problems inherent in those "easy" solutions. Some would say that there may be times when the psychiatrist has a stronger duty to override his duty of dispassionate professionalism in order to do the "right thing." Thus, Alan Stone, writing about this issue, tells of his experience as an army psychiatrist asked to testify at the court-martial of soldier being tried for theft. The issue was whether the soldier was kleptomanical or, at least, sufficiently kleptomanical to absolve him of responsibility for the thefts. Stone's professional clinical judgment at the time was that the soldier was, in fact, responsible according to the army's legal criteria. However, his interviews with the soldier, who was black and had suffered because of that, convinced Stone (in retrospect, at least) that the societal causes might morally excuse the soldier's behavior. And, in retrospect, Stone wonders if he ought to have had lied about the evaluation in order to get the soldier off.[11]

Stone's quandary was not unique. Psychiatrists who do not believe in capital punishment face the same issue if they are asked to evaluate prisoners for competency to be executed. They face the issue if they are asked to treat incompetent prisoners to make them competent to be executed. When Dr. Overholzer lied to government prosecutors about his own professional opinion that Ezra Pound was competent so that he could to save Pound from trial, he was not acting with evil intent. On the contrary, he thought he was acting morally. The forensic psychiatrist asked to medicate a very psychotic defendant who intends to make an insanity defense may face the same problem. He may believe that a medicated and competent but calm, passive, and unemotional defendant will have less chance of convincing a jury. The psychiatrist morally opposed to military conscription who is asked by a emotionally stable 19-year-old for an exempting psychiatric diagnosis faces the same issue, as may the forensic psychiatrist who believes there were morally but not legally mitigating circumstances that justified or excused an elderly defendant's criminal act. If his psychiatric judgment is that the defendant is competent or legally sane, and he further believes that the court will not be lenient with the defendent, he may face the same dilemma that Stone faced.

The moral conflict that can arise from seeming conflicting moral requirements of a person's various roles is often called a problem of "dual agency" or "divided loyalties." It is an issue that occurs, in one form or another, more commonly than any other in forensic psychiatry.

In his role as a treating psychiatrist, the forensic psychiatrist is obliged to do no harm to those in his care and more than simply doing no harm, he is obliged to help them. However, a psychiatrist is also a person and a citizen and he may have obligations to individuals, or the public at large which may go contrary to the best interests of his patient. In his role as an evaluating psychiatrist working for the courts he may have obligations that conflict with his role as healer.

Confidentiality

These conflicts often emerge as issues relating to confidentiality. The forensic psychiatrist evaluating a detainee for competency to stand trial, or evaluating a person for fiduciary competency, is not doing treatment. Thus, he may not be bound by the Hippocratic obligations of confidentiality. It has already been suggested[12] that it follows that he has a duty to issue a sort of Miranda warning to those he is evaluating. It would be a warning that, although he has all the appearance of a psychiatrist and sometimes is a psychiatrist, as an evaluator he is, in some sense, not a psychiatrist. Yet we know that even that sort of warning may not work to protect the interests of an arrestee. The arrestee may not be mentally capable either of understanding the warning or understanding the implications of the process of evaluation. Or, the evaluee may be capable of understanding the warning, but transference issues may operate in a way that make him unable to act on the warning. That is, in spite of the evaluator's warning to "keep your distance," the evaluee may still create a positive transference with the psychiatrist and unwittingly lose that distance. Such, for example, may have been the case with John Hinckley, who became enraged in court when a state psychiatrist whom he came to trust testified against him. The elderly defendant may be cowed by the system, confused by the process, may be afraid and may, despite the psychiatrist's warnings, still see him as a treating physician and thus a source of help to cling to. The psychiatrist, trained to be compassionate and, perhaps, flattered by the trust, may forget to reestablish the distance.

My mother was never arrested, and never reached the point at which a fiduciary competency evaluation was necessary. But I am certain that if she had, she would have revealed every peccadillo she had ever done to a psychiatrist evaluating her. Her awe of physicians and fear of the system would have overwhelmed any belief that she would be better off keeping silent. A Miranda warning almost certainly would not have stopped her.

Moreover, whether the psychiatrist is treating or evaluating, whether the evaluee keeps his guard up or not, there are still other problems about confidentiality that may not be easily resolvable with a Miranda warning. The Tarasoff decision[13] has refocused attention on the conflict between the role of healer and the role of citizen. We know some of the difficulties with Tarasoff—the difficulty of making accurate predictions about the seriousness of a threat, and the probable destruction of the therapeutic alliance once confidentiality has been breached. There are also difficulties about the scope of the duty to warn intended victims and, hence, difficulties about the scope and specificity of any proposed warning to patients or evaluees about the limits of confidentiality. For example, how does one determine whether the evaluee's or patient's intended crime is serious enough to warrant reporting? A recent court case in the state of Vermont has extended the Tarasoff duty to warn victims of intended personal violence to victims of intended property damage.[14] Should then the psychiatrist warn the owner of an empty building that his homeless patient or evaluee has been breaking in every night to sleep there and cook his meals on an unguarded sterno flame? And should the psychiatrist warn his patient or evaluee that he will report such behavior?

And what if the behavior in question has been intentionally brought up by the patient (or evaluee) as a problem he would like to work on in therapy?*

First, those who propose a Miranda warning to patients or evaluees may face a great deal of difficulty creating a statement that stipulates what will and what will not be held in confidence.†

Second, although the psychiatrist may formally separate his role as evaluator from his role as treating psychiatrist by a warning statement to the evaluee, he still may have a problem morally separating roles. For example, suppose that, as evaluator, he sees the evaluee's mental status worsening during an interview? Suppose too that the interview is proceeding quite well in gathering information for the courts. Should he continue the interview?

CONCEPTUAL ISSUES IN FORENSIC PSYCHIATRY

The conceptual issues that surround definitions of mental health/mental illness are not new, as we can see in the following quote from John Stuart Mill:

> There is something contemptible and frightful in the sort of evidence on which, of late years, any person can be judicially declared unfit for the management of his affairs; and after his death, his disposal of his property can be set aside if there is enough of it to pay the expenses of litigation—which are charged on the property itself. All the minute details of his daily life are pried into. and whatever is found which, seen through the medium of the perceiving and describing faculties of the lowest of the low, bears an appearance unlike absolute commonplace, is laid before the jury as evidence of his insanity, and often with success.[15]

Mill's concern remains with us. Certainly it is an issue in our imaginary case. The physician who called in the psychiatrist very likely believes that his patient is (in lay DSM-III terms) "crazy" for refusing the proposed procedure. Most likely, he has called in the psychiatrist for confirmation of his belief. It is a common occurrence, and liaison psychiatrists often face enormous peer pressure from attending physicians to find such patients mentally incapacitated.

I have said that we seem to be moving toward accepting the concept of "internal autonomy" as the basis for definitions of mental health and

---

*In a recent case in California an individual in family therapy with his wife revealed that he had been sexually abusing his step-daughter. The therapist duly followed the law and reported the information.

†Nursing staff on a forensic psychiatric ward face still another confidentiality issue—should they warn detainees that all their behavior in the ward is being observed and reported to the service's psychiatrists?

mental illness. Speaking very roughly, an individual is internally autonomous if there are no internal constraints on his ability to make rational choices. Optimally, a healthy individual actually makes use of that ability. That is, he attempts to be an active participant in his life rather than allowing himself to be a piece of wood moved only by outside forces.

As noble and as convincing as that approach is in principle, it is not easy to put into practice. It is not a simple matter to determine if and to what degrees an individual possesses or lacks autonomy. That difficulty emerges time and time again in the practice of forensic psychiatry. It emerges in insanity defenses, where the very issue at hand is to determine whether the defendant was autonomous when he committed the crime. It emerges in competency evaluations, particularly when dealing with the elderly. Whether the evaluation is for competency to stand trial, or for fiduciary competency, or for competency to refuse treatment, the issue is the same—Is the individual able to make rational, responsible, and voluntary decisions?

CONCLUSIONS

None of these issues have pat and easy solutions. There are many who believe some of the issues have no solution. There are also those, like myself, who are more sanguine. These are optimists who believe that the claim that there are no solutions is an unproven claim. It may be that solutions are attainable, but require long and hard work—sometimes a labor of decades rather than of hours. Moreover, some solutions may arrive in increments rather than in fell swoops. For example, I believe that progress has been made in unraveling the still problematic concept of informed consent. Conceptual issues, such as those relating to the criteria for autonomy, have facets that catch and reflect light from the sciences as well as from medicine and the humanities. The solutions will depend on input from many disciplines.

What follows here are not solutions but comments. The reader should be warned that they are comments that may show a bias in favor of an autonomy-theory approach.

The Conscious Ceding of Autonomy

There may be both strong utilitarian and autonomy arguments that go against the practice of immediate acquiescence with a patient's desire to be passive. In terms of the consequences of patient passivity, it is an empirical question whether the patient who is a passive recipient of treatment will really flourish more than the patient who is actively aware of what is happening to him. It may be that the patient who is active in his own care decisions may be healthier and get healthier in spite of the anxiety that goes along with active participation in decision making.

Moreover, in many cases there are available treatment options that are made unavailable to the patient if he does not know the truth about his condition. This is particularly the case with terminally ill patients where options such as hospice care and withholding of life-sustaining measures have now become acceptable.

As a result of those available options, there may be a question as to whether a patient who has voluntarily waived the right to information about his diagnosis and prognosis can give a binding informed consent to any major procedure proposed from that point on.

The temptation to go along with a geriatric patient's desire to be "taken care of" must be weighed against those factors just mentioned.

There is a sense in which anyone who contracts to get help from another waives some autonomy. If I bring my car to a mechanic for a repair that I could, but do not want to do myself, I have ceded some autonomy. That is, I have passively allowed another to do something I could do myself. After all, there are constraints on our ability to control actively all the aspects of our lives that we are capable of controlling. We sometimes must create priorities, and we believe that we are entitled to be passive at times. Full autonomy is an ideal that is approachable, but probably not fully attainable. The dividing line between being a responsible person and being a passive onlooker is often blurry. It would seem that the higher the stakes, the more active the person should be. The stakes in our context of forensic psychiatry are usually high. If the psychiatrist believes in the ethical ideal of autonomy, and if he believes in a psychotherapeutic goal of autonomy, then he has an obligation to encourage strongly patients to be autonomous.

The issues relating to transference and autonomy are more complicated. There is much literature on the ethical issue of truth telling and truth withholding in medicine, but most of it seems to deal with disclosing facts that patients could be expected to accept immediately as true. Very little work has been done on the theoretical relationship between telling truths to patients and convincing patients of the more complicated truths of psychotherapy. It is another thread that must be unwound before we have a full and adequate concept of autonomy.

Double-Agency

The pervasive issue of double-agency is complicated in other ways. Certainly the points raised by Stone are cogent. They seem to be supported by the post-Nuremburg belief that health professionals may have moral obligations that override their legal obligations. A strict autonomy theorist would say that Stone might actually have done the soldier a service by testifying that he was responsible for his thefts. To claim that ideas and actions have irrational causes is to claim that they are irrational ideas and actions. It is to imply that they are the outpourings of a maniac and thus not to be considered seriously, much less justified. Thus, the autonomy theorist might claim, a "humane diversion" of someone from prison might provide him with happiness at the expense of demeaning the essence of his humanity—his standing as a person responsible for his ideas and actions.

Those unconvinced by that argument might also consider how fairly such personal justice tends to be dispensed. It may be that it will tend to save those with whom the psychiatrist feels a prejudiced empathy—the verbal middle-class draft resistor, or the erudite poet, and sacrifice those who he cannot relate to.

In issues relating to confidentiality, it seems to me that in spite of the many difficulties about warning evaluees and patients, there still remains a strong duty to warn them. Although the specificity and range of the warnings present great difficulties, an ongoing attempt should be made to tune and retune the warnings rather than dispense with them entirely.

Moreover, when the psychiatrist, ever wary, sees the evaluee forgetting the warning, he has a duty to warn him again. No doubt this will often interfere with the evaluatory process, but the forensic psychiatrist should not forget that his role must be considered adversarial. He may be tempted to rationalize and thus try to view the evaluation as being in the interest of the evaluee. But here, as in all these matters, he should be guided by the Delphic rule that Socrates said should guide us all—Know thyself.

REFERENCES

1. Macklin R: A philosophical perspective on ethics in forensic psychiatry, in Rosner R (ed): *Critical Issues in American Psychiatry and the Law*, New York, Plenum Press, 1985, vol 2, p 19.
2. Gorovitz S: Could Oliver Wendell Holmes, Jr, and Sigmund Freud work together if Socrates were watching?, in Rosner R (ed): Critical Issues in American Psychiatry and the Law, Springfield, Ill, Charles C Thomas, 1982, vol 1.
3. Kant I: *Foundations of the Metaphysics of Morals*, Wolff R (ed). Indianapolis, Md, Bobbs-Merrill, 1969, pp 18–19.
4. Kant I: *Lectures on Ethics*, Infield L (trans). New York, Harper & Row, 1963, p. 240.
5. Aristotle: *Nicomachean Ethics*, Ostwald M (trans). New York, Bobbs-Merrill, 1962.
6. Thomas: *Summa Theologiae*, English Dominican Fathers (trans). London, Dominican Fathers, 1936.

7. *Diagnostic and Statistical Manual of Mental Disorders,* ed 3. Washington DC, American Psychiatric Association, 1980, p 6.
8. Boorse C: What a theory of mental health should be, in Edwards R (ed): *Psychiatry and Ethics.* Buffalo, NY, Prometheus, 1982, p 29.
9. Culver C, Gert B: *Philosophy in Medicine.* New York, Oxford University Press, 1982.
10. Moore M: *Law and Psychiatry.* New York, Cambridge University Press, 1984.
11. Stone A: *Law, Psychiatry, and Morality.* Washington DC, American Psychiatric Press, 1984, pp 254–255.
12. *Ethical Guidelines for the Practice of Forensic Psychiatry.* The American Academy of Psychiatry and the Law (second draft: July, 1985) p 1.
13. *Tarasoff v Regents of University of California,* P.2d 553, 118 Cal. Rptr. 129 (1974).
14. Peck v Counseling Service of Addison County, Inc., 499 A.2d 422.
15. Mill JS: *On Liberty.* Indianapolis, Md, Bobbs-Merrill, 1956, pp 84–85.

VI

Ethnic Considerations: Overcoming Potential Bias

20

Issues in the Psychiatric-Legal Evaluation of the Black Elderly

MICHAEL BELL AND RICHARD ELLISON

INTRODUCTION

The psychiatrist examining the black geriatric patient for legal purposes must, as in any other forensic psychiatric evaluation, have a systematic method of conducting the evaluation. It is important that the examiner be aware of factors that can influence his evaluation, his collection of the relevant data, and his application of that data to the psychiatric-legal issue at hand, such as determination of competency, guardianships, civil commitment, etc.

In order to facilitate an understanding of the elderly black patient, we will consider several important issues: the historical perspective of blacks in America; demographic variables of the black elderly population; health problems specific to the black elderly; and specific aspects related to the forensic psychiatric evaluation of the black elderly.

Over the years, relatively little empirical research has been conducted with regards to black psychiatric patients of any type. Whereas blacks tend to be overrepresented in certain clinical populations (e.g.,

MICHAEL BELL • University of Medicine and Dentistry of New Jersey, Newark, NJ 07103.   **RICHARD ELLISON** • Department of Psychiatry and Behavioral Sciences, School of Medicine, State University of New York at Stony Brook, Stony Brook, NY 11790; Department of Psychiatry, Veterans Administration Hospital, Northport, NY 11768.

state mental hospitals, prisons, etc.), they are grossly underrepresented in the medical profession, and in psychiatry in particular. Of all the graduates of United States medical schools in 1985, only 4.8% were black, and at least 25% of these were from predominantly black medical schools, such as Howard and Meharry.[1] As of September 1, 1984, only 4.4% of the psychiatric residents on duty in the United States were black.[1] The percentage of forensic psychiatrists who are black is slimmer still.[1] Undoubtedly, the problem of the scarcity of black related psychiatric research is due, at least to some degree, to this severe shortage of black mental health professionals.

Needless to say, there is a dearth of literature concerning psychiatric problems of the black elderly, and literally none addressing forensic psychiatric issues of the black elderly. However, there are several issues peculiar to the black elderly that the forensic psychiatrist should be aware of, and that we will discuss in this chapter.

HISTORICAL PERSPECTIVE

In trying to understand any patient, it is usually helpful to have some knowledge of the patient's cultural background. For black people in America today, particularly for the elderly black person, who may be only one generation removed from slavery, it is imperative that the clinician have an historical perspective.

Black people were brought to America in the early 1600s almost solely for the purpose of slavery. They were quickly and legally distinguished from their European "indentured servant" counterparts, and made slaves for life. Deliberate attempts were made to separate incoming slaves from their families, friends, and tribesmen, effectively annihilating the various cultures that were, for the most part, deeply rooted in traditions in which the family was central.[2] This evil was compounded by the common practice, during the generations of slavery that followed, of separating children from parents, and husbands from their wives, usually by the sale or trade of a slave to another owner.[2] The slave was considered to be less than fully human. Marriage was not legally recognized among black slaves.[2] Birth and death records of slaves were only informally maintained, if at all, by the slave owner, usually as part of property records.[2]

In the post-Civil War, "Jim Crow" South, blacks were effectively eliminated from most economic, political, and social gains. Black men wandered from place to place looking for work, and many abandoned families they could not support. Many were imprisoned or killed. The so-called matriarchal system of family life developed, with the grandmother as the head of an extended family.[2]

The black elderly person of today is likely to have been involved in some way, either directly, or through family members, with the massive, unprecedented waves of migration of blacks from the Jim Crow South to the North and West during and after the two world wars. Opportunities were not much greater for blacks in these parts of America, and they found themselves concentrated in large urban ghettoes.[2]

It is imperative that the clinician keep this cultural perspective in mind. The average elderly black patient is likely to have been the victim of Jim Crow-like practices of racial discrimination at some point during their lifetime. The average elderly black patient is unlikely to have had much, if any experience in dealing with white people as peers. He or she is likely to experience a visit to a doctor's or a lawyer's office as intimidating. In the lifetime of many elderly blacks, a mistrust of whites and/or white dominated institutions develops, in some cases as a necessary and helpful tool for survival, both physically and emotionally. Grier and Cobbs, in their classic work *Black Rage,* describe a "healthy cultural paranoia" that blacks tend to develop, based on an inherent suspicion of American institutions that have a tradition of racism.[3] This paranoia of the elderly black can lead to misdiagnosis by the psychiatrist, who may mistake as pathology something that is a cultural characteristic.

DEMOGRAPHIC VARIABLES

As of 1983, there were more than 2.2 million black people in the United States who were 65 years of age or older.[4] Elderly blacks accounted for 8% of all people 65 years of age or older in this country.[4] The majority of elderly blacks are females (60.2%). It is projected that by the year 2030, the black elderly population will have more than tripled, to approximately 7.3 million people.[4]

The black geriatric population is heavily dependent on Social Security payments for their income. This is particularly true of the black elderly person who lives alone, for whom it can be expected that almost 75% of his or her income will be derived from Social Security.[4] Forty-one percent of elderly black males, and 70.2% of elderly black females had incomes below $5,000 in 1983.[4] The elderly black male who is single is three times more likely than the elderly white single male to have a gross annual income of below $5,000.[4] In 1983, one third of all elderly blacks in America were living below the poverty level.[4]

With regard to living situations, elderly whites are more likely (72.1%) to own their homes than are elderly blacks (57.8%).[4] In addition, whites are almost twice as likely to reside in nursing homes. Five percent of whites, and 3.2% of blacks reside in nursing homes.[4]

With regard to education, there are further striking differences to be noted. Elderly whites are more than 2.5 times more likely to have completed high school than are elderly blacks.[4] In 1984, the median level of education for blacks age 65 to 69 was 8.7 years, compared to a median of 12.3 years of education completed by elderly whites in the same age group.[4] In the 75 years old and above age group, the median number of years of education completed is 6.8 for blacks, and 10.1 for whites.[4]

Of the myriad of distinctions that can be made between black and white Americans with regards to health care delivery and the epidemiolgy of diseases, one area in particular for the psychiatrist to note is that of cerebrovascular disease, which is responsible for dementia and stroke. Three quarters of all strokes occur in the elderly. The incidence of stroke in the elderly black is much higher than in the elderly white, undoubtedly because hypertension is two times more common in blacks than in whites.[5] The death rate from stroke is 66% higher for blacks than for whites.[5] American blacks as a group have the world's second highest death rate for stroke. The white death rate for stroke is 98.8/ 100,000. The death rate for American blacks is 164.6/100,000.[5](pp 17—37) The psychiatrist who is evaluating the elderly black patient should be mindful of the increased incidence of hypertension, cardiovascular disease, and cerebrovascular disease in this population. Screening for the possible presence of these problems, and assessing their effect on the mental status of the patient, is essential. And in the presence of any of these conditions, the physician must encourage the patient to receive adequate medical treatment, if they are not already doing so.

DIAGNOSTIC CONSIDERATIONS

Based on his life history and the experience of being black in America, the elderly black patient is likely to be culturally quite different from the psychiatrist examining him. The physician is more likely to be white, with a far more extensive educational experience, and a much higher socioeconomic status. Much has been written of the tendency for blacks to be misdiagnosed.[6,7] We have already alluded to the possibility of mistaking cultural paranoia as a pathological condition. There is also a tendency to diagnose schizophrenia and to avoid the diagnosis of affective illness in blacks. It is essential that the psychiatrist be aware that cultural differences can influence the accuracy of the conclusions that he draws from his psychiatric-legal evaluation. The psychiatrist must be able to communicate effectively with his patient, so that the appropriate diagnosis can be made. For example, Jones and Gray, in a 1986 article published in *Hospital and Community Psychiatry,* state:

> Language not understood is often considered evidence of thought disorder, styles of relating are sometimes misinterpreted as disturbance of affect, and unfamiliar mannerisms are considered bizarre. These mistakes are avoidable.[8]

An ethical issue arises if the psychiatrist, unable to communicate with an elderly black patient, assumes this lack of communication to be evidence of a thought disorder, cognitive deficits, or affective disturbance. It is essential to gather all of the relevant data needed to arrive at a conclusion that is not biased by cultural differences between the psychiatrist and the patient, nor biased by any racial stereotypes that the psychiatrist may assume to be true.[9]

The elderly black patient may not have the funds to obtain a private psychiatric evaluation, should a psychiatrist working for an institution or legal system have findings that the client considers incorrect and/or not in his best interests. On the other hand, the psychiatrist who is aware of the history and cultural background of the black elderly patient may overidentify with him or her, and produce a sympathetic, yet biased report. The physician must avoid this pitfall as well. It is therefore incumbent on the psychiatrist to be absolutely certain that he or she has made as fair and careful an assessment as is possible under the circumstances.

THE EVALUATION PROCESS

The most obvious and traditional manner in which to proceed in a psychiatric-legal evaluation is to obtain as thorough a history as is possible from the patient. In addition, information from as many collateral sources as possible should be obtained. These sources would include relatives, children, grandchildren, anyone living with the patient, neighbors, friends, health care providers who know the patient, etc. This type of investigation is basic to the process of a forensic evaluation, and will not be elaborated further here.

Another, less common method of gaining greater understanding of the elderly black patient is by means of making a house call as part of the evaluation. Most patients become anxious during a visit to a doctor's office. This anxiety can be even worse for the elderly black patient because of the racial and cultural differences previously discussed. Such anxiety can surely interfere with the forensic evaluation, perhaps to the patient's detriment. An examination in an environment that is familiar and comfortable can reduce this anxiety considerably.[10]

A more compelling argument for the house call as part of the forensic psychiatric evaluation of the black elderly patient is what such a visit may add to the evaluation itself, and to the knowledge and experience of the physician. Direct knowledge of the neighborhood and the immediate environment of the patient, of the actual physical characteristics of the patient's apartment or home can go a long way toward bridging the wide gulf between the life of the patient on the one hand, and the physician on the other. In addition, learning how the patient functions in his or her environment adds a great deal to the objectivity of the assessment of the patient's competency to manage their own affairs—a matter frequently at issue in forensic psychiatric determinations of the elderly.

If a house call can be safely and conveniently arranged, it is likely to be particularly useful in the evaluation of the black elderly. The physician's ability to empathize with his patient will be greatly enhanced.

THE EFFECTS OF RACISM

Racism is a fact of life in America and elsewhere in the world today. There are two general aspects of racism that are relevant to the forensic evaluation of the black elderly that will be discussed here. The first involves personal attitudes that have been colored by the effects of racism. The second involves the psychiatrist's role in procedures for legal redress for emotional damages wrought by overt racism.

We have already briefly alluded to the necessity for the physician to possess a minimum awareness of the history of racism in America and how it may have impacted on the patient. The black elderly patient of today is likely to have had limited and confined relationships with whites during his or her lifetime. The physician must be cognizant of how the past racial experiences of the patient may affect their current presentation.

Racism can have more current, overt effects on a patient. The landmark case of *Dillon v Legg* firmly established the principle that psychological pain and suffering caused by negligence is a compensable damage.[10] Courts have accepted a causal connection between discriminatory conduct and emotional distress. (*Gore v Turner*).[10] Courts have even been willing to infer emotional distress once a determination of racial discrimination has been made, equating outrageous racist conduct with psychological injury (*Seaton v Sky Realty Co., Inc.*).[9]

The exact psychological effects of racist behavior or racial epithets on an individual have not been studied. Neither would anyone suggest that blacks are more likely than anyone else to suffer any damages from racism. Nevertheless, the psychiatrist may be called on to assess these effects. Based on a knowledge of the patient's past exposure to racism, the physician may be able to determine the patient's sensitivities to racial slurs or racist behavior, and to judge the severity of the ensuing psychological injury or disability. Such considerations can pose both practical and ethical challenges for the physician. However, it is important to keep in mind that in this post-civil rights era, courts will consider psychological damages that are the result of overtly racist conduct as compensable, and knowledge of his patient will be helpful to the psychiatrist in performing an evaluation in such a case.[10] The psychiatrist must also be aware of his own attitudes toward racial differences, and not allow his own biases to interfere with an objective evaluation.

CONCLUSION

In this chapter, we have attempted to shed some light on a topic in which little has been studied or written previously. It is hoped that this discussion will be of assistance to the psychiatrist who evaluates black elderly patients. It is also hoped that psychiatrists will be mindful of the relevance of profound cultural differences between themselves and their black patients, and use that awareness to produce fair, accurate, and unbiased reports.

REFERENCES

1. American Medical Association: *JAMA* 1985; 254.
2. Bass B, Wyatt G, Powell G: *The Afro-American Family: Assessment, Treatment, and Research Issues.* Grune and Stratton, 1982, pp 3–12.
3. Grier WH, Cobbs PM: *Black Rage.* New York, Bantam Books, Inc., 1968, pp 102–129.
4. National Caucus and Center on Black Aged, Inc.: *A Profile of Elderly Black Americans.* A publication prepared and distributed by the National Caucus from Washington, D.C., February, 1985.
5. Hall WD, Saunders E, Shulman NB: *Hypertension in Blacks.* Yearbook Medical Publishing, 1985, pp 3–16.
6. Thomas A, Sillen S: *Racism and Psychiatry.* New York, Brunner/Mazel Inc, 1972.
7. Williams DH: The epidemiology of mental illness in Afro-Americans. *Hosp Community Psychiatry* 1986; 37:42–49.
8. Jones BE, Gray BA: Problems in diagnosing schizophrenia and affective disorders among blacks. *Hosp Community Psychiatry* 1986; 37:61–66.
9. Mayo JA: The significance of sociocultural variables in the psychiatric treatment of black outpatients. *Compr Psychiatry* 1974; 15:471–482.
10. Griffith WH, Griffith EJ: Racism, psychological injury, and compensatory damages. *Hosp Community Psychiatry* 1986; 37:71–75.

21

Special Considerations in the Psychiatric-Legal Evaluation of the Hispanic Elderly

SHELDON TRAVIN AND
HILDA BREWER-GIZZARELLI

This chapter is concerned with special considerations in the psychiatric-legal evaluation of a practically unknown sector of the American population, namely the Hispanic elderly. Even more than the general elderly population, minority elderly's mental health needs are unknown; moreover, they face enormous cultural and linguistic barriers to obtaining services. Their situation constitutes a "triple jeopardy—old, poor, and minority."[1] Minorities must cope not only with the normal problems of aging and its attendant physical decline, but also with socioeconomic discrimination and often with poverty, factors that greatly contribute to the difficulties of old age. The subject population of this chapter, the Hispanic elderly, an almost invisible minority within a minority, exemplifies this predicament.

The usual purpose of any psychiatric-legal evaluation of the elderly is to determine the subject's competency to participate in his or her own decision making. Evaluating the decision-making capacity of any older mental patient, who may have varying degrees of organic impairment, is

SHELDON TRAVIN • Department of Psychiatry, Bronx-Lebanon Hospital Center, 1285 Fulton Avenue, Bronx, NY 10456; Albert Einstein College of Medicine, Bronx, NY 10456.   HILDA BREWER-GIZZARELLI • Forensic Psychiatry Clinic, Bronx-Lebanon Hospital Center, 1285 Fulton Avenue, Bronx, NY 10456.

critical to many forensic issues, including informed consent for treatment and research, guardianship and substituted judgments, living wills, the right to die, and do-not-resuscitate (DNR) participation. These issues are, by their very nature, extremely complex. When the complexity that necessarily attends such decisions is compounded by differences in culture and language, the difficulty of adequately and accurately assessing the decision-making capacity of such a patient requires and demands attention to the special considerations that are addressed in this chapter. Therefore, our purpose is not to focus on the competency evaluation *per se,* but rather to explore these special considerations. We intend to explore the attitudes, ignorances, and misunderstandings of both evaluator and evaluatee in the evaluation and treatment situations when the doctor–patient relationship involves differences in culture and language.

Cultural and language concerns are of paramount importance to the Hispanic elderly. For all minority-group people, Butler and Lewis[2] consider two major issues to be extremely important: the unique cultural attributes of each minority group, and the majority culture's effect on older minority people. The authors feel that most of the available research does not focus enough on cultural factors. They believe that this shortcoming in our extraordinarily diverse society should be addressed by transcultural studies. Attention must also be given to the importance of language, particularly to the use of interpreters. Language barriers to adequate psychiatric evaluation and treatment plans must be recognized and overcome. In addition to the presence of a minority staff and physical accessibility of the services to the elderly, increased sensitivity of psychiatrists and other mental health workers toward minorities increases the likelihood that adequate services can be provided to the minority elderly.[1] Providing anything less than the full range of adequate psychiatric services has profound ethical and legal implications. We would add that without an adequate evaluation and understanding of the individual Hispanic elderly, appropriate treatment cannot take place.

DEFINING AND IDENTIFYING HISPANICS

A major problem in planning public services for minority populations in general, and for elderly Hispanics in particular, is the difficulty of defining and identifying persons belonging to the group. In the past, Hispanics have been included in demographic statistics among blacks, whites, and others. They have been distinguished imprecisely from the general population by attributes such as Spanish surnames, Spanish language use, and birthplace of the person and his parents. Even when separated into an Hispanic category, the problem remains of distinguishing between the multitude of ethnic groups, including Mexican-Americans, Puerto Ricans, Cubans, Dominicans, Colombians, and so forth. Butler and Lewis,[2] although they find no single satisfactory label, prefer to use the term *Latino,* the term they consider to be both recognizable and less offensive than others. Hayes-Bautista[3] argues, to the contrary, that imprecise definitions of differences among Hispanic groups could result in operational differences that could prevent the results of the populations studied from being compared with one another. He considers the terms *Hispanic* and *Spanish Origin* to be "misleading, stereotypical, and racist." Lampe[4] is also highly critical of the various ethnic labels bestowed on those individuals of Mexican descent. He argues that the use of such labels makes the study of each ethnic group more problematic and constitutes a hindrance to scientific research.

In this context, however, it is interesting to note that Dowd and Bengston[5] found that, although differences among older blacks, Mexican-Americans, and whites were found to exist, especially on income and self-assessed health, measurement of life satisfaction and frequency of interaction with relatives among these three groups were much more homogeneous. This study, which emphasizes similarities as well as differences, supports the "age as a leveler" hypothesis. This same hypothesis is supported by another study suggesting that older Mexican-Americans may have more in common with their Anglo and black peers than with younger Mexican-Americans.[6]

Another important reason for the tendency to overlook the Hispanic elderly is the common perception of Hispanic Americans as a rapidly increasing and relatively youthful population. The 1980 census counts Hispanics at 14.6 million, or 6.4% of the United States population. Compared with the general population's increase of 50% between 1950 and 1980, Hispanics have increased by 270%. If this trend continues, Hispanics will constitute at least 15% of the total population by the year 2020, which will make them the largest American minority group.[7] In 1980, 20% of Hispanics were younger than 10 years of age compared to 14% for the general population. Among Hispanics, the median age was 23 years. Hispanics over 70 years of age, however, constituted only 3% of their population, compared with 7.7% of the nonminorities.[8] Other estimates place the older Latino population at 4%, and for the Cuban elderly (over 65 years) at 8.6%.[2] Because death certificates have not in the past listed Hispanics separately, it is impossible to tabulate accurately the death rate. There are some indications in the data for 1950 and 1960 that the general Hispanic category had a higher mortality rate than whites, but not as high as blacks.[9] Although

Hispanics in general appear to suffer somewhat less coronary heart disease, they are just as prone as whites to develop cerebrovascular disease. Diabetes mellitus has increasingly been recognized as a major health problem among Hispanics, especially for Mexican-Americans and Puerto Ricans.[8]

There is a tremendous shortage of Hispanic care providers. In medical schools, for example, only 1.9% of the enrolled students are Hispanic. Even in neighborhood health centers, Hispanics make up only 19.3% of the registrants.[9] Additional demographic data relevant to our subject reveal that although Hispanics live in every state of the union, over 60% live in California, Texas, and New York. Most Hispanics are bilingual; only 10 are monolingual in Spanish.[9] As is true of other minorities, Hispanics are disproportionately located below the poverty level in this country.

UNDERUTILIZATION OF PSYCHIATRIC RESOURCES

Older people in general underutilize community mental health services,[10] and it is likely that this underutilization is even more pronounced among Hispanic elderly. Lacayo[11] has reported on the needs of Hispanic elderly for professional social and health care services that are not being met in the community. In a recent study of older persons enrolled in a Community Services System (CSS), a comprehensive care management system in Tucson, Arizona, Green and Monahan[12] found that Hispanic elderly used considerably fewer agency services than did Anglo elderly, even though the Hispanic elderly demonstrated higher levels of impairment. However, the Hispanic elderly utilized considerably higher levels of informal support, a fact Green and Monahan felt could explain their lower use of agency services. Torrey[13] has enumerated the reasons he believes traditional mental health services are irrelevant to urban Mexican-Americans. Torrey argues that traditional mental health services are inaccessible, problematic regarding language considerations, class bound, culture bound, caste bound; moreover, he points out that Mexican-Americans have their own system of mental health services in San Jose and Santa Clara County: curanderos or folk healers and mental health "ombudsmen." Similarly, Acosta[14] singles out social class and cultural dissimilarity, the tendency to rely on general physicians first for psychological problems, and language differences as significant barriers to mental health services for Mexican-Americans. However, he feels that the situation can be remedied and he makes specific recommendations which will be discussed later. He also admonishes the dominant Anglo-American culture for having maintained erroneous assumptions about Mexican-Americans, which has enabled it to minimize the significance of the underutilization of psychiatric services by this minority population. Unlike Torrey, Acosta does not believe that Mexican-Americans are prejudiced against seeing Anglo therapists,[15] nor does he believe that *curanderos* (folk healers) are widely used.

There is every indication that the Hispanic elderly's utilization of nursing homes is low. There has never been significant usage of nursing homes by minorities.[16] In a study of the utilization of nursing homes by Mexican-Americans in the southwest, Eribes and Bradley-Rawls[17] found nursing homes to be significantly underused by Mexican-Americans. Every attempt was made by Mexican-American families to keep their elderly out of institutional facilities. This reflects, in part, the tendency of Hispanic-Americans to have strong familial and community connections and to respect their elders and value their support in times of crises.[8] Nursing home placement is not considered a culturally acceptable choice; instead it is viewed as the last possible choice.[17]

ATTITUDES TOWARD MENTAL ILLNESS

Perceptions of mental illness and attitudes about psychological problems profoundly influence whether or not the individual and the family will seek help and the kind of treatment they would be willing to accept. Among the elderly, a general bias and fear of mental disorders and mental health professionals causes them to prefer to be treated for mental problems by general physicians.[18] Undoubtedly, Hispanic elderly would also be expected to seek first professional help for mental disorders from general physicians, especially because the Hispanic elderly clearly underutilize community mental health resources.

Although there is a paucity of data about Hispanic elderly's perceptions and attitudes toward mental illness, there have been some interesting studies concerning this issue involving general Hispanic groups, especially Mexican-Americans. In a large group of Mexican-Americans studied in Los Angeles, the majority were found to prefer being referred to general physicians for mental problems.[19] Regarding perception of severe mental illness, systematic interviewing of residents in Anglo and Mexican-American households in two East Los Angeles communities of similar socioeconomic status revealed, in an initial analysis, that there were few statistically significant differences between these two major groups.[20]

In an effort to determine the connection between attitudes toward mental illness and the usage of language, Edgerton and Karno[21] further evaluated data derived from interviewing 224 Anglo-Americans and 444

Mexican-Americans, 260 of whom spoke mainly or only Spanish. Although they found no statistically significant differences in over 75% of the responses, they did find meaningful differences in the perception of mental illness between the two Mexican-American groups in six response categories: (a) depression (English speakers believed the depression was due to the described woman's "loneliness," whereas Spanish speakers felt it was due to her "nervous" condition), (b) juvenile delinquency (English speakers felt this resulted from faulty parent–child interaction, but the Spanish speakers tended to blame the child, (c) schizophrenia (English speakers described it as "mental" and Spanish speakers as "nerves"), (d) inheritance of mental illness (Spanish speakers, more than English speakers, believed that mental illness is inherited), (e) relationship of prayer and cure (Spanish speakers believed prayer to be more important than did the English speakers), and (f) role of the family (English speakers felt a mentally ill person would better recover away from his family, but Spanish speakers believed the person would improve more at home). Edgerton and Karno concluded that English-speaking Mexican-Americans and Anglo-Americans were basically the same in their perception of mental illness; that they spoke English was itself a reflection of their acculturation. Mexican-Americans who spoke mainly or only Spanish, on the other hand, differed meaningfully in their perception of mental illness; these differences, in turn, reflected their cultural distinctions. Keefe's[22] assertion that Mexican-Americans underutilize public mental health services, as well as psychiatric care in general, primarily results from their perception of mental illness and its required treatment supports the findings of the study cited earlier.

Obviously, much remains to be learned about such differences. As Acosta,[14] whose study of significant barriers to Mexican-American access to mental health services was mentioned earlier, points out, the differences between Anglos and Mexican-Americans' perceptions of conditions that do not constitute severe mental illness has not been settled. He believes that research should further examine the relative capabilities of the two groups to identify psychological problems and their readiness to seek psychiatric intervention.

FOLK HEALING AND SPIRITUALISM

As independent alternatives or at least as complements to conventional psychotherapists, especially for the older Hispanic generation, folk healers and spiritualists should not be overlooked. Spiritualism is accepted by all levels of Hispanic society, though it is more commonly found in the lower classes.[23] Cancela and Martinez[24] write that the exact extent of these beliefs and practices among different Hispanic groups has never been systematically researched, although available reports suggest different estimates. The authors cite various reports that indicate *curanderismo* (Mexican folk healing) is decreasing among Mexican-Americans[25,26]; *santeria* (a healing cult that regards African deities much like Catholics regard saints, and conducts rituals in which they offer food and live animals to these deities) has been gaining popularity among Cuban-Americans in certain sections of the country[27]; and *espiritismo* (the belief in spirits who can attach themselves to human beings) is increasing among Puerto Ricans.[28] Cancela and Martinez[24] acknowledge the effectiveness of these folk healing systems, an effectiveness they feel that results from a combination of the following attributes of these systems: a concrete, commonsense, action-oriented treatment; flexibility in offering individual, family, or group therapy; the absence of impersonal and bureaucratic procedures; a holistic approach based on the unity of mind and body; a familiar literal and symbolic language; an attempt to determine the precise cause of the difficulty; a lack of stigmatizing the client; a placebo effect; and the power of suggestion. But although these authors acknowledge the benefits of folk healing systems, they believe the other side of the issue has not been adequately addressed, namely the "potentially limiting and possibly oppressive aspects of folk beliefs" and healing systems.

Espiritismo occurs particularly though certainly not exclusively among Puerto Ricans. Lubchansky *et al.*,[29] impressed by figures suggesting frequent visits by Puerto Rican New Yorkers to spiritualists, feel that the more serious the mental illness, the more likely the Puerto Rican is to consult with a spiritualist. These investigators found that spiritualists have highly idiosyncratic conceptions of mental illness, and they tend to describe primarily disorganized thought processes. Also, because spiritualists tend to expect an improvement in the illness, chronicity may be averted. They therefore share certain aspects of contemporary psychiatric thinking. It also appears that, for all practical purposes, the Puerto Rican spiritualist occupies a position of indigenous nonprofessional healer. Rogler and Holingshead[23] point out that Puerto Rican mentally ill persons frequently visit spiritualists before, during, and after the course of their treatment with psychiatrists. Purdy *et al.*[30] emphasize the widespread use of spiritualism as a supplement to psychiatry or as a substitute method of treatment. Interestingly, these investigators found that when patients who consulted spiritualists at the same time they were receiving psychiatric treatment improved, they attributed their improvement solely to the spiritualists.

INTERPRETERS AND LANGUAGE BARRIERS

The role of the interpreter in interviewing non-English-speaking patients has only in the last two decades become a serious topic of psychiatric study. This relatively new interest coincides with the awareness of the diversity of the recent immigrant groups coming to the United States. The different cultures and languages frequently constitute barriers that prevent these groups from receiving adequate psychiatric evaluations and treatments. Obviously, this language concern applies to Hispanics, and especially to the older members of these groups.

In *The Psychiatric Interview in Clinical Practice,* published in 1971, MacKinnon and Michels[31] pointed out the lack of psychiatric literature on the subject of interviewing through an interpreter. According to those authors, the ideal interpreter should be thoroughly familiar with both cultures and be able to discern the "subtleties," "nuances," and "finer emotional tones" in the interview. They recommend that the interpreter translate the interview sentence by sentence, rather than approximating or summarizing his or her understanding of the interview. They also suggest that the interviewer face the patient and speak slowly in a normal tone of voice in order to express feelings by his voice, gesture, and facial expression and so be able to develop a favorable working relationship. Cannon,[32] writing about using an interpreter in cross-cultural counseling, points out that attitudes toward the need for interpretation invariably affect the success in interpretation; she argues, in fact, that unless the client's language is accepted as being just as valid as that of the counselor's, effective communication using an interpreter cannot be attained. Cultural differences in gender patterns may also bear on the effectiveness of communication. Because of "machismo" and "sumisa" in Hispanic culture, it may be necessary for a counselor of the opposite sex to use an interpreter of the same sex as the client.

Drawing on his careful study of interviews conducted through interpreters, Marcos[33] has enumerated what he believes to be the major interpreter-related distortions that lead to misevaluation of a patient's mental status: (a) The interpreter's lack of language competence and translation skills, especially obvious in omissions, additions, substitutions, and condensations. (b) The interpreter's lack of psychiatric knowledge. Most common was a tendency to normalize and make sense of the patient's disorganized thoughts such as circumstantiality, tangential thinking, loose associations, and blockings. (c) The interpreter's attitudes. Some relatives used as interpreters tended to distort the patient's psychopathology in a positive or negative way, or even to answer the clinician's questions without really asking the patient.

In an effort to minimize these distortions, Marcos makes the following recommendations to clinicians: (a) Meet with the interpreter before the interview to discuss the goals and the areas that will be covered; (b) endeavor to make clear to the psychiatrically inexperienced interpreter the distinctions between formal and content aspects of the patient's statements, and (c) meet with the interpreter after the evaluation for further clarification of the substance of the interview and the dynamics of the transaction.

The question of whether the same kind of psychopathology can be elicited in an interview of a patient when he or she speaks in a foreign language (English) as when he or she speaks in the native language has also been under considerable discussion. In an earlier study, Del Castillo,[34] whose observations were made at the New Jersey State Hospital in Trenton with "criminally insane" patients, described several cases in which psychotic symptoms were detected when the patients were interviewed in their native languages but not when interviewed in foreign languages. Del Castillo reasoned that expressing oneself in a foreign language is psychologically equivalent to the administration of certain stimuli disposed to awaken a sleeping person or to arouse his unconscious toward a vigilance that is absent during sleep or moments of relaxation. He thus suggests that speaking in a foreign language can stimulate the patient to think, and may put him or her in better contact with reality. Marcos *et al.*,[35] on the other hand, report the exact opposite. These investigators used standardized mental status interviews recorded on closed-circuit television to evaluate the recent admissions of schizophrenic patients at Bellevue Psychiatric Hospital. They found that when those patients who were sufficiently fluent in English were interviewed in that language, the patients disclosed more psychopathology in English than when the same patients were interviewed in their native Spanish. Like Del Castillo, the investigators concluded that changes related to the difficulties in speaking in a foreign language were clinically important; but their interpretations of the source and effects of these changes differs radically from Del Castillo's. Marcos *et al.*[35] observed that when speaking in English, the patient may become tense, and he or she may even stop trying, and thus appear emotionally detached and uncooperative. In addition, the clinician's frame of reference, though relevant to native English-speaking patients, cannot be directly applied in evaluating the psychopathology of persons with different native languages and cultures.

After a fuller scrutiny of their data, Marcos *et al.*[36] stress that the language barriers that often exist when evaluating Spanish-speaking patients who are interviewed in English are sufficient to make possible misunderstandings and misdiagnoses. If the patient has a poor command of English and is obliged to speak in that language, this alone can distress the patient enough to lead to a worsening of his or her cognitive functioning. The patients may become less verbally expressive, give briefer responses, and appear more emotionally withdrawn compared to when they spoke in Spanish. The patient's flow of thought may be interfered with, causing him or her to display what could be mistakenly interpreted as cognitive decline or illogical, concrete, and impoverished thinking. The patient may exhibit the kinds of speech disturbances that are usually associated with anxiety, as well as the slowed up speech with long silent periods commonly seen in depression.

Yet other studies differ in their findings and conclusions from those of Marcos *et al.*[36] For example, Gonzelez[37] found that bilingual Mexican-American schizophrenic males, who were interviewed in both Spanish and English, showed considerably more discomfort in the Spanish-language interview. Gonzalez interpreted this difference as one of degree or flavor rather than kind; that is, he found psychopathology to be presented more intensely in Spanish than in English. He also points out that an English-speaking psychiatrist could easily miss this discomfort revealed in the native language interview, and therefore may not be able to accurately appraise the case. Price and Cuellar's[38] study of 32 Mexican-American patients supports Del Castillo's finding that, when speaking Spanish, the patients exhibited more symptomatology indicative of psychopathology. Price and Cuellar point out that the interviews in their study were rated by bilingual, bicultural raters. Although the researchers previously cited agree that speech disturbances may contribute to the misdiagnosis of the case, they differ on the reasons and results of these disturbances. Certainly, they contribute to the difficulty and complexity of accurate diagnosis.

TRANSCULTURAL CONSIDERATIONS IN PSYCHOTHERAPY

There has been increasing recognition that older adults can benefit from a variety of psychotherapy interventions.[39] As in other areas related to the Hispanic elderly, there are few references specifically mentioning this population in the literature. Szapocznik *et al.*[40] describe a demonstration project in Florida where life enhancement counseling (consisting of life review and ecological assessment and intervention) serves as an alternative to psychopharmacology for depressed Cuban-American elderly. For the purpose of this chapter, we are using the term *psychotherapy* in a generic sense, meaning psychosocial therapy excluding somatic and biological treatment. In referring to psychotherapy of minority groups in general, the past literature underscores the numerous pitfalls involved, and only more recently includes reports of success in intercultural treatment approaches.

A major shortcoming to evaluating and treating a non-English-speaking minority patient through a translator is the possibility that one aspect of the patient's psychopathology may be recognized but not another. Sabin[41] reported on two cases of suicide by Spanish-speaking patients who were being treated by English-speaking psychiatrist through translators. The author believes that the various psychiatrists who, using translators, evaluated and treated these patients routinely misjudged the extent of the affective aspects of their illnesses. The psychiatrists recognized that both patients were seriously ill, but the patients' psychotic thinking was focused on at the expense of full appreciation of the patients' despair. One of the cases involved a 45-year-old South American woman who was diagnosed as suffering from schizoaffective illness and had been psychiatrically hospitalized four times within the previous 18 months. Earlier that day she talked about suicide but the emergency room doctor decided that she was not then suicidal. The second case involved a 71-year-old former businessman from Puerto Rico who, following a physical illness, had become chronically depressed. Both the patient and his family refused hospitalization. One day he slipped away from his sister and hung himself. Sabin cites a study by Schwab *et al.*[42] of the sociocultural aspects of depression in general medical inpatients, a study in which the investigators discovered that physicians, who generally have attained high social status, tend to ignore depression in lower-class patients. Sabin[41] goes on to note that this physician tendency to miss depression in lower-class patients is compounded by the language factor that leads to the physician's making direct contact with the translator but not with the patient.

The notion of language as an emotional barrier in psychotherapy has been of particular concern to Marcos,[43,44] who describes an "emotional-detachment effect associated with conversing in a second language."[43] This phenomenon occurs in "subordinate bilinguals"—bilinguals who are not nearly as proficient in the second language as they are in their native tongue. Marcos argues that subordinate bilinguals are forced to "perform a more elaborate and difficult encoding work (the mechanism whereby a speaker's ideas, feelings, images, and so forth, become coded into intelligible sounds in a given language) in their second language." This additional process accounts for the emotional detachment seen in subordinate bilinguals. Marcos also points out that, conversely, the language barrier may allow the patient to verbalize highly charged material in the second language (when the patient is emotionally detached) that ordinarily could not be expressed in his or her native language. It is incumbent on psychotherapists treating a subordinate bilingual to be aware of these linguistic factors in order to minimize the possibility of misevaluating the patient's emotional responses and resistances.[44]

Regarding bilingualism and psychotherapy, it is instructive to consider the concept of information processing. Kolers[45] has explored the phenomenon of bilingual learning. Among his conclusions, he observed that information learned in one context may be unavailable in another context. This suggests the extent to which the language used in interviewing a bilingual patient may limit and/or determine the information elicited in the interview. Marcos and Alpert[46] point out that if only one language is used, some areas of the bilingual's intrapsychic world may remain untouched and unavailable to treatment. This can be one result of the phenomenon of language independence that allows the bilingual to "maintain two separate language codes—each with its own lexical, syntactic, phonetic, and ideational components—and to use them independently with minimal interaction." During psychotherapy, if both languages are used, the patient may shift from one language to another as a form of resistance to experiencing affectively charged material, which becomes unavailable in the other language.

Another unresolved question concerns the extent of the intermixture of culture with psychopathology as it relates to psychotherapy. Bluestone and Vela,[47] pointing out the difficulty in differentiating between what is cultural and what is psychopathological, suggest that there is a continuum between the two that may be impossible to demarcate precisely. These authors draw attention to the difficulties Puerto Rican adults have in modulating their aggressive feelings as a result of child-rearing practices that include the fostering of dependence and the inhibition of aggression. They conclude that the psychotherapist treating Puerto Rican patients must become "a kind of cultural anthropologist" in order to distinguish the psychopathological from the culturally determined. At the very least, psychiatrists who intend to evaluate or treat Hispanics must be familiar with at least some basic cultural determinants. But here, too, conflicting and even opposing views can greatly complicate the psychiatrist's attempt to do so. For example, Diaz Guerrero[48] hypothesized in 1955 that in the traditional Mexican family the two basic propositions of family life—the absolute supremacy of the father and the absolute self-sacrifice of the mother—were sufficiently stressful to produce neurotic conflicts. Maldonado-Sierra *et al.*[49] challenged this hypothesis, and instead found that the normal family adhered more to these traditional family values than did the neurotically troubled group. Fernandez-Pol's[50] study of the association between lower-class New York Puerto Ricans' family values and the development of psychopathology backed up Maldonado-Sierra's findings on the subject. Fernandez-Pol concluded that there was an inverse relationship between maintaining traditional family beliefs and psychopathology. A 1958 study[51] attempted to assess the extent to which changes have occurred in Puerto Rican family values that differentiate New York Puerto

Rican families from those in Mexico. This study arrived at the following descriptive conclusions: the Puerto Rican mother is held in high esteem, yet the concept of male superiority and dominance in the family prevails, and the male child is accorded greater status. Although the middle-class Puerto Rican family's basic value system is still more similar to that of Mexicans than to that of mainland Americans, the Puerto Rican families are in the process of change, shifting toward more Americanization of beliefs, and there are indications of some questioning of male authority. The well-known *ataque*, which has been labeled the Puerto Rican syndrome, has been studied by Fernandez-Marina.[52] According to his hypothesis, the origin of the syndrome has both psychodynamic and cultural determinants.

Most of the past literature emphasized the difficulties minorities faced when they sought psychiatric treatment. Black patients, for example, were less likely to be seen in individual therapy and were seen fewer times than whites.[53] Investigators in a psychiatric outpatient clinic at the Los Angeles County General Hospital found in 1968 that minority patients received the least intensive therapy and were more often discharged than nonminority patients.[54] Another study in 1966 also concluded that minority patients are less likely to be accepted for psychotherapy, and if accepted receive less of it than nonminority patients of the same social class.[55] These kinds of studies, in turn, led to an increasing recognition of the widespread existence and impact of institutional racism[56] and the failure of training programs to produce psychiatrists prepared to deal with the special needs of black patients.[57]

Added to the general discrimination and poverty experienced by minorities in an Anglo-dominated society, it is not surprising that these obstacles of institutional racism and of insensitivity to minority groups on the part of many psychiatrists led to the situation Kline[58] described when, in 1969, he wrote that Spanish-Americans perceived Anglo psychiatrists as "cold, exploitative, and insincere." This perception, in turn, contributes to the underutilization of psychiatric services among this group, and suggests the special problems that must be surmounted in order to treat those minorities who do seek help.

The more recent literature indicates an emerging sensitivity towards minorities, and expresses an increased optimism about our capacity to provide adequate mental health services to minorities. The results of a questionnaire given to 21 Spanish-speaking patients who needed interpreters, 40 bilingual patients, and the 16 psychiatric residents who conducted the initial therapy interviews at a clinic located in Los Angeles, revealed wide discrepancies in their perceptions of the interviewers. The patients who were interviewed through interpreters felt understood, helped, and wanted to return. The psychiatric residents, on the other hand, felt that the patients who used interpreters felt less understood and helped than the bilingual patients, and that these patients did not want to return. The researchers concluded that in our efforts to develop more bilingual-bicultural therapists, we should not minimize our present therapists' efforts. These researchers argue that the fact that the therapists misjudged the Spanish-speaking patients' responses merely illustrates the difficulty we all have in surmounting cultural and linguistic barriers.[59] Padilla *et al.*[60] reviewed three kinds of emerging models—professional adaptation model, family adaptation model, and barrio service center model—that may help to improve community mental health services to Hispanics. And Acosta and Cristo[61] described an innovative bilingual interpreter program created to increase the professional mental health services available to Spanish-speaking patients in East Los Angeles. The model training program, primarily for community aides to be trained as bilingual interpreters, was started in 1975. Since 1975 the percentage of Spanish-speaking patients comprising new admissions to the clinic has risen from 1% to 15%. The great majority of these cases were treated by non-Spanish-speaking therapists in collaboration with the trained bilingual interpreters.

Moreover, most recent discussions of psychiatric training acknowledge the importance of addressing issues involved in cross-cultural psychotherapy. Kinzie's[62] assertion that cross-cultural psychotherapy demands "the sensitive use of the medical model, awareness of the nonverbal forms of communication, and awareness of the subjective phenomenology of each individual patient," points to the ways in which psychiatric training must change if the psychiatrist is to be capable of successfully evaluating and treating his or her patients in our ethnically and culturally diverse society. Furthermore, these changes in psychiatric training promise not only more accessible mental health care, but also as Wen-Shing and McDermott[63] note, learning about the universal elements common to all psychotherapeutic activity and becoming aware of cultural factors that make psychotherapy relevant in different sociocultural settings can improve our conceptualization of psychotherapy and treatment skills.

RACE AND ETHNICITY IN GERIATRIC PSYCHIATRIC RESEARCH

The need for psychiatric research with geriatric patients is abundantly clear. As a subgroup of older mental patients, Hispanic elderly present additional challenges to researchers. Although considerable discussion has addressed the relationship between sociocultural factors and manifest psychiatric symptomatology among various cultural groups,[64] the possibility that racial and ethnic differences account for certain differences in symptom presentation, and the racial and ethnic differences manifested in biological markers and psychotropic medication responses,[65] a review of this literature emphasizes, above all, the need for more research in these areas.

In a review chapter on transcultural psychiatry, Wittkower and Prince[64] discuss the cultural dimension of psychiatry. Regarding the question of whether or not the symptomatology of mental disorders are specific "culture-bound syndromes," the authors comment that "most observers stress similarities rather than differences" in mental disorders, but they go on to add that variations exist in "intensity, duration, verbalized content and affective coloring of disorders." Sometimes the differences in symptomatology have seemed significant enough to suggest a different category of disorder, that is, a culture-bound syndrome. On further study, however, it becomes apparent that it is not, after all, a separate syndrome but one of the standard syndromes described in the official American Psychiatric Association diagnostic manual.

In these authors' view, transcultural studies offer both theoretical and practical benefits. One important practical application is that the sensitivity and understanding gained from transcultural studies can more adequately prepare mental health workers to work with different ethnic groups. Further, we believe that transcultural studies with Hispanic elderly subjects, for example, could yield important information about possible differences in cultural expression of the psychiatric symptomatology of standard mental disorders. Obviously, such information could have important diagnostic and treatment implications for this population.

There are enormous deficiencies in our knowledge of Hispanic elderly. Although there is a small body of literature on some segments of the Hispanic elderly, such as the Mexican-American population, comparatively less is known about the Puerto Rican-American, Cuban-American, and other Hispanic elderly groups. Because Hispanic elderly share a common language and some cultural similarities, information accumulated from the diverse Hispanic groups have been presumed applicable to all of them. However, Costa-Robles Newton[66] points out that even the research that has been done with the Mexican-American elderly did not take into consideration the heterogeneous aspects of this population, a fact that may have accounted for many of the conflicting results. He enumerates the following differences on which future research should focus: age-cohort differences, acculturation level differences, sex differences, rural-urban differences, and regional differences.

Differences in psychiatric symptomatology, in biological markers, and in psychotropic medication responses associated with race and ethnicity are indeed being reported by researchers. Adebimpe *et al.*[67] found differences in symptoms between black and white schizophrenic patients, and between rural and urban black schizophrenic patients. These researchers feel that their data on differences in symptom presentations have "implications for researchers whose samples are drawn from multiethnic and multinational populations, and for the use of diagnostic manuals derived from one group for patients of other cultural groups." And, in a comprehensive review article, Lawson[65] cites numerous research findings showing differences in symptom presentations between blacks and whites. Discussing bipolar affective disorders in black and Hispanic patients, Lawson argues that their increased likelihood of having hallucinations may contribute to their being misdiagnosed as schizophrenics. He surveys the various studies that have revealed racial differences in biological markers, including serum creatinine phosphokinase, platelet serotonin, and HLA antigen A2. Lawson also notes research findings that point to racial and ethnic differences in psychotropic medication response; he cites research findings of Asians requiring lower dosages than whites for a variety of psychotropic medications, including neuroleptics, antidepressants, and lithium, and a study indicating that Hispanic patients require less antidepressant medication and show more side effects at dosages half that of Anglo therapeutic dosages.[68] This latter study in particular would have important implications in the treatment of depression in Hispanic elderly patients. Clearly, psychiatric research with Hispanic elderly patients would have to take these racial and ethnic factors into consideration, and would work to expand our knowledge and awareness of such factors.

DISCUSSION

Despite the limited information available about Hispanic elderly, the psychiatrist evaluating any Hispanic older person must be sensitive to issues of culture and language; without this sensitivity an adequate evaluation cannot be made. In conducting examinations on elderly Mexican-Americans, Costa Robles Newton[66] emphasizes the importance of *respeto* (respect). This means the proper honoring of the other's *dignidad* (dignity). As Costa-Robles Newton points out, *respeto* (respect), *orgullo* (pride), and *dignidad* (dignity) comprise the cultural system of *personalismo*—the pattern of social interaction encouraging people to relate to each other as whole persons, in a human and humane way. In order to establish rapport with an elderly Hispanic patient (as is true with all patients), the evaluator must treat him or her with friendliness and respect.

It is particularly important to state the obvious in a chapter that emphasizes, as this one does, the significance of cultural differences, that although generalities about groups of people should be considered, they may not apply to the specific case. It has been clinically observed that among lower-class groups psychological suffering generally tends to be expressed in psychophysiological symptoms.[69] In older patients, many of whom have accompanying physical ailments, psychophysiological complaints would undoubtedly confound the clinical picture. Regarding specific Puerto Rican symptomatology, Abad and Boyce[70] are impressed by the similarities seen in many patients. They found that the common complaints were depression (expressed mostly as bodily symptoms), anxiety (expressed as heart palpitations, dizziness, and fainting), somatic concerns, hallucinations, and actual or feared loss of control (a familiar example is the *ataque de nervios* or Puerto Rican syndrome). Unfortunately, nothing specific about Hispanic elderly symptomatology was mentioned.

Some additional points about bilingualism should be brought to bear on its consideration in the psychiatric and legal evaluation of the Hispanic elderly. Marcos and Alpert[46] point out that "the encoding and decoding mechanisms associated with a primary language have greater resistance to language pathology than do those associated with a secondary language." They cite some of the existing literature suggesting that bilingual patients demonstrate different degrees of impairment of their two languages after suffering strokes or during worsening chronic organic brain syndromes. (However, one of these citations, Brain,[71] emphasizes the extreme complexity of the subject.) Marcos and Alpert[46] also refer to the case of a bilingual patient, who after receiving electroconvulsive therapy experienced a language regression and was unable to speak English but started to speak her native French. The situation lasted for about 2 weeks and then she began to speak English again. During the time she was unable to speak English, the patient was reported to be alert, oriented, and without any change in her personality.[72] However, in accordance with the notion of content differences between the two languages bilinguals speak, reverting back to one's native language conceivably could entail changes in personality and verbal behavior. That is, "it is possible that a shift in language is associated with a shift in social roles and emotional attitudes."[73] This shift in attitude could compel the patient to change his or her mind about further treatment and invalidate informed consent. The psychiatrist could be faced, for example, with the dilemma of making a new determination of competency based on the personality speaking the primary (native) language. Admittedly, there are tremendous gaps of knowledge here and a concomitant need for further investigation in this clinical area. Laski and Taleporos[74] comment that, generally, bilingualism and its relation to psychopathology has been a poorly researched area. Their description of a case of bilingualism involving a schizophrenic adolescent male who suffered an anticholinergic delirious psychosis, one of the symptoms of which were hallucinations in his native language, vividly illustrates our need for further research on bilingualism.

CONCLUSION

In evaluating and treating older Hispanic mental patients, the psychiatrist assumes legal and ethical obligations. Kapp[75] has focused attention on the fiduciary and contractual aspects of the relationship between doctor and older patient that begin when the doctor undertakes voluntarily to treat the patient. The doctor is obligated to maintain quality of care-related standards that are befitting the older patient. These should include being knowledgeable about and sensitive to special age-based characteristics and needs, being fully informed about the issues of aging in contemporary America, and being able and willing to develop appropriate clinical skills. Anything less than this deviates from the legally acceptable standard of care. These exhortations of commitment to render the highest quality of professional service to the elderly apply to the evaluation and treatment of older Hispanic mental patients with great force. The need to improve our diagnostic and treatment abilities to meet the challenge posed by the unique cultural and linguistic situation of the Hispanic elderly is equally great. Only when the complexity and cultural specificity of the situation of the elderly Hispanic patient is taken into consideration, can the evaluating doctor meet the demands of law and ethics.

REFERENCES

1. Hooyman NR, Fallcreek SJ: Programs for the elderly, in Austin MJ, Hershey WE (eds): *Handbook on Mental Health Administration.* San Francisco, Washington, London, Jossey-Bass Publishers, 1982, p 539.
2. Butler RN, Lewis MI: *Aging and Mental Health,* ed 3. St. Louis, Mo, The C.V. Mosby Company, 1982.
3. Hayes-Bautista DE: Identifying "Hispanic" populations: The influence of research methodology upon public policy. *Am J Public Health* 1980; 70(4):353–356.
4. Lampe PE: Mexican Americans: Labeling and mislabeling. *Hispanic J Behavioral Sciences* 1984; 6(1):77–85.
5. Dowd J, Bengtson V: Aging in minority populations: An examination of the double jeopardy hypothesis. *J Gerontol* 1978; 33:427–436.
6. Gaitz CM, Scott J: Mental health of Mexican-Americans: Do ethnic factors make a difference? *Geriatrics* 1974; 29:103–110.

7. Burgoyne E: Hispanics: The sleeping giant wakes, editors introduction, in *Race and Politics*, New York, The H. W. Wilson Company, 1985, p 65.

8. *Report of the Secretary's Task Force on Black and Minority Health*. Margaret M. Heckler, Secretary. U.S. Government Printing Office, 1985.

9. *Hispanic Mosaic. A Public Health Service Perspective*. A report of the 1st Annual Forum of the Status of Hispanic Health. Satomayor M (ed): Washington, DC The Public Health Service Hispanic Employee Organization, 1983.

10. Shapiro S, Skinner EA, Kessler LG, *et al.*: Utilization of health and mental health services. *Arch Gen Psychiatry* 1984; 41:971–978.

11. Lacayo C: Triple jeopardy: Underserved Hispanic elders. Generations 1982; 6:25, 58.

12. Greene VL, Monahan DJ: Comparative utilization of community based long term care services by Hispanic and Anglo elderly in a case management system. *J Gerontol* 1984; 39:730–735.

13. Torrey EF: The irrelevancy of traditional mental health services for urban Mexican-Americans. *Am J Orthopsychiatry* 1970; 40:240–241.

14. Acosta FX: Barriers between mental health services and Mexican-Americans. An examination of a paradox. *Am J Community Psychol* 1979; 7:503–520.

15. Acosta FX, Sheehan JG: Preferences toward Mexican American and Anglo American Psychotherapists. J Consult Clin Psychol 1976; 44:272–279.

16. Pollack W: Utilization of alternative care settings by the elderly, in Lawton MP, Newcomer RJ, Byerts TO (ed): *Community Planning for an Aging Society: Designing Services and Facilities*. Stroudsburg, Pa, Dowden, Hutchinson and Ross, 1976.

17. Eribes RA, Bradley-Rawls M: The underutilization of nursing home facilities by Mexican-American elderly in the southwest. *Gerontologist* 1978; 18:363–371.

18. Waxman HM, Carner EA, Klein M: Underutilization of mental health professionals by community elderly. *Gerontologist* 1984; 24:23–30.

19. Edgerton RB, Karn M, Fernandez I: Curanderismo in the metropolis. *Am J Psychotherapy* 1970; 24:124–134.

20. Karno M, Edgerton RB: Perception of mental illness in a Mexican-American community. *Arch Gen Psychiatry* 1969; 20:233–238.

21. Edgerton RB, Karno M: Mexican-American bilingualism and the perception of mental illness. *Arch Gen Psychiatry* 1971; 24:286–290.

22. Keefe SE: Mexican-American's underutilization of mental health clinics: An evaluation of suggested explanations. *Hispanic J Behavioral Sciences* 1979; 1:93–115.

23. Rogler LH, Hollingshead AB: The Puerto Rican Spiritualist as a psychiatrist. *Am J Sociology* 1961; 67:17–21.

24. Cancela VDL, Martinez IZ: An analysis of culturalism in Latino mental health: Folk medicine as a case in point. *Hispanic J Behavioral Sciences* 1983; 5:251–274.

25. Torrey EF: Psychotherapists in a Mexican-American subculture, in Fuller-Torrey E (ed): *The Mind Game: Witchdoctors and Psychiatrists*. New York, Emerson Hall Publishers, 1972.

26. Keefe SE: Folk medicine among urban Mexican-Americans: Cultural persistence, change, and displacement. *Hispanic J Behavioral Sciences* 1981; 3:41–58.

27. Sandoval MC: Santeria as a mental health care system: An historical overview. *Soc Sci Med* 1979; 13B:137–151.

28. Garrison V: The "Puerto Rican syndrome" in psychiatry and espiritismo, in Crapanzano V, Garrison V (eds): *Case Studies in Spirit Possession*. New York, John Wiley & Sons, 1977.

29. Lubchansky I, Egri G, Stokes J: Puerto Rican spiritualists view mental illness: The faith healer as a paraprofessional. *Am J Psychiatry* 1970; 127:88–97.

30. Purdy B, Flores S, Bluestone H, *et al.*: Mellaril or medium, Stelazine or seance? A

study of spiritism as it affects communication, diagnosis, and treatment of Puerto Rican people. *Am J Orthopsychiatry* 1970; 40:239–240.

31. MacKinnon RA, Michels R: *The Psychiatric Interview in Clinical Practice.* Philadelphia, W.B. Saunders Company, 1971.

32. Cannon CB: using an interpreter in cross-cultural counseling. *The School Counselor* 1983; September:11–16.

33. Marcos LR: Effects of interpreters on the evaluation of psychopathology in non-English-speaking patients. *Am J Psychiatry* 1979; 136:171–174.

34. Del Castillo JC: The influence of language upon symptomatology in foreign-born patients. *Am J Psychiatry* 1970; 127:160–162.

35. Marcos LR, Alpert M, Urcuyo L, *et al.:* The effect of interview language on the evaluation of psychopathology in Spanish-American schizophrenic patients. *Am J Psychiatry* 1973; 130:549–553.

36. Marcos LR, Urcuyo L, Kesselman M, *et al.:* The language barrier in evaluating Spanish-American patients. *Arch Gen Psychiatry* 1973; 29:655–659.

37. Gonzalez JR: Language factors affecting treatment of bilingual schizophrenics. *Psychiatric Annals* 1978; 8:356–358.

38. Price CS, Cuellar I: Effects of language and related variables on the expression of psychopathology in Mexican American patients. *Hispanic J Behavioral Sciences* 1981; 3:145–160.

39. Lazarus LW (ed): *Clinical Approaches to Psychotherapy with the Elderly.* Washington, DC, American Psychiatric Press, Inc., 1984.

40. Szapocznik J, Santisteban D, Kurtines WM: Life enhancement counseling and the treatment of depressed Cuban American elders. *Hispanic J Behavioral Sciences* 1982; 4:487–502.

41. Sabin JE: Translating despair. *Am J Psychiatry* 1975; 132:197–199.

42. Schwab JJ, Bialow M, Holzer CE, *et al.:* Sociocultural aspects of depression in medical inpatients. *Arch Gen Psychiatry* 1967; 17:533–538.

43. Marcos LR: Bilinguals in psychotherapy: language as an emotional barrier. *Am J Psychotherapy* 1976; 30:552–560.

44. Marcos LR: Linguistic dimensions in the bilingual patient. *Am J Psychoanalysis* 1976; 36:347–354.

45. Kolers PA: Bilingualism and information processing. *Scientific American* 1968; 218:78–86.

46. Marcos LR, Alpert M: Strategies and risks in psychotherapy with bilingual patients: The phenomenon of language independence. *Am J Psychiatry* 1976; 133:1275–1278.

47. Bluestone H, Vela RM: Transcultural aspects in the psychotherapy of the Puerto Rican population in New York City. *J Am Acad Psychoanal* 1982; 10:269–283.

48. Diaz-Guerrero R: Neurosis and the Mexican family structure. *Am J Psychiatry* 1955; 112:411–417.

49. Maldonado-Sierra ED, Trent RD, Fernandez-Marina R: Neurosis and traditional family beliefs in Puerto Rico. *Int J Soc Psychiatry* 1960; 6:237–246.

50. Fernandez-Pol B: Culture and psychopathology: a study of Puerto Ricans. *Am J Psychiatry* 1980; 137:724–726.

51. Fernandez-Marina R, Maldonado-Sierra ED, Trent RD: Three basic themes in Mexican and Puerto Rican family values. *J Soc Psychol* 1958; 48:167–181.

52. Fernandez-Marina R: The Puerto Rican syndrome: its dynamics and cultural determinants. *Psychiatry* 1961; 24:79–82.

53. Rosenthal D, Frank JD: The fate of psychiatric clinic outpatients assigned to psychotherapy. *J Nerv Ment Dis* 1958; 127:330–343.

54. Yamamoto J, Quinton JC, Palley N: Cultural problems in psychiatric therapy. *Arch Gen Psychiatry* 1968; 19:45–49.

55. Karno M: The enigma of ethnicity in a psychiatric clinic. *Arch Gen Psychiatry* 1966; 14:516–520.

56. Sabshin M, Diesenhaus H, Wilkerson R: Dimensions of institutional racism in psychiatry. *Am J Psychiatry* 1970; 127:787–793.

57. Jones BE, Lightfoot OB, Palmer D, *et al.:* Problems of black psychiatric residents in white training institutes. *Am J Psychiatry* 1970; 127:798–803.

58. Kline LY: Some factors in the psychiatric treatment of Spanish-Americans. *Am J Psychiatry* 1969; 125:1674–1681.

59. Kline F, Acosta FX, Austin W, *et al.:* The misunderstood Spanish-speaking patient. *Am J Psychiatry* 1980; 137:1530–1533.

60. Padilla AM, Ruiz RA, Alvarez R: Community mental health services for the Spanish-speaking/surnamed population. *Am Psychol* 1975; 30:892–905.

61. Acosta FX, Cristo MH: Bilingual-bicultural interpreters as psychotherapeutic bridges: A program note. *J Community Psychology* 1982; 10:54–56.

62. Kinzie JD: Lessons from cross-cultural psychotherapy. *Am J Psychotherapy* 1978; 32:510–520.

63. Wen-Shing T, McDermott JF: Psychotherapy: historical roots, universal elements and cultural variations. *Am J Psychiatry* 1975; 132:378–384.

64. Wittkower ED, Prince D: A review of transcultural psychiatry, in Arieti S (ed): *American Handbook of Psychiatry*, ed 2. New York, Basic Books, Inc., 1974, vol 2, p 535–550.

65. Lawson WB: Racial and ethnic factors in psychiatric research. *Hosp Community Psychiatry* 1986; 37:50–54.

66. Costa-Robles Newton F: Issues in research and service delivery among Mexican-American elderly: A concise statement with recommendations. *Gerontologist* 1980; 20:208–213.

67. Adebimpe VR, Chu CC, Klein HE, *et al.:* Racial and geographic differences in the psychopathology of schizophrenia. *Am J Psychiatry* 1982; 139:888–891.

68. Marcos LR, Cancro R: Pharmacotherapy of Hispanic depressed patients: Clinical observations. *Am J Psychotherapy* 1982; 36:505–512.

69. Crandell DL, Dohrenwend BP: Some relations among psychiatric symptoms, organic illness, and social class. *Am J Psychiatry* 1967; 123:1527–1537.

70. Abad V, Boyce E: Issues in psychiatric evaluations of Puerto Ricans: A socio-cultural perspective. *J Operational Psychiatry* 1979; 10:28–39.

71. Brain L: Speech Disorders: *Aphasia, Apraxia and Agnosia.* London, Butterworth & Co. Publishers, 1965.

72. Lipsius LH: Electroconvulsive therapy and language. *Am J Psychiatry* 1975; 132:459.

73. Ervin SM: Language and TAT content in bilinguals. *J Abnormal and Social Psychology* 1964; 68:500–507.

74. Laski E, Taleporos E: Anticholinergic psychosis in a bilingual: A case study. *Am J Psychiatry* 1977; 134:1038–1040.

75. Kapp MB: Medical ethics and humanities. Legal and ethical standards in geriatric medicine. *J Am Geriatr Soc* 1985; 33:179–183.

Index